Fruit and vegetable biotechnology

Related titles from Woodhead's food science, technology and nutrition list

Cereal biotechnology (ISBN: 1 85573 498 2)

Genetic modification is one of the more important and controversial issues facing the food industry, nowhere more so than in cereal production and processing. This major work explains the techniques involved and their enormous potential for food producers and consumers. It also explains how this new technology is regulated, the methods for assessing its potential risks, and the ways that cereal biotechnology can add value. Much has been written on this significant issue but until now there has been no single guide for both academia and industry. By examining both sides of the coin, this book bridges the gap between these groups, giving each a greater awareness of the other's role, a more rounded picture of the business and an increased understanding of all the issues at stake.

Fruit and vegetable processing: improving quality Second edition (ISBN: 1 85573 548 2)

Fruit and vegetables are both major food products in their own right and key ingredients in many processed foods. This new book provides an authoritative survey of the latest research on improving the sensory, nutritional and functional qualities of fruit and vegetables whether as fresh or processed products for the consumer.

Lockhart and Wiseman's crop husbandry eighth edition (ISBN: 1 85573 549 0)

Lockhart and Wiseman's crop husbandry is widely recognised as the standard introduction to its subject for both students and practitioners. This major new edition has been comprehensively revised. The book has been totally reorganised and includes new chapters on the influence of climate, cropping techniques, integrated crop management and quality assurance, seed production and selection.

Details of these books and a complete list of Woodhead's food science, technology and nutrition titles can be obtained by:

- visiting our web site at www.woodhead-publishing.com
- contacting Customer services (e-mail: sales@woodhead-publishing.com; fax: +44 (0) 1223 893694; tel.: +44 (0) 1223 891358 ext.30; address: Woodhead Publishing Limited, Abington Hall, Abington, Cambridge CB1 6AH, England)

If you would like to receive information on forthcoming titles in this area, please send your address details to: Francis Dodds (address, tel. and fax as above; e-mail: francisd@woodhead-publishing.com). Please confirm which subject areas you are interested in.

Fruit and vegetable biotechnology

Edited by
Victoriano Valpuesta

CRC Press
Boca Raton Boston New York Washington, DC

WOODHEAD PUBLISHING LIMITED
Cambridge England

Published by Woodhead Publishing Limited, Abington Hall, Abington
Cambridge CB1 6AH, England
www.woodhead-publishing.com

Published in North America by CRC Press LLC, 2000 Corporate Blvd,
NW Boca Raton FL 33431, USA

First published 2002, Woodhead Publishing Limited and CRC Press LLC

British Library Cataloguing in Publication Data
A catalogue record for this book is available from the British Library.

Library of Congress Cataloging-in-Publication Data
A catalog record for this book is available from the Library of Congress.

Woodhead Publishing Limited ISBN 1 85573 467 2
CRC Press ISBN 0-8493-1436-4
CRC Press order number: WP1436

Cover design by The ColourStudio
Project managed by Macfarlane Production Services, Markyate, Hertfordshire (e-mail: macfarl@aol.com)
Typeset by MHL Typesetting Limited, Coventry, Warwickshire
Printed by TJ International, Padstow, Cornwall, England

Contents

Contributors

Chapter 1

Professor Victoriano Valpuesta
Departmento de Biología Molecular y Bioquímica
Facultad de Ciencias
Universidad de Málaga
Campus de Teatinos
29071 Málaga
Spain

Tel: +34 95-213-1932
Fax: +34 95-213-1932
E-mail: valpuesta@uma.es

Chapter 2

Javier Pozueta-Romero
Centro de Biotecnología Agraria Vegetal
Universidad Pública de Navarra
Ctra. Mutilva s/n
31192 Mutilva Baja
Navarra
Spain

Tel: +34 948-242-834
Fax: +34 948-232-191
E-mail: javier.pozueta@unavarra.es

Chapter 3

Dr Luciani Baldoni and Professor Eddo Rugini
IR Miglioramento Genetico Piante Foraggere CNR
Via Madonna Alta
130 – 06128
Perugia
Italy

Tel: +39 075-501-4878
Fax: +39 075-501-4869
E-mail: L.Baldoni@irmgpf.pg.cnr.it

Chapter 4

Dr M. A. Gomez-Lim
Departmento de Ingeniería Genetíca CINVESTAV – Irapuato
KM 9.6 Carretera Irapuato-León
Apartado Postal 629
Irapuato
GTO
Mexico
36500

Tel: +52 462-396-00
Fax: +52 462-396-50 / 462-458-49
E-mail: mgomez@ira.cinvestav.mx

Chapter 5

Dr J. R. Botella
Department of Botany
University of Queensland
Brisbane
Qld 4072
Australia

Tel: +61 7-3365-1128
Fax: +61 7-3365-1699
E-mail: J.Botella@botany.uq.edu.au

Chapter 6

Professor V. Valpuesta, Dr M. A. Botella and Dr I. Amaya
Departmento de Biología Molecular y Bioquímica
Facultad de Ciencias
Universidad de Málaga
Campus de Teatinos
29071
Spain

Tel: +34 95-213-1932
Fax: +34 95-213-1932
E-mail: valpuesta@uma.es

Chapter 7

David G. Lindsay
CEBAS-CSIC
Apartado de Correos 4195
30080
Murcia
Spain

Tel: +34 908-39-63-34
Fax: +34 968-27-47-93
E-mail: dlindsay@terra.es

Chapter 8

Dr Ann L. T. Powell and Professor Alan B. Bennett
Mann Laboratory
University of California
Davis
CA 95616

Tel: +530 752 9096
Fax: +530 752 4554
E-mail: alpowell@ucdavis.edu
E-mail: abbennett@ucdavis.edu

Chapter 9

Professor H. Davies
Head of Cellular and Environmental Physiology Department
Scottish Crop Research Institute
Invergowie
Dundee
DD2 5DA

Tel: +44 (0)1382-568-513
Fax: +44 (0)1382-568-503
E-mail: h.davies@scri.sari.ac.uk

Chapter 10

Dr A. Bernadac, Dr A. Latché, Dr J.-P. Roustan, Dr M. Bouzayen and Dr J.-C. Pech
Avenue de l'Agrobiopole
BP 107
Auzeville Tolosane 31320
Castanet Tolosan Cedex
France

E-mail: pech@ensat.fr

Chapter 11

Dr Lynn Frewer
Head, Consumer Science Group
Institute of Food Research
Norwich Research Park
Colney Lane
Norwich
NR4 7UA

Tel: +44 (0) 1603-255-000
Fax: +44 (0) 1603-507-723
E-mail: lynn.frewer@bbsrc.ac.uk
www.ifr.bbsrc.ac.uk

Chapter 12

Dr Jeremy Sweet
National Institute of Agricultural Botany
Huntingdon Road
Cambridge
CB3 0LE

Tel: +44 (0) 1223-276-381
E-mail: jeremy.sweet@niab.com

1

Introduction

V. Valpuesta, Universidad de Málaga

Biotechnology can be seen as an imprecise term since the harnessing of any biological process could justifiably be called biotechnology. In food processing it could reasonably be applied to processes as long established as bread making and brewing. However, the revolution in our understanding of the molecular mechanisms underlying the processes of life, in particular our understanding of DNA, has resulted in the potential to manipulate those mechanisms for our requirements. This new-found knowledge and ability is loosely termed biotechnology.

There are two main applications of biotechnology to fruit and vegetable production:

1. as an aid to conventional breeding programmes
2. its ability to transfer genes between different organisms.

Physiological or morphological traits are governed by genes carried on chromosomes. The ability to monitor the presence or absence of such genes in plants is a great aid to plant breeders. This is done through the use of molecular markers, characteristic DNA sequences or fragments that are closely linked to the gene or genes in question. Molecular biological methods allowing the monitoring of such markers in many independent individuals, for example those arising from a cross between two plant varieties. This is a great aid to the selection process.

The ability to transfer genes means that specific genes can be added to a crop variety in one step, avoiding all the back-crossing that is normally required, providing a major saving of time and effort. Furthermore, those genes that are added need not come from a species that is sexually compatible with the crop in question. Conventional breeding is, of course, limited to the introduction of genes from plants of the same species or very near relatives. By employing the science of genetic engineering, it is possible to bring into a crop plant different genes from other plants or even bacteria, fungi or animals. Genes are, simplistically, made up of two parts: the coding region which determines what the gene product is, and the promoter, a set of instructions specifying where, when and to what degree a gene is expressed. Coding regions and promoters from different genes can be spliced together in the laboratory to provide genes with new and useful properties (recombinant DNA). These foreign or recombinant genes can then be introduced back into crop plants through the techniques of plant genetic transformation. The introduced genes integrate into the plant genome and will be passed on to the offspring in the normal way. In this way it is possible to enhance existing characteristics and introduce new attributes into a crop.

This book explores the application of biotechnology in this second area of fruit and vegetable cultivation and their subsequent use in food processing. Chapter 2 describes the basic tools and methods of genetic manipulation, from the selection and isolation of genes to safety issues such as the stability of transgenes. Part I then considers the range of target properties for genetic enhancement, starting with two chapters on how biotechnology can improve quality and productivity in fruit and vegetable cultivation. Chapter 3 looks at the genetic modification of agronomic traits in fruit crops such as herbicide resistance, resistance to plant pests and environmental stresses, increasing yield and fruit quality. Chapter 4 looks in more detail at improving plant defences against pathogens. A group of three chapters then discusses the enhancement of traits which affect final product quality. Chapter 5 considers how biotechnology can help in extending the post-harvest life of fruit and vegetables, an increasingly important issue given the complexity of modern supply chains. Chapter 6 reviews the use of molecular genetics to improve food properties such as nutritional quality and sensory characteristics such as colour and flavour. Given its importance, Chapter 7 looks in more detail at the nutritional enhancement of plant foods.

Part II includes three case studies on the application of biotechnology to particular crops. Tomato was the subject of the first commercial release of a transgenic food product, the Flavt Savr tomato with extended shelf life of the ripe fruit, and has subsequently been a particular focus for research in this field. Chapter 8 reviews the range of work. Chapter 9 considers current commercial developments with transgenic potato whilst Chapter 10 reviews work on a range of other vegetables and fruit from melon and cucumber to cabbage, broccoli, cauliflower and lettuce. Finally, Part III looks at the all-important issues of consumer attitudes and risk assessment, with chapters on these issues and identifying GMOs in foods.

2

Tools of genetic engineering in plants

J. Pozueta-Romero, Universidad Pública de Navarra

2.1 Introduction

Transfer and expression of foreign genes in plant cells, now routine practice in several laboratories around the world, has become a major tool to carry out gene expression studies and to obtain plant varieties of potential agricultural interest. The capacity to introduce and express diverse foreign genes in plants, first described for tobacco in 1984,[1] has been extended to many species. Transgenic crops such as tomato, cotton, maize, soybean, etc., are now available for human consumption and by complementing traditional methods of crop improvement (and thus becoming an integral part of agriculture), they will have a profound impact on food production, economic development and on the development of a sustainable agricultural system during the 21st century.

Although the capacity to introduce and manipulate specific gene expression in plants provides a powerful tool for fundamental research, much of the support for plant transformation research has been provided because of the generation of plants with useful and rapidly discernible phenotypes which are unachievable by conventional plant breeding, i.e., resistance to viruses, insects, herbicides, or post-harvest deterioration.[2–9] Plants useful for production of materials ranging from pharmaceuticals[10] to biodegradable plastics.[11] have been obtained using this new technology. Remarkably also, plant biotechnology techniques have been used to create plants overexpressing genes from human pathogens, the resulting plants accumulating proteins with immunogenic properties. These plants have been proved to be effective in causing oral immunization against diseases such as hepatitis B, cholera and rabies[12–14] which demonstrate the feasibility of using transgenic plants as expression and delivery systems for oral vaccines. In this chapter the technical aspects of the state of the art in plant engineering are described. It also identifies technical problems remaining in the development of systems of plant transformation applicable to crop improvement.

2.2 Selection and isolation of genes

Genetic information is carried in the linear sequence of nucleotides in DNA. Its expression involves the translation of the linear sequence of specific regions of DNA existing in the nucleus of the cell (called coding regions or genes) into a colinear sequence of amino acids (proteins). As an intermediate step, however, DNA must be copied into a different type of polynucleotide known as ribonucleic acid (RNA) which retains all the information of the DNA sequence from which it was copied. Single-stranded RNA molecules are synthesized by a process known as DNA transcription which is regulated by interactions between DNA sequences located upstream of the gene (promoters) and proteins (transcription factors). Thousands of RNA transcripts can be made from the same DNA segment in a given cell. Many of these RNA molecules undergo major chemical changes before they leave the nucleus to serve as the messenger RNA (mRNA) molecules that direct the synthesis of proteins in the cytosol.

Fragments of DNA can be amplified by a process called DNA cloning which consists in inserting the DNA into a plasmid or a bacterial virus and then growing these in bacterial (or yeast) cells. Plasmids are small circular molecules of DNA that occur naturally in bacteria, where they replicate as independent units. As these bacteria divide, the plasmid also replicates to produce an enormous number of copies of the cloned DNA fragment. Although restricted genomic DNA fragments can be cloned to produce genomic libraries, cDNA libraries are most frequently used to isolate and characterize genes necessary for the production of genetically engineered plants. cDNA libraries represent the information encoded in the mRNA of a particular tissue or organism. mRNA molecules are exceptionally labile and difficult to amplify in their natural form. For this reason, the information encoded by the mRNA is converted into a stable DNA duplex (cDNA) via enzymatic reactions catalyzed by reverse transcriptase and DNA polymerase I, and then is inserted into a self-replicating plasmid. The resulting heterogeneous population of cDNA molecules collectively encodes virtually all of the mRNAs sysnthesized by the cell. Once the information is available in the form of a cDNA library, individual processed segments of the original genetic information can be isolated and examined with relative ease.

A representative cDNA library should contain full-length copies of the original population of mRNA. cDNA libraries provide a method by which the transcription and processing of mRNA can be examined and interpreted to produce models for the flow of information responsible for the fundamental characteristics of each organism and tissue type. Comprehensive cDNA libraries can be routinely established from small quantities of mRNA, and a variety of reliable methods are available to identify cDNA clones corresponding to extremely rare species of mRNA. As the enzymatic reactions used to synthesize cDNA have improved, the sizes of cloned cDNAs have increased, and it is often possible to isolate cloned full-length cDNAs corresponding to large mRNAs.

Screening of recombinant clones for the search of agronomically interesting genes can be carried out effectively with only two types of reagents: antibodies and nucleic acid probes. In those instances when both types of reagents are available, nucleic acid probes are preferred because they can be used under a variety of different stringencies that minimize the chance of undesired cross-reactions. Furthermore, nucleic acid probes will detect all clones that contain cDNA sequences, whereas antibodies will react only with a subset of these clones (in some cases one in six at best) in which the cDNA has been inserted into the vector in the correct reading frame and orientation.

The higher the concentration of the sequences of interest in the starting mRNA, the easier the task of isolating relevant cDNA clones becomes. It is therefore worthwhile investing some effort to make sure that the richest source of mRNA available is being used. Whenever possible, estimates should be obtained of the frequency with which the mRNA of interest occurs in the starting preparation. mRNAs that represent less than 0.5% of the total mRNA population of the cell are classified as 'low-abundance' mRNAs. Using the protocol to generate cDNA libraries explained above, the isolation of cDNA clones from low-abundance mRNAs presents two major problems, first, construction of a cDNA library whose size is sufficient to ensure that the clone of interest has a good chance of being represented and secondly, identification and isolation of the clone(s) of interest. These problems have been overcome by the possibility of amplifying specific segments of DNA by the polymerase chain reaction (PCR) which is an *in vitro* method for the enzymatic synthesis of specific DNA sequences, using two oligonucleotide primers that specifically hybridize to opposite strands and flank the region of interest in the target DNA.[15] Starting from minute amounts of DNA, repetitive series of cycles involving template denaturation, primer annealing, and the extension of the annealed primers by thermostable DNA polymerase results in the exponential accumulation of a specific fragment. *In vitro* amplification systems have the advantage of being specific, rapid, but above all they allow the detection and amplification of low-abundance transcripts from total RNA.[16] PCR can be also used to produce probes, DNA sequencing and *in vitro* generation of mutations in DNA molecules.

2.3 Transformation and regeneration of plants

Development of procedures in cell biology to regenerate plants from single cells and the discovery of techniques to transfer and express foreign genes to plant cells provided the prerequisite for the practical use of genetic engineering in crop improvement. The essential requirements in a gene transfer system for production of transgenic plants are the availability of a target tissue having cells competent for both plant regeneration and transformation, a method to introduce

DNA into cells, a procedure to select transformed cells and a system to regenerate plants from the transformed cells at a satisfactory rate.

2.3.1 DNA delivery systems

Agrobacterium tumefaciens

This bacterium is a natural transformer of somatic host cells of plants into tumorous crown gall cells. Its ability to transform cells with a piece of DNA was exploited by plant biologists, and now *Agrobacterium* plays a prominent role in transgenesis of plants. This natural gene transfer system is highly efficient, frequently yielding transformants containing single copies of the transferred DNA which have a relatively uncomplicated integration pattern compared with other transformation procedures.

Its utility was developed from the understanding of the molecular basis of the crown gall disease, namely, the transfer of DNA from the bacterium to the plant nuclear genome during the tumor-formation process. Only a small discrete portion of the ca. 200 kbp tumor-inducing plasmid (Ti) existing in the bacterium is transferred to the plant genome. The transferred DNA, now familiarly referred to as T-DNA, is surrounded by two 25-bp imperfect direct repeats and contains oncogenes encoding enzymes for the synthesis of the plant growth regulators auxin and cytokinin and for the synthesis of novel amino acid derivates called opines. The DNA transfer is mediated by a set of bacterial proteins encoded by genes (*vir* genes) existing in the Ti-plasmid, which become induced by phenolic compounds released upon wounding of the plant tissue. The key aspect in regard to gene transfer is that none of the T-DNA genes are involved in the transfer process and therefore, any or all of these genes can be removed, mutated, or replaced by other genes, and the T-DNA region can still be transferred to the plant genome.

Direct gene transfer

For some time there was good reason to believe that *Agrobacterium tumefaciens* was the vector system with the capacity for gene transfer to any plant species and variety. As this was not the case, numerous alternative approaches of 'direct gene transfer' have been tested. Most methods of direct gene transfer, such as the introduction of DNA via electroporation,[17–19] PEG-mediated DNA uptake,[20–1] protoplast fusion with liposomes containing DNA,[22] biolistics[23] or microinjection,[24] require the regeneration of plants from protoplasts. The recalcitrance of many plant species for efficient regeneration from protoplasts, elaborate protocols and prolonged tissue culture phases, are a disadvantage. Other methods for direct gene transfer in which DNA is introduced directly into tissue or whole plants[25–9] do not require protoplasts.

Biolistics, or acceleration of heavy microparticles coated with DNA, has been developed into a technique that carries genes into virtually every type of cell and tissue. Without too much manual effort, this approach has advantages such as easy handling, regeneration of multiple transformants in one shot and utilization of a broad spectrum of target cells, i.e., pollen, cultured cells, meristematic cells, etc. Using this technique, a number of transgenic crops have been produced. Remarkably some of them correspond to recalcitrant species not readily amenable to infection by *Agrobacterium* such as oat,[30] sugarcane,[31] maize,[32–3] wheat,[34–5] barley,[36] cotton,[37] banana,[38] and soybean.[39] It is not unreasonable to expect that additional major crops will be engineered using this technology. However, although biolistics has impacted significantly on agricultural biotechnology, it is certainly not a panacea. This technique is inefficient in yielding stable integrative events and most of the transformation events are transient. This makes recovery of large numbers of independently derived transformation events labor intensive and expensive.

Electroporation is one of several standard techniques for routine and efficient transformation of plants from protoplasts.[17, 40–1] This technique refers to the process of applying a high-intensity electric field to reversibly permeabilize bilipid membranes and it may be applicable to all cell types. Discharge of a capacitor across cell populations leads to transient openings in the plasmalemma which facilitates entry of DNA molecules into cells if the DNA is in direct contact with the membrane. Transgenic plants recovered using this technique contain from one to few copies of the transfected DNA, which is generally inherited in a Mendelian fashion.

2.3.2 The selection and analysis of transformants

Using either *Agrobacterium* or direct gene transfer systems, it is now possible to introduce DNA into virtually any regenerable plant cell type. However, only a minor fraction of the treated cells become transgenic while the majority of the cells remain untransformed. It is therefore essential to detect or select transformed cells among a large excess of untransformed cells, and to establish regeneration conditions allowing recovery of intact plants derived from single transformed cells.

Selectable genes

Selectable marker genes are essential for the introduction of agronomically important genes into important crop plants. The agronomic gene(s) of interest are invariably cointroduced with selectable marker genes and only cells that contain and express the selectable marker gene will survive the selective pressure imposed in the laboratory. Plants regenerated from the surviving cells will contain the selectable marker joined to the agronomic gene of interest.

The selection of transgenic plant cells has traditionally been accomplished by the introduction of an antibiotic or herbicide-resistant gene, enabling the transgenic cells to be selected on media containing the corresponding toxic compound. The antibiotics and herbicides selective agents are used only in the laboratory in the initial stages of the genetic modification process to select individual cells containing genes coding for agronomic traits of interest. The selective agents are not applied after the regeneration of whole plants from those cells nor during the subsequent growth of the crop in the field. Therefore, these plants and all subsequent plants and plant products will neither have been exposed to, nor contain the selective agent.

By far, the most widely used selectable gene is the neomycin phosphotransferase II (*NPTII*) gene[42] which confers resistance to the aminoglycoside antibiotics kanamycin, neomycin, paromomycin and G-418.[43–4] A number of other selective systems has been developed based on resistance to bleomycin,[45] bromoxynil,[46] chloramphenicol,[47] 2,4-dichlorophenoxy-acetic acid,[48] glyphosate,[49] hygromycin,[50] or phosphinothricin.[51]

The increasing knowledge of modes of action of herbicides, and rapid progress in molecular genetics have led to the identification, isolation and modification of numerous genes encoding the target proteins for herbicides. Engineering herbicide tolerance into crops has proved useful not only as a selection system, but also as a valuable trait for commercial agriculture. To be useful in agriculture, herbicides must distinguish between crop plant and weed. Although they are designed to affect significant processes in plants such as photosynthesis and amino-acid biosynthesis, these processes are common to both crops and weeds. Consequently, at present, selectivity is based on differential herbicide uptake between weed and crop, or controlled timing and site of application of the herbicide by the crop plant. As to the different strategies employed to introduce herbicide tolerance in crops, the overexpression or modification of the biochemical target of the herbicide[52–4] and detoxification-degradation of the herbicide before it reaches the biochemical target[55–6] are the general routes by which this trait is engineered in plants.

Reporter genes

Reporter genes are 'scoreable' markers which are useful for screening and labeling of transformed cells as well as for the investigation of transcriptional regulation of gene expression. Furthermore, reporter genes provide valuable tools to identify genetic modifications. They do not facilitate survival of transformed cells under particular laboratory conditions but rather, they identify or tag transformed cells. They are particularly important where the genetically modified plants cannot be regenerated from single cells and direct selection is not feasible or effective. They can also be important in quantifying both transformation efficiency and gene expression in transformants. The reporter gene should show low background activity in plants, should not have any detrimental effects on plant metabolism and should come with an assay system that is quantitative, sensitive, versatile, simple to carry out and inexpensive.

The gene encoding for the enzyme β-glucuronidase, *GUS*, has been developed as a reporter system for the transformation of plants.[57–8] The β-glucuronidase enzyme is a hydrolase that catalyzes the cleavage of a wide variety of β-glucuronides, many of which are available commercially as spectrophotometric, fluorometric and histochemical substrates. There are several useful features of *GUS* which make it a superior reporter gene for plant studies.

Firstly, many plants assayed to date lack detectable *GUS* activity, providing a null background in which to assay chimaeric gene expression. Secondly, glucuronidase is easily, sensitively and cheaply assayed both *in vitro* and *in situ* in gels and is robust enough to withstand fixation, enabling histochemical localization in cells and tissue sections. Thirdly, the enzyme tolerates large amino-terminal additions, enabling the construction of translational fusions.

The gene encoding firefly luciferase has proven to be highly effective as a reporter because the assay of enzyme activity is extremely sensitive, rapid, easy to perform and relatively inexpensive.[59] Light production by luciferase has the highest quantum efficiency known of any chemiluminescent reaction. Additionally, luciferase is a monomeric protein that does not require post-translational processing for enzymatic activity.[60]

The use of green fluorescent protein (GFP) from the jellyfish *Aequorea victoria* to label plant cells has become an important reporter molecule for monitoring gene expression *in vivo*, *in situ* and in real time. GFP emits green light when excited with UV light. Unlike other reporters, GFP does not require any other proteins, substrates or cofactors. GFP is stable, species-independent and can be monitored noninvasively in living cells. It allows direct imaging of the fluorescent gene product in living cells without the need for prolonged and lethal histochemical staining procedures. In addition, GFP expression can be scored easily using a long-wave UV lamp if high levels of fluorescence intensity can be maintained in transformed plants. Another advantage of GFP is that it is relatively small (26 kDa) and can tolerate both N- and C-terminal protein fusions, lending itself to studies of protein localization and intracellular protein trafficking.[61] It has been reported that high levels of GFP expresion could be toxic to plant growth and development.[62] Solution to this problem comes from the utilization of GFP mutant genes. Among the various GFP mutations, the S65T (replacement of the serine in position 65 with a threonine) is one of the brightest chromophores characterized by its faster formation and greater resistance to photobleaching than wild-type GFP photobleaching. Furthermore, this mutant is characterized by having a single excitation peak ideal for fluorescin isothiocyanate filter sets[63] and also by its harmless action to the plant cell.[64]

2.3.3 Plant regeneration systems

The introduction of foreign genes by genetic engineering techniques as a means of plant improvement requires the development of an efficient regeneration system for the desired plant species. Such a system must be rapid, reliable and applicable to a broad range of genotypes. Until the early 1980s, efficient regeneration of plants from cultured cells and tissues of most of the important food crops had proven to be very difficult. The problem was solved by the culture of explants from immature tissues, which retain their morphogenetic potential, on nutrient media containing potent plant-growth regulators. Development of the leaf disk transformation system by Horsch and colleagues[65] and the use of regenerable embryogenic cell cultures (so-called because they form somatic embryos) represented a technological breakthrough allowing almost routine transfer of foreign genetic material into a number of recalcitrant plant species. These techniques overcame many of the problems inherent in the protoplast transformation systems, particularly the extended culture period required and the limited regeneration of plants from protoplasts. However, the lack of efficient tissue culture systems generally applicable to agriculturally important crops is a major obstacle in the application of genetic engineering technology.

In tissue culture systems, it is important that a large number of *in vitro* culturable cells are accessible to the gene transfer treatment and they retain the capacity for regeneration of fertile plants during gene transfer and selection treatments. In some circumstances, especially in the design of gene transfer programs to produce desired commercial traits into elite vegetatively propagated cultivars, the need to avoid undesirable random genetic variation (somaclonal variation[66]) becomes the overriding consideration in the choice of tissue culture system. Minimizing the phase of tissue culture leading to the adventitious regeneration of plants is a factor favorably contributing to reduce the risk of somaclonal variation and morphological abnormality. This goal has been approached in several crops by particle bombardment into meristematic tissues, shoot proliferation and screening for transformed sexual progeny.[67] The limiting factors remain the ability to prepare the explants, transfer genes into regenerable cells, and select or screen for transformants at an efficiency sufficient for practical use in crop improvement.

2.4 Stability of the transgenes

Desirable new phenotypes created by genetic engineering of plants are frequently unstable following propagation, leading to a loss of the newly acquired traits.[68] This genetic instability is due not to mutation or loss of the transgene but rather to its inactivation. A widely accepted factor causing the variation in transgene expression is the difference in genomic integration sites (position effects). Chromosomal regions with distinct levels of transcriptional activity, adjacent enhancers, or silencing elements may differentially influence the expression of the transgene. Besides the integration site, the copy number of the transgene[69–70] and its configuration[71] may induce gene silencing. As proposed by Finnegan and McElroy,[68] transgene inactivation is a consequence of events including chromatin restructuring, DNA methylation and the inhibition of mRNA processing, transport, export or translation. Silencing phenoma may also result from the introduction of transgenes expressed under the control of strong promoters. It may affect the expression of the transgene alone, leading to a plant devoid of its original interest. Silencing may also affect the expression of homologous host genes, a phenomenon referred to as co-supression that can have dramatic consequences for the survival of the plant if it involves a housekeeping gene or a defence-related gene. Therefore, the limiting process in the application of plant transformation to biotechnology is generally not the production of transformants but the screening required to eliminate transformants with collateral genetic damage or silenced transgene expression that would interfere with meaningful physiological analysis or commercial use.

2.5 Environmental risk assessment

Despite the scientific advantages made in crop improvement, the commercialization of genetically engineered plants has been slowed by public concerns on the issue of the environmental safety of genetically engineered organisms. The assumption underlying regulations is that all transgenic plants are potentially hazardous because of the gene transfer method(s) used. However, as public experience and understanding of plant transformation increase, it is hoped that regulatory process to assess environmental risk will focus on products of the transgene expression rather than on the method of gene transfer.

Regulatory agencies and commercial interests are concerned about the environmental impact, distribution uncertainty, and public perception of widespread release of organisms expressing genes that confer resistance to antibiotics or herbicides. Although products of expression of such genes are not necessarily harmful[72] these concerns can be alleviated by removing selection markers from the host genome. Selectable markers can be eliminated by a Cre/Lox site-specific recombination.[73] However, to suggest that it should be used to remove marker genes is to fail to appreciate the implications of applying the method to agronomically important crops. For vegetatively propagated crops, the Cre/Lox system would be particularly cumbersome since the necessary sexual crosses and seed production scramble the elite genome. Therefore, if regulatory agencies decided that selectable markers should be removed, crops such as potato, apple and strawberry would be much more difficult to improve using plant biotechnology.

Selectable marker genes not only are essential to those constructing genetically modified plants but also are useful to plant breeders, legislative bodies, and monitoring agencies. Plant breeders can use selectable markers to identify progeny of crosses which contain the gene of agronomic interest because the two are linked. This saves the breeder having to assay the gene of commercial interest by more complex and expensive methods such as Southern and PCR analyses based on the utilization of specific probes and primers. Very importantly, selectable markers can be used by breeders, and by regulatory and monitoring agencies to distinguish transgenic from non-transgenic plants by a simple test which does not involve advanced molecular biology.

2.6 Future trends

Methods for DNA delivery into plant cells are now sufficiently developed to allow transformation of essentially any plant species in which regenerable cell can be identified. However, what currently limits the practical transformation of many plant species is the combination of high frequency of undesired genetic damage or unpredictable transgene expression with low frequency of transformation. These problems necessitate expensive large-scale transformation and screening programs to produce useful transformants.

2.6.1 Gene targeting

In plants there is a preference for random integration of the introduced DNA, which frequently leads to the accidental inactivation of important genes and to variable and unpredictable expression of the transgene itself. In some plants, over 90% of T-DNA insertions may disrupt transcriptional units leading to transformants with visible mutant phenotypes.[74] These observations, together with the silencing phenomena described above, sound an alarm for direct production of improved cultivars in highly selected crops, where most phenotypic changes from random mutations are likely to be adverse. Therefore, there is an urgent need to develop techniques for the directed integration of transgenes at specific locations in the genome.

Homologous genetic recombination is the transfer of genetic information between regions of similar sequence composition. Gene targeting, that is the directed integration of introduced DNA into the genome via homologous recombination, can be a valuable tool to solve the problems of genetic damage and gene silencing in genetically engineered plants. As an alternative tool to antisense strategies, gene targeting can be also a valuable tool in both fundamental and applied research to down-regulate gene expression by reverse genetics approaches. Nevertheless, the main route used by somatic plant cells for integration of transgenes is via non-homologous recombination, irrespective of the transformation procedure used for the introduction of the genes.[75–7] The efficiency of homologous recombination is in the range of 10^{-3} to 10^{-5} compared with non-homologous recombination. In contrast to the case of mammalian cells in which several factors have been shown to influence homologous recombination frequencies[78–80] factors such as vector type, homology and isogenicity of the delivered DNA, do not affect gene targeting in plants. However, analysis of the recombination enzymes and mechanisms operating in plant cells, and their possibly different prevalence in different cell types, will hopefully shed more light on the different recombination events that take place in plants.[81]

Knowledge of the enzymes participating in recombination reactions may favorably contribute to the development of strategies for gene targeting. Most of such enzymes have been purified directly or have been identified through the molecular analysis of recombination mutants in *E. coli* and *S. cerevisiae*. In *E. coli* the RecA single-stranded DNA binding protein plays a key role in homologous recombination. Remarkably, a plant homolog of the *E. coli recA* gene has been isolated from *Arabidopsis thaliana* on the basis of sequence conservation.[82] In yeast, Rad51 has a role in recombinational repair of DSBs whereas Dmc1 has a function in DSB repair and formation of synaptonemal complexes. Recently, in lily (*Lilium longiflorum*), as well as in *Arabidopsis thaliana*, plant homologs of the yeast Dmc1 and Rad51 proteins were identified.[83–5] Further progress in plant recombination is envisaged by the isolation of interesting mutants with altered recombinational behavior.

In plants, homologous recombination is performed in tissues or cells that are highly competent for non-homologous recombination, which is not necessarily the best choice. It would also be very interesting to test the capacity of meiotic or meristematic cells for homologous recombination of foreign DNA in plants.

2.6.2 Transformation of recalcitrant species

Cereals, legumes, and woody plants are commonly categorized as recalcitrant to transformation. However, the hypothesis that some plants lack the biological capacity to respond to essential triggers for integrative transformation, or have cellular mechanisms preventing integrative transformation, can effectively be rejected. Broadly applicable selection methods are well established and the key to transform recalcitrant species appears to be the development of methods to expose many regenerable cells to nondestructive gene transfer treatments.

Knowledge of the relative susceptibility of different cells and tissues to transformation by *Agrobacterium tumefaciens*, would be helpful in devising strategies for transformation experiments for recalcitrant plant species. Although we know much about the contribution of the bacterium, we know little about its interaction with the plant cell and about the events surrounding gene transfer. It is known that *Agrobacterium* DNA transfer is highly regulated and is triggered only in the presence of susceptible cells of the plant host. However, does *Agrobacterium* select between cell types? What features determine favored cells for gene transfer? Are there physiological requirements for efficient T-DNA integration? Can wound response of recalcitrant plant species efficiently induce the expression of *vir* genes existing in the Ti plasmid of *Agrobacterium*?

A clear understanding of the factors determining the amenability of the transformed cells for regeneration will also favorably contribute to overcome the problem of transforming recalcitrant species. Despite a vast lore of information on hormonal control, largely arrived at through trial and error, knowledge of the fundamental biology underlying induction of plant regeneration and organogenesis remains scanty. For example, gene expression associated with organ-specific inductive events is poorly characterized and the mechanism(s) by which growth factors such as auxins and cytokinins act to induce organogenesis is still a mystery. In a developmental perspective, it has been suggested that plant tissues are composed of cell populations with different states of developmental competence.[86] Although this implies that cells belonging to different populations have different fates, the major issue remains as to the molecular characterization of the different developmental states of the cell and the determination of organogenic 'markers'. Additionally, what makes a cell competent for dedifferentiation, proliferation and regeneration?

Protocols aimed to avoid long tissue culture- and hormone-dependent regeneration processes have been developed which are based on the natural capability of plants for spontaneous regeneration. These protocols, which are characterized by the requirement of a limited number of plant manipulations, proved to be successful for the stable transformation of plants acting as important model systems in fundamental research (ie. *Arabidopsis thaliana*,[87] and for the transformation of crops such as tomato.[88] These protocols should be applicable for the genetic engineering of recalcitrant plant species such as bell pepper where transformation,[89–91]) has been limited because of the difficulties of developing an efficient and universal plant regeneration system. The regeneration of bell pepper has been performed using empirically determined combinations of growth regulators.[92–6] However, protocols for spontaneous plant regeneration have been applied to different cultivars of bell pepper which proved to be efficient.[97–9] Some of these protocols, combined with *Agrobacterium tumefaciens* mediated gene transfer and selection, have been shown to be effective in regenerating stable transformed plants of tomato and they are also promising tools to transform bell pepper.

2.6.3 More 'friendly' selectable markers: the positive selection method

In some instances there are disadvantages in using antibiotic or herbicide resistant genes in a selection system, such as toxicity or allergenicity of the gene product and interference with antibiotic treatment.[72, 100] Other problems are linked to the capacity for cross-fertilization of some domestic crop species with wild varieties. Oat, for instance, is cross-fertile with wild oat species and transference of phosphinothricin resistance from transgenic oat to weedy wild oats has been reported.[30] The concerns are that phosphinothricin-resistant wild oat would eliminate control of wild oats using phosphinothricin and compromise the usefulness of transgenic crops resistant to this herbicide such as wheat.[34] Therefore, the use and release of selectable genes into the environment has been the cause of concern among environmental authorities. While many of such concerns may prove unfounded[101] they may nevertheless lead to governmental restrictions on the use of selectable genes in transgenic plants, and it is therefore desirable to develop new selection methods.

In contrast to the traditional selection where the transgenic cells acquire the ability to survive on selective media while the non-transgenic cells are killed (negative selection), the positive selection method, first developed by Joersbo and Okkels,[102] favors regeneration and growth of the transgenic cells while the non-transgenic cells are starved but not killed. The positive selection method exploits the fact that cytokinin must be added to plant explants in order to obtain optimal shoot regeneration rates. By adding cytokinin as an inactive glucuronide derivate, cells which have acquired the *GUS* gene by transformation are able to convert the cytokinin glucuronide to active cytokinin while untransformed cells are arrested in development. In this system, *GUS* serves the dual purpose of being both a selectable and screenable marker gene. Another interesting system of positive selection uses the xylose isomerase gene from *Thermo-anaerobacterium thermosulforogenas* as a selectable gene, which expression allows effective selection of transgenic plan cells using D-xylose as the selection agent.[103] The transformation frequencies obtained by positive selection appear to be higher than using the negative selection method. This could be related to the fact that during negative selection the majority of the cells in the explants die. Such dying cells may release toxic substances which in turn may impair regeneration of the transformed cells. In addition, dying cells may form a barrier between the medium and the transgenic cells preventing uptake of essential nutrients.

2.6.4 Use of more appropriate promoters

Silencing phenomena may result from the introduction of transgenes expressed under the control of strong promoters. The most commonly used promoter has been the constitutive *35S-CaMV* promoter which has been used to engineer herbicide- and pathogen-resistant plants. In many instances however, the efficient manipulation of other agronomically or commercially interesting traits would require the expression of the transgene in a predictable and suitable manner which, in turn, would avoid undesired genetic damage and unpredictable transgene expression. In this context, inducible promoters provide an ideal tool to express heterologous genes. However, use of these promoters is limited because the naturally occurring levels of signal molecules may vary according to the environmental and developmental factors. Furthermore, these signals generally alter the expression of many endogenous genes. To circumvent these problems, the production of synthetic promoters responding to chemical inducers would be of great value.[104]

2.7 Sources of further information and advice

Development of plant transformation systems and their potential application are topics comprehensively addressed in excellent reviews[23, 81, 105–6] to which the reader is referred for background information. For further details about molecular aspects on T-DNA transfer, readers are referred to several excellent reviews.[107–8] For those interested in *Agrobacterium*-based vectors available for DNA transfer to plant cells, numerous useful methodologies have been reported.[109–10]

2.8 References

1 DE BLOCK M, HERRERA-ESTRELLA L, VAN MONTAGU M, SCHELL J and ZAMBRYSKI P, 'Expression of foreign genes in regenerated plants and their progeny'. 1984 *EMBO J* **3** 1681–1689.

2 NELSON R S, McCORMICK S M, DELANNAY X, DUBE P, LAYTON J, ANDERSON E J, KANIEWSKA M, PROKSCH R K, HORSCH R B, ROGERS S G, FRALEY R T and BEACHY R N, 'Virus tolerance, plant growth, and field performance of transgenic tomato plants expressing coat protein from tobacco mosaic virus' 1988 *Bio/Technology* **6** 403–409.
3 VAECK M, REYNAERTS A, HOFTE H, JANSENS, DE BEUCKELLER M, DEAN C, ZABEAU M, VAN MONTAGU M and LEEMANS J, 'Transgenic plants protected from insect attack' 1987 *Nature* **328** 33–37.
4 THEOLOGIS A, 'Control of ripening' 1994 *Curr Opin Biotechnol* **5** 152–157.
5 SHAH D M, HORSCH R B, KLEE H J, KISHORE G M, WINTER J A, TUMER N E, HIRONAKA C M, SANDERS P R, GLASSER C S, AYKENNT S, SIEGEL N R, ROGERS S G and FRALEY R T, 'Engineering herbicide tolerance in transgenic plants' 1986 *Science* **233** 478–481.
6 STASKAWICZ B J, AUSUBEL F M, BAKER B J, ELLIS J G and JONES J D G, 'Molecular genetics of plant disease resistance' 1995 *Science* **268** 661–667
7 MOFFAT A S, 'Plants as chemical factories' 1995 *Science* **268** 659–661.
8 HARTMAN C L, LEE L, DAY P R and TUMER N E, 'Herbicide resistant turfgrass (Agrostis palustris Huds.) by biolistic transformation' 1994 *Bio/Tech* **12** 919–923.
9 DUAN X, LI X, XUE Q, ABO-EL-SAAD M, XU D and WU R, 'Transgenic rice plants harboring an introduced potato proteinases inhibitor II gene are insect resistant' 1996 *Nat Biotech* **14** 494–498.
10 HAQ T A, MASON H S, CLEMENTS J D and ARNTZEN C J, 'Oral immunization with recombinant bacterial antigen produced in transgenic plants' 1995 *Science* **268** 714–716.
11 NAWRATH C, POIRIER Y and SOMERVILLE C, 'Plant polymers for biodegradable plastics: cellulose, starch and polyhydroxyalkanoates' 1995 *Mol Breed* **1** 105–122.
12 MASON H S, LAM D M-K and ARNTZEN C J, 'Expression of hepatitis B surface antigen in transgenic plants' 1992 *Proc Natl Acad Sci USA* **89** 11745–11749.
13 MODELSKA A, DIETZSCHOLD B, SLEYSH N, FU Z F, STEPLEWSKI K, HOOPER D C, KOPROWSKI H and YUSIBOV V, 'Immunization against rabies with plant-derived antigen' 1998 *Proc Natl Acad Sci USA* **95** 2481–2485.
14 ARAKAWA T, CHONG D K X and LANGRIDGE W H R, 'Efficacy of a food plant-based oral cholera toxin B subunit vaccine' 1998 *Nat Biotech* **16** 292–297.
15 SAIKI R K, SCHARF S J, FALOONA F A, MULLIS K B, HORN G T, ERLICH H A and ARNHEIM N, 'Enzymatic amplification of β-globin genomic sequences and restriction site analysis for diagnosis of sickle cell anemia' 1985 *Science* **230** 1350–1354.
16 GOBLET C, PROST E, BOCKHOLD K J and WHALEN R, 'One-tube versus two-step amplification of RNA transcripts using polymerase chain reaction' 1992 *Methods Enzymol* **216** 160–168.
17 RIGGS C D and BATES G W, 'Stable transformation of tobacco by

electroporation: evidence for plasmid concatenation' 1986 *Proc Natl Acad Sci USA* **83** 5602–5606.

18 CHRISTOU P, MURPHY J E and SWAIN W F, 'Stable transformation of soybean by electroporation and root formation from transformed callus' 1987 *Proc Natl Acad Sci USA* **84** 3962–3966.

19 SHIMAMOTO K, TERADA R, IZAWA T and FUJIMOTO H, 'Fertile transgenic rice plants regenerated from transformed protoplasts' 1989 *Science* **338** 274–276.

20 HAYASHIMOTO A, LI Z and MURAI N, 'A PEG-mediated protoplast transformation system for production of fertile transgenic rice plants' 1990 *Plant Physiol* **93** 857–863.

21 TORRES M, SIEMENS J, MEIXNER M and SACRISTÁN M D, 'An improved method for direct gene transfer and subsequent regeneration of *Arabidopsis thaliana* Landsberg erecta and two marker lines' 1997 *Plant, Cell, Tissue and Organ Culture* **47** 111–118.

22 CABOCHE M, 'Liposome-mediated transfer of nucleic acids into plant cells' 1990 *Physiol Plant* **79** 173–176.

23 CHRISTOU P, 'Genetic transformation of crop plants using microprojectile bombardment' 1992 *Plant J* **2** 275–281.

24 SCHNORF M, NEUHAUS-URI G, GALLI A, IIDA S and POTRYKUS I, 'An improved approach for transformation of plant cells by microinjection: molecular and genetic analysis' 1991 *Transgenic Res* **1** 23–30.

25 CHRISTOU P, FORD T and KOFRON M, 'Production of transgenic rice (Oriza sative L.) plants from agronomically important indica and japonica varieties via electric discharge particle acceleration of exogenous DNA into immature zygotic embryos' 1991 *Bio/Tech* **9** 957–962.

26 D'HALLUIN K, BONNE E, BOSSUT M, DE BEUCKELEER M and LEEMANS J, 'Transgenic maize plants by tissue electroporation' 1992 *Plant Cell* **4** 1495–1505.

27 CHOWRIRA G M, AKELLA V and LURQUIN P F, 'Electroporation-mediated gene transfer into intact nodal meristems *in Planta*' 1995 *Molecular Biotechnology* **3** 17–23.

28 KLEIN T M, WOLF E D, WU R and SANFORD J C, 'High-velocity microprojectiles for delivering nucleic acids into living cells' 1987 *Nature* **327** 70–73.

29 BARCELO P, HAGEL C, BECKER D, MARTIN A and LÖRZ H, 'Transgenic cereal (tritordeum) plants obtained at high efficiency by microprojectile bombardment of inflorescence tissue 1994 *Plant J* **5** 583–592.

30 SOMERS D A, RINES H W, WEINING G, KAEPPLER H F and BUSHNELL W R, 'Fertile transgenic oat plants' 1992 *Bio/Tech* **10** 1589–1594.

31 BOWER R and BIRCH R G, 'Transgenic sugarcane plants via microprojectile bombardment' 1992 *Plant J* **2** 409–416.

32 FROMM M E, MORRISH F, ARMSTRONG C, WILLIAMS R, THOMAS J and KLEIN T M, 'Inheritance and expression of chimeric genes in the progeny of transgenic maize plants' 1990 *Bio/Tech* **8** 833–833.

33 GORDON-KAMM W J, SPENCER T M, MANGANO M L, ADAMS T R, DAINES R J, START W G, O'BRIEN J V, CHAMBERS S A, ADAMS W R, WILLETTS N G, RICE T B, MACKEY C J, KRUEGER R W, KAUSCH A P and LEMAUX P G, 'Transformation of maize cells and regeneration of fertile transgenic plants' 1990 *Plant Cell* **2** 603–618.

34 VASIL V, CASTILLO M A, FROMM M E and VASIL I K, 'Herbicide resistant fertile transgenic wheat plants obtained by microprojectile bombardment of regenerable embryogenic callus' 1992 *Bio/Technology* 10 667–674.

35 WEEKS J T, ANDERSON O D and BLECHL A E 'Rapid production of multiple independent lines of fertile transgenic wheat (Triticum aestivum)' 1993 *Plant Physiol* **102** 1077–1084.

36 WAN Y and LEMAUX P G 'Generation of large numbers of independently transformed fertile barley plants' 1994 *Plant Physiol* **104** 37–48.

37 UMBECK P, JOHNSON G, BARTON K A and SWAIN W F, 'Genetically transformed cotton (*Gossypium hirsutum* L.) plants' 1987 *Bio/Tech* **5** 263–266.

38 SÁGI L, PANIS B, REMY S, SCHOOFS H, DE SMET K, SWENNEN R and CAMMUE B P A, 'Genetic transformation of banana and plantain (*Mus spp.*) via particle bombardment' 1995 *Bio/Technolgy* **13** 481–485.

39 CHRISTOU P, McCABE D E, MARTINELLI B J and SWAIN W F, 'Soybean genetic engineering – commercial production of transgenic plants' 1990 *Trends Biotech* **8** 145–151.

40 FROMM M, CALLIS J, TAYLOR L P and WALBOT V, 'Electroporation of DNA and RNA into plant protoplasts' 1987 *Methods Enzymol* **153** 351–366.

41 FROMM M E, TAYLOR L P and WALBOT V, 'Stable transformation of maize after electroporation' 1986 *Nature* **319** 791–793.

42 FRALEY R T, ROGERS S G and HORSCH R B, *Crit Rev Plant Sci* 1986 **4** 1–45.

43 BEVAN M, FLAVELL R B and CHILTON M D, 'A chimeric antibiotic resistance gene as selectable marker for plant cell transformation' 1983 *Nature* 1983 **394** 184–187.

44 GUERCHE P, BELLINI C, LEMOULLEC J M and CABACHE M, 1987 *Biochimie* **69** 621–628.

45 HILLE J, VERHEGGEN F, ROELVINK P, FRANSSEN H, KAMMEN A V and ZABEL P 1986 *Plant Mol Biol* **7** 171–176.

46 STALKER D M, McBRIDE K E and MALVJ L D, 'Herbicide resistance in transgenic plants expressing a bacterial detoxification gene' 1988 *Science* **242** 419–423.

47 FRALEY R T, ROGERS S G, HORSCH R B, SANDERS P R, FLICK J S, ADAMS S P, BITTNER M L, BRAND L A, FINK C L, FRY J S, GALLUPPI G R, GOLDBERG S B, HOFFMANN N L and WOO S C, 'Expression of bacterial genes in plant cells' 1983 *Proc Natl Acad Sci* **80** 4803–4807.

48 STREBER W R and WILLMITZER L 1989 *Bio/Tech* **7** 811–816.

49 SHAH D M, ROMMENS C M T and BEACHY R N, 'Resistance to diseases and insects in transgenic plants: progress and applications to agriculture' 1995 *Trends Biotech* **13** 362–368.

50 WALDRON C, MURPHY E B, ROBERTS J L, GUSTAFSON G D, ARMOUR S L and MALCOLM S K 1985 *Plant Mol Biol* **5** 103–108.

51 DE BLOCK M, BOTTERMAN J, VANDEWIELE M, DOCKY J, TOEN C, GOSSELE V, MOVVA N R, THOMPSON C, VAN MONTAGU M and LEEMANS J, 'Engineering herbicide resistance in plants by expression of a detoxifying enzyme' 1987 *EMBO J* **6** 2513–2518.

52 COMAI L, FACCIOTTI D, NIATT W R, THOMPSON G, ROSE R E and STALKER D M, 'Expression in plants of a mutant aroA gene from Salmonella typhimurium confers tolerance to glyphosate' 1985 *Nature* **317** 741–744.

53 CHEUNG A Y, BOGORAD L, VAN MONTAGU M and SCHELL J, 'Relocating a gene for herbicide tolerance: a chloroplast gene is converted into a nuclear gene' 1988 *Proc Natl Acad Sci USA* **85** 391–395.

54 MAZUR B J, CHUI C-F and SMITH J K, 'Isolation and characterization of plant genes coding for acetolactate synthase, the target enzyme for two classes of herbicides' 1987 *Plant Physiol* **85** 1110–1117.

55 WOHLLEBEN W, ARNOLD W, BROER I, HILLEMANN D, TRAUCH E and PHULER A, 'Nucleotide sequence of the phophinothricin N-acetyltransferase gene from *Streptomyces viridochromogens* Tu 494 and its expression in *Nicotiana tabacum*' 1988 *Gene* **70** 25–37.

56 LYON B R, LLEWELLYN D J, HUPPATZ J L, DENNIS E S and PEACOCK W J, 'Expression of a bacterial gene in transgenic tobacco plants confers resistance to the herbicide 2,4-dichlorophenoxyacetic acid' 1989 *Plant Mol Biol* **13** 533–540.

57 JEFFERSON R A, BURGESS S M and DAVID H, 'β-glucuronidase from *Escherichia coli* as a gene-fusion marker' 1986 *Proc Natl Acad Sci USA* **83** 8447–8451.

58 VANCANNEYT G, SCHMIDT R, O'CONNOR-SANCHEZ A, WILLMITZER L and ROCHA-SOSA M, 'Construction of an intron-containing marker gene: splicing of the intron in transgenic plants and its use in monitoring early events in Agrobacterium-mediated plant transformation' 1990 *Mol Gen Genet* **220** 245–250.

59 OW D 1986 *Science* **234** 856.

60 DE WET J, WOOD K V, HELINSKI D R and DELUCA M, 'Cloning of firefly luciferas cDNA and the expression of active luciferase in *Escherichia coli*' 1985 *Proc Natl Acad Sci USA* **82** 7870–7873.

61 KAETHER C and GERDES H H, 'Visualization of protein transport along the secretory pathway using green fluorescence protein' 1995 *FEBS Lett* **369** 267–271.

62 ROUWENDAL GJ, MENDES O, WOLBERT E J and DOUWE DE B, 'Enhanced expression in tobacco of the gene encoding green fluorescent protein by modification of its codon usage' 1997 *Plant Mol Biol* **33** 989–999.

63 HEIM R, CUBITT A B and TSIEN R Y, 'Improved green fluorescence' 1995 *Nature* **373** 663–664.

64 NIWA Y, HIRANO T, YOSHIMOTO K, SHIMIZU M and KOBAYASHI H, 'Non-invasive quantitative detectin and applications of non-toxic, S65T-type

green fluorescent protein in living plants' 1999 *Plant J* **18** 455–463.

65 HORSCH R B, FRY J E, HOFFMAN N L, EICHHOLTZ D, ROGERS S G and FRALEY R T, 'A simple and general method for transferring genes into plants' 1985 *Science* **227** 1129–1231.

66 LARKIN P J and SCOWCROFT W R, 'Somaclonal variation – a novel source of variability from cell cultures for plant improvement' 1981 *Theor Appl Genet* **60** 197–214.

67 CHRISTOU P, 'Strategies for variety-independent genetic transformation of important cereals, legumes and woody species utilizing particle bombardment' 1995 *Euphytica* **85** 13–27.

68 FINNEGAN J and McELROY D, 'Transgene inactivation: plants fight back!' 1994 *Bio/Tech* **12** 883–888.

69 MATZKE A J M, NEUHUBER F, PARK Y, AMBROS P F and MATZKE M A, 'Homology-dependent gene silencing in transgenic plants: epistatic silencing loci contain multiple copies of methylated transgenes' 1994 *Mol Gen Genet* **244** 219–229.

70 FLAVELL R B, 'Inactivation of gene expression in plants as a consequence of specific sequence duplication' 1994 *Proc Natl Acad Sci USA* **91** 3490–3496.

71 ASSAAD F F, TUCKER K L and SIGNER E R, 'Epigenetic repeat-induced gene silencing (RIGS) in Arabidopsis' 1993 *Plant Mol Biol* **22** 1067–1085.

72 FUCHS R L, REAM J E, GAMMOND B G, NAYLOR M W, LEIMGRUBER R M and BERBERICH S A, 'Safety assessment of the neomycing phosphotransferase II (NPTII) protein' 1993 *Bio/Tech* **11** 1543–1547.

73 DALE E C and OW D W, 'Gene transfer with subsequent removal of the selection gene from the host genome' 1991 *Proc Natl Acad Sci USA* **88** 10558–10562.

74 FELDMANN K A, 'T-DNA insertion mutagenesis in *Arabidopsis*: mutational spectrum' 1991 *Plant J* **1** 71–82.

75 PASZKOWSKI J, BAUR M, BOGUCKI A and POTRYKUS I, 'Gene targeting in plants' 1988 *EMBO J* **7** 4021–4026.

76 KEMPIN S A, LILJEGREN S J, BLOCK L M, ROUNSLEY S D and YANOFSKY M F, 'Targeted disruption in *Arabidopsis*' 1997 *Nature* **389** 802–803.

77 SCHAEFER D G and ZRŸD J P, 'Efficient gene targeting in the moss *Physcomitrella patens*' 1997 *Plant J* **11** 1195–1206.

78 HASTY P, RIVERA-PÉREZ J and BRADLEY A, 'The length of homology required for gene targeting in embryonic stem cells' 1991 *Mol Cell Biol* **11** 5586–5591.

79 NEGRITTO M T, WU X, KUO T, CHU S and BAILIS A M, 'Influence of DNA sequence identity on efficiency of targeted gene replacement' 1997 *Mol Cell Biol* **17** 278–286.

80 BRONSON S K and SMITHIES O, 'Altering mice by homologous recombination using embryonic stem cells' 1994 *J Biol Chem* **269** 27155–27158.

81 VERGUNST A C and HOOYKAAS P J J 'Recombination in the plant genome and its application in biotechnology' 1999 *Crit Rev Plant Sci* **18** 1–31.

82 CERUTTI H, OSMAN M, GRANDONI P and JAGENDORF A T, 'A homolog of *Escherichia coli* RecA protein in plastids of higher plants' 1992 *Proc Natl Acad Sci USA* **89** 8068–8072.

83 SATO S, HOTTA Y and TABATA S 'Structural analysis of a recA-like gene in the genome of *Arabidopsis thaliana*' 1995 *DNA Res* **2** 89–93.

84 DOUTRIAUX M, COUTEAU F, BERGOUNIOUX C and WHITE C, 'Isolation and characterization of the Rad51 and Dmc1 homologs from *Arabidopsis thaliana*' 1998 *Mol Gen Genet* **257** 283–291.

85 KOBAYASHI T, KOBAYASHI E, SATO S, HOTTA Y, MIYAJIMA N, TANAKA A and TABATA S 'Characterization of cDNAs induced in meiotic prophase in lily microsporocytes' 1994 *DNA Res* **1** 15–26.

86 HICKS G S, 'Shoot induction and organogenesis in vitro: a developmental perspective' 1994 *In Vitro Cell Dev Biol* **30P** 10–15.

87 CHANG S S, PARK S K, KIM B C, KANG B J, KIM D U and NAM H G, 'Stable genetic transformation of *Arabidopsis thaliana* by *Agrobacterium* inoculation *in planta*' 1994 *Plant J* **5** 551–558.

88 CHYI Y-S and PHILLIPS G C, 'High efficiency Agrobacterium-mediated transformation of Lycopersicon based on conditions favorable for regeneration' 1987 *Plant Cell Rep* **6** 105–108.

89 ZHU Y-X, OU-YANG W-J, ZHANG Y-F and CHEN Z-L, 'Transgenic sweet pepper plants from *Agrobacterium* mediated transformation' 1996 *Plant Cell Rep* **16** 71–75.

90 KIM S J, LEE S J, KIM B D and PAEK KH, 'Satellite-RNA-mediated resistance to cucumber mosaic virus in transgenic plants of hot pepper (*Capsicum annumm* cv. Golden Tower)' 1997 *Plant Cell Rep* **16** 825–830.

91 MANOHARAN M, VIDYA C S S and SITA G L, '*Agrobacterium*-mediated genetic transformation in hot chili (*Capsicum annuum* L. var. *Pusa jwala*)' 1988 *Plant Sci* 131 77–83.

92 DÍAZ I, MORENO R and POWER J B, 'Plant regeneration from protoplasts of *Capsicum annuum*' 1998 *Plant Cell Rep* **7** 210–212.

93 BINZEL M L, SANKHLA N, JOSHI S and SANKHLA D, 'Induction of direct somatic embryogenesis and plant regeneration in pepper (*Capsicum annuum* L.)' 1996 *Plant Cell Rep* **15** 536–540.

94 ARROYO R and REVILLA A, '*In vitro* plant regeneration from cotyledon and hypocotyl segments in two bell pepper cultivars' 1991 *Plant Cell Rep* **10** 414–416.

95 EBIDA A I A and HU C-Y, '*In vitro* morphogenesis responses and plant regeneration from pepper (*Capsicum annuum* L. cv. Early California Wonder) seedling explants' 1993 *Plant Cell Rep* **13** 107–110.

96 SZÁSZ A, NERVO G and FÁRI M, 'Screening for *in vitro* shoot-forming capacity of seedling explants in bell pepper (*Capsicum annuum* L.) genotypes and efficient plant regeneration using thidiazuron' 1995 *Plant Cell Rep* **14** 666–669.

97 EZURA H, NISHIMIYA S and KASUMI M, 'Efficient regeneration of plants independent of exogenous growth regulators in bell pepper (*Capsicum*

annuum L.)' 1993 *Plant Cell Rep* **12** 676–680.
98 RAMÍREZ-MALAGÓN R and OCHOA-ALEJO N, 'An improved and reliable chili pepper (*Capsicum annuum* L.) plant regeneration method' 1996 *Plant Cell Rep* **16** 226–231.
99 POZUETA-ROMERO K, HOULNÉ G, CAÑAS L, SCHANTZ R and CHAMARRO J, 'Procedimiento simple y eficaz para la transformación de tomate sin utilizar hormonas exógenas' 1999 XIII reunión de la Sociedad Española de fisiología vegetal.
100 NAP J-P, BIJVOET J and STIEKEMA W J, 'Biosafety of kanamycin-resistant transgenic plants' 1992 *Transgenic Res* **1** 239–249.
101 FLAVALL R B, DART E, FUENS R L and FRALEY R T, 'Selectable marker genes: safe for plants?' 1992 *Bio/Tech* **10** 141–144.
102 JOERSBO M and OKKELS F T, 'A novel principle for selection of transgenic plant cells: positive selection' 1996 *Plant Cell Rep* **16** 219–221.
103 HALDRUP A, PETERSEN S G and OKKELS F T, 'The xylose isomerase gene from *Thermoanaerobacterium thermosulforogenes* allows effective selection of transgenic plant cells using D-xylose as the selection agent' 1998 *Plant Mol Biol* **37** 287–296.
104 WARD E R, RYALS J A and MIFLIN B J, 'Chemical regulation of transgene expression in plants' 1993 *Plant Mol Biol* **22** 361–366.
105 ZUPAN J R and ZAMBRYSKI P, 'Transfer of T-DNA from *Agrobacterium* to the plant cell'. *Plant Physiol* 1995 **107** 1041–1047.
106 BIRCH R G, 'Plant transformation: problems and strategies for practical application' 1997 *Annu Rev Plant Physiol Plant Mol Biol* **48** 297–326.
107 ZAMBRYSKI P C, 'Chronicles from the Agrobacterium-plant cell DNA transfer story' 1992 *Annu Rev Plant Physiol Plant Mol Biol* **43** 465–490.
108 HOOYKAAS P J J and BEIJERSBERGEN A, 'The virulence system of *Agrobacterium tumefaciens*' 1994 *Ann Rev Phytopathol* **32** 157–179.
109 HOEKEMA A, HIRSCH P R, HOOYKAAS P J J and SCHILPEROORT R A, 'A binary vector strategy based on separation of vir- and T-region of the *Agrobacterium tumefaciens* Ti-plasmid' 1983 *Nature* **303** 179–180.
110 LICHENSTEIN C P and FULLER S L, 'Vectors for the genetic engineering of plants' 1987 *Genetic Engineering* **6** 103–183 New York: Academic.

Part I

Targets for transformation

3

Genetic modification of agronomic traits in fruit crops

L. Baldoni and E. Rugini, IR Miglioramento Genetico Piante Foraggere CNR, Perugia

3.1 Introduction

The genetic improvement of fruit crops has a range of objectives, including:

- selecting cultivars or rootstocks which tolerate biotic and abiotic stresses, allowing reduced pesticide use and controlling damage from, for example, plant diseases and pests, frost or drought
- reducing the size and altering the shape (apical dominance) of the plant in order to increase orchard plant density, lower harvesting and pruning costs, shorten the unproductive period and improve radiation of the canopy
- selecting self-fertile genotypes, both to eliminate pollinators in the orchards which, in some cases, do not produce marketable fruit, and to maintain a more consistent yield over time
- achieving simultaneous fruit ripening for mechanical harvesting, supplying cultivars with a different ripening season
- selecting genotypes with higher nutritional value of the fruit (sugar, oil, vitamins, functional components such as flavonoids)
- improving the organoleptic qualities and shelf-life of fruits.

In meeting these and other objectives, conventional genetic improvement of most species of fruit crops faces a range of obstacles. These include the long juvenility period of some species, seedlessness, frequent inter- and intra-species incompatibility, high heterozygosity, sterility and the presence of specific traits only in wild species. These characteristics make conventional breeding techniques difficult, expensive and time consuming (Mehlenbacher 1995). Common techniques used to reduce juvenility, for example, such as grafting scions on adult plants, are not always effective in all species. This explains why some fruit crops have been improved almost exclusively with clonal selection, using variability from spontaneous mutations or selecting plants derived from natural hybrydisation. Recent developments, such as induced mutations by ionising irradiation, have given few promising results both for cultivars and rootstocks, as in the example of olive, almond and cherry. However, few of these mutations have been commercialised, partly because stable mutations and significant improvement are rare.

Recent molecular and biotechnological approaches such as somaclonal variation or gene transformation, which are the main subject of this chapter, offer an attractive alternative to conventional genetic improvement, since they make possible a greater range of improvements to commercial varieties in a relatively short period of time with minimum or no change to other characteristics. Protoplast technology in fruit crops, for example, provides the potential for making significant changes to varieties, since it can be used for:

1. somatic hybridisation: fusion of cells belonging to different species or genera not sexually compatible, both in making symmetric and asymmetric hybrids (cybrids) to create stable new variations;
2. transferring alien genes by the technique of recombinant DNA:
   (a) co-cultivation protoplasts with *Agrobacterium*
   (b) direct DNA uptake with fusogen agents or by electroporation
   (c) fusion of bacterial spheroplasts with protoplasts, and
   (d) uptake of liposome carrying DNA into protoplasts;
3. selection by selective agents (toxin, culture filtrate of pathogens, and others).

In particular, cybrids may make a good impact on genetic improvement since some important characteristics are governed by organelle genome. Among fruit crops, cybrids are reported almost exclusively in *Citrus* spp by several authors (Vardi and Galum 1988; Grosser *et al.* 1996; Saito *et al.* 1993). Studies on inheritance of organelle genomes in citrus somatic hybrids have been carried out by Moreira *et al.* (2000). Somatic hybrids have been obtained between species of *Citrus* (Moriguchi *et al.* 1997; Gou and Deng 2001) and from different genera (Motumura *et al.* 1995). Hybrids have been used to improve rootstocks to control tree size (Gmitter *et al.* 1992; Moriguchi *et al.* 1997; Deng *et al.* 1992); improve resistance to diseases (Deng *et al.* 1995); and to improve the scion (Grosser *et al.* 1998) to strengthen resistance to viruses, nematodes and *Phytophthora*, as well as confer cold hardiness, drought and salt resistance (Louzada *et al.* 1992; Guo and Deng 2001). Using direct gene transfer to protoplasts, transgenic plants have already been recovered from *Citrus sinensis* (Kobayashi and Uchimiya 1989; Vardi *et al.* 1990) and strawberry (Nyman and Wallin 1992).

3.2 Somaclonal variation

This technique has been described in detail by Larkin and Scowcroft (1981) and specifically in fruit crops by Hammerschlag (1992). It arises when plant explants are subjected to a tissue culture cycle. The cycle includes establishment of a de-differentiated cell or tissue culture under defined conditions and the subsequent regeneration of plants. Variation at cellular level occurs either in cells before explant excision or during the tissue culture cycle (Skirvin 1978; Jain 2001; Remotti 1998; Rami and Raina 2000). The degree of variation depends on many factors, including:

- the origin of the explant used (organ, age, genotype) (Murashige 1974; D'Amato 1975; Barbier and Dulieu 1980)
- the time that cells or tissues are maintained *in vitro* (Barbier and Dulieu 1980)
- the time and intensity of the mutagenic agents used (Burk and Matzinger 1976)

Reduction of somaclonal variation is achieved by using appropriate culture media and by shortening subculture intervals. Somaclonal variation can be 10,000 times higher than spontaneous mutation rates in whole plants (Larkin and Scowcroft 1981). Many phenotypic variations reported in the regenerated fruit crop plants were extensively reviewed by Hammerschlag (1992). Important changes include growth rate and reproductive apparatus modification (sterility, precocious flowering and flower abnormalities, internodal length), and leaf (variegation, albino, chlorotic, etc.), thornlessness, isoenzymatic activity changes, and increased salt resistance, fruit colour, etc. An increased ploidy level has been reported in kiwi subcultures (Rugini *et al.* 2000b) and in grape (Kuksova *et al.* 1997). Some changes are not hereditable, since they have epigenetic origin. These changes include:

- cytokinin and auxin habituation (Meins and Binns 1977)
- chilling resistance (Dix and Street 1976)
- changing susceptibility to fungal attack (Potter 1980)
- susceptibility to certain pathogens, due maybe to virus elimination during regeneration, which can also alter plant habit.

3.3 Gene transformation

The technique of recombinant DNA is promising in fruit crops because, more than other biotechnological techniques, it seems to be more precise in correcting deficiencies in commercial cultivars or rootstocks without disrupting their otherwise desirable genetic make-up (Schuerman and Dandekar 1993). At present the insertion of foreign (alien) genes into the plant DNA, which could alter the functionality of neighbour genes and the induction of somaclonal variation, cannot yet be fully controlled. These problems can be overcome by producing a high number of plants from many transformation events, selecting the best genotype among a large number of transformants. The procedure used to transfer genes to fruit crops has been described, for example, in Dandekar (1992) and is discussed in more detail in Chapter 2 of this book, and reviews of fruit crop transformation are reported by Singh and Sansavini (1998).

3.4 Genetic stability

When a gene is transferred or induced to change by physical or chemical agents in long-lived perennials such as fruit trees, it is essential that stable patterns of gene expression are maintained for long periods of time and, although fruit trees are normally vegetatively propagated, the T-DNA should also be heritable in the progeny. Several studies have been carried out on genetic stability and inheritability with marker genes (James *et al.* 1996) or with important genes for agronomic performance, such as *rolABC* in transgenic kiwi plants of both cvs staminate GTH and pistillate Hayward (Rugini *et al.* 1997; 2000a). After 12 years the staminate *rolABC* plants still maintain the same morphology and the offspring (transgenic staminate X normal pistillate) was transgenic in 50% of plants. The cherry rootstock Colt, transgenic for *RiT-DNA* which seems able to modify the scion vigour (Rugini and Gutierrez-Pesce 1999), showed the same stability after four years in the field (Rugini, pers. com.). Transgenic apricot for virus coat protein still maintains its tolerance to viruses after several years in the field (Laimer, pers. com.). A lot of work has been done in the USA by Scorza and co-workers on transgenic plants of *Prunus domestica* carrying plum pox virus coat protein (*PPV-CP*), *gus* and *npt*II genes. The expression has been stable in the greenhouse for over five years and the progeny produced from hybridisation of transgenic plants carrying plum pox virus coat protein inherited the transgenes and expressed it (Ravelonandro *et al.* 1997). One should note that in some cases the transgenic plants may require different agronomic management in the field to optimise the performance of the plants, that is, vigorous growth observed in kiwi *rolABC* plants may require less N_2 fertilisation to avoid pathogen attacks (Balestra *et al.* 2001) and maybe less water.

3.5 Plant development and reproduction

Gene modification to produce plants more suitable for high-density orchards can be performed both on cultivars and rootstocks (Table 3.1). Plants with an extensive root system and/or with reduced water consumption or changes in canopy architecture, dwarf and semi-dwarf canopy, with short and numerous shoots, could increase orchard density and improve plant performance. At present, reduction of plant size is achieved by using mainly dwarfing rootstocks and, in a few cases, by using spur varieties, selected by clonal selection or among seedling population. Both dwarfing rootstocks and spur varieties are available for only a few species and graft compatibility often presents a problem. Biotechnology techniques may contribute to the creation of dwarfing rootstocks and dwarf varieties either by somaclonal variation (better if in combination with gamma irradiation treatments) or by genetic engineering to modify hormone activity or light receptors. *In vitro* cultures treated with ionising radiation frequently produce shoots modified in their growth which maintain this characteristic also in the field. More interesting, however, is a transformation approach with some already available genes. Both phytohormone, or phytohormone-like and phytochrome genes seem to be good candidates in modifying plant architecture, particularly in response to different light conditions. Phytohormones are recognised as modulators of growth and differentiation in plants, since their levels can influence the growth rate such as branching, apical dominance, flowering, sex determination, regrowth and rest period. Since their synthesis is linked to light quality, modifying genes for hormone synthesis or for light receptors (phytocromes), some interesting modifications should be expected.

3.5.1 Phytohormones modification

Genes encoding enzymes for phytohormone production (*ipt*, *iaaH*, *iaaM*) or other related genes such as *rol A, B, C, D* have been isolated from strains of *Agrobacterium rhizogenes* and *A. tumefaciens* and *Pseudomonas* (Tepfer 1984; Slightom *et al.* 1986; Spena *et al.* 1987; Cardarelli *et al.* 1987a; Schmulling *et al.* 1988; Capone *et al.* 1989). The *ipt* gene codes are for the isopentyl transferase, the first enzyme in the cytokinin biosynthetic pathway, while both *iaa*/M and *iaa*/H genes are codifying for enzymes (tryptophan-2 monoxigenase and indoleacetamide hydrolase respectively) involved in the pathway of IAA synthesis. Regarding *rol* genes, their functions as transcription products are not completely clear. *RolB* and *rolC* are probably responsible for the beta-glucosidase activity and are able to release active auxines (*rolB*) and active cytochinines (*rolC*) from conjugated glucoside (Estruch *et al.* 1991a, 1991b). In addition a tirosine-phosphatase activity associated to *rolB* has been demonstrated, explaining the strong morphogenic action in root organogenesis (Filippini *et al.* 1996). Transgenic plants with chimeric construct with *gus+rolB*, revealed that the proteins of *rolB* and *rolC* are localised in the plasmatic membrane and in the cytosol respectively.

The stable integration of these *genes* in plants which include fruit crops, under constitutive promoter control, showed altered phenotype in morphology (usually except for *rolB*) and could alter resistance to diseases, positively or negatively, according to the prevalence of expression of auxins or cytokinins respectively (see Section 3.7). Fruit crops, such as pear and trifoliate orange, transgenic for *rolABC,* showed a reduction in size, in internode length and in leaf area (Kaneyoshi and Kobayashi 1999), and in active gibberellin synthesis. The association of more *rol* genes modifies morphology and biotic and abiotic stress resistance. Kiwi fruit expressing *rolABC* (Rugini *et al.* 1997, 2000b), as well as cherry rootstock Colt (Gutierrez-Pesce *et al.* 1998; Rugini and Gutierrez-Pesce 1999), apple (Lambert and Tepfer 1992), papaya (Rugini *et al.* 1994), expressing the T-DNA of *A. rhizogenes*, showed 'hairy root' phenotype and morphological similarity to tobacco transgenic for the oat *phyA*, having in common the internode length reduction, reduced apical dominance, late vegetative period and increased chlorophyll content (Wanger *et al.* 1991; Cherry *et al.* 1991a; Whitelam and Harberd 1994). Furthermore, the *rol* genes

Table 3.1 Genetic modification of fruit crops for plant development and reproduction

Fruit crop	Technique	Alien Gene(s)	System/Plasmid or selective agents	Origin of plant material	Modification in *planta*	Authors
Apple (*Malus X domestica*) M26 rootstock	T	*RiT-DNA*	*A.rh.*	Microcutting	Increased rooting ability and altered morphology	Lambert and Tepfer 1992
Apple (*Malus X domestica*) cv Granny Smith	T	*Ipt*	*A.t.*	Leaf segment	Bushy phenotype	Trifonova *et al.* 1994
Apple (*Malus X domestica*) (M26) rootstock	T	*rolA*	*A.t.*	Leaf	Altered morphology	Holefors *et al.* 1998
Apple (*Malus X domestica*) (M26) rootstock	T	*rolB*	*A.t.*	Leaf	Rooting capacity	Welander *et al.* 1998; Zhu *et al.* 2001
Banana (*Musa* spp AAA group)	S.V	–	–	Meristem	Dwarfism, abnormal leaves colour of pseudostem; ploidy change	Hawang 1986; Hawang and Ko 1987; Reuveni *et al.* 1985; Stover 1987; Stover and Buddenhagen 1986
Banana (*Musa* spp AAB group)	S.V	–	–	Meristem	Flower and leaf abnormalities	Ramcharan *et al.* 1985; Vuylsteke *et al.* 1988
Blackberry (*Rubus laciniatus*)	S.V.		–	Shoot tips	Thornyness, dwarf phenotype	Swartz *et al.* 1983

Blackberry (*Rubus fruticosus*)	S.V.	–	Tissue culture	Buds	Thornless	Hall *et al.* 1986
Citrange troyer (*C. sinensis X Poncirus trifoliata*) and Orange (*C. sinensis*) cv Tarocco	T	*rolABC*	*A.t.*	Internodes	Altered morphology	Gentile *et al.* 1999
Clementine (*Citrus clementine*)	S.V	–	–	Nucellus	Thornlessness	Navarro *et al.* 1985
Colt rootstock (*P. avium X P. pseudocerasus*)	T	*RiT-DNA*	*A.rh.*	Roots	Hairy root phenotype	Gutierrez-Pesce *et al.* 1998; Rugini and Gutierrez-Pesce 1999
Colt rootstock (*P. avium X P. pseudocerasus*)	T	*PhyA*	*A.t.*	Stem	Altered tree habit and light perception	Negri *et al.* 1998; Muleo and Iacona 1998
Grape (*Vitis vinifera*) cv Koshusanjaku	T	*RiT-DNA*	*A.rh.*	Leaf embryogenic calli	Ri phenotype Increase root mass	Nakano *et al.* 1994
Grape (*Vitis vinifera*) cv Parodok Magaracha	S.V.	–	Gamma irradiation		Tetraploids	Kuksova *et al.* 1997
Kiwi fruit (*Actinidia deliciosa*) male (cv GTH) and female (cv Hayward)	T	*rolABC*	*A.t.*	Leaf discs	Altered morphology (Hairy root phenotype)	Rugini *et al.* 1991; Rugini *et al.* 2000b

Table 3.1 Continued

Fruit crop	Technique	Alien Gene(s)	System/Plasmid or selective agents	Origin of plant material	Modification in *planta*	Authors
Kiwi fruit (*Actinidia deliciosa*) female (cv Hayward)	T	*rolB*	*A.t*	Leaf discs	Normal phenotype	Rugini and Mariotti 1992
Kiwi fruit (*A. deliciosa*), cvs: Hayward, Abbot, Matsua and Bruno	T	*RiT-DNA*	*A.rh.*IFO14555, A5, ArM123, A13	Petiole	(Adventitious buds)	Yamakawa and Chen 1996
Kiwi fruit (*Actinidia deliciosa)*	T	*RiT-DNA*	*A.rh.* NIAES *1724*	Hypocotils	Hairy root phenotype	Yazawa *et al.* 1995
Kiwi fruit (*Actinidia deliciosa*	T	*OSH1*	*A.t.*	Leaf	Dwarf	Kusaba *et al.* 1995
Kiwi fruit (*A. kolomikta*)	T	*rolC*	*A.t.*	Leaf	Altered morphology	Firsov and Dolgov 1997
Mexican lime (*C. aurantifolia*)	T	*RiT-DNA*	*A.rh.*	Internode	Altered morphology	Perez and Ochoa 1998
Papaya (*Carica papaya* L.)	T	*Ri-TDNA*	*A.rh.*	Zygotic embryos	Hairy root phenotype	Rugini *et al.* 1994
Papaya (*Carica papaya* L.)	T	*rol* genes	*A.rh.*	Petiole leaf	Hairy root phenotype	Cabrera-Ponce *et al.* 1996
Peach (*Prunus persica*)	T	*Ipt*	*A.t.*	Zygotic embryos	Compact habit	Hammerschlag and Smigocki 1998

Pear (*Pyrus communis*)	S.V	–	–	Protoplasts	Leaf morphology, rootability	Ochatt 1987
Persimmon (*Diospiros kaki*)	T	*RiT-DNA*	*A.rh.*	Stem of micropropagated shoots	Altered morphology	Tao *et al.* 1994
Plum rootstock (MRS2/5)	T	*RiT-DNA*	*A. rh*	Transgenic roots	Altered morphology	Rugini and Gutierrez-Pesce 1999
Red raspberry (*Rubus ideaus*)	T	*Hpt, SAMase*	*A.t.*	Leaf and petiole	Altered morphology	Mathews *et al.* 1995
(*Rubus laciniatus* and *Robus ursinus* loganobaccus)	S.V.	–	–	Meristem callus	Thornlessness	McPheeters and Skirvih 1983; Hall *et al.* 1986
Strawberry (*Fragaria X Ananassa*) cv Calypso	T	*rolC*	*A.t.*	Leaves	Compact habit	Mazzara *et al.* 1998
Strawberry (*Fragaria X ananassa*)	T	*rolABC*	*A.t.*	Leaf stipule	Compact habit	Lolletti 1999
Trifoliate Orange (*P. trifoliata*)	T	*RiT-DNA*	*A.rh.* 1724	Epicotyl	Altered morphology (reduced geotropism)	Kaneyoshi and Kobayashi 1999
Trifoliate Orange (*P. trifoliata*)	T	*rolC*	*A.t.*	Epicotyl	Altered morphology	Kaneyoshi and Kobayashi 1999

can determine alterations in floral morphology, probably due to the polyamine content variation and modifications in the architecture of the root system, reduced pollen and seed production, abundant and partially geotropic root system, increased rooting ability and juvenility reduction (Cardarelli *et al.* 1987; Jouanin *et al.* 1987; Spena *et al.* 1987; Vilaine and Casse-Delbart 1987; Vilaine *et al.* 1987). In addition, when used as rootstock, *rol* transgenes seem to influence scion by reducing growth, indicating that some of the products of those genes (primary product transcripts or translated or secondary products, induced from their expression) can migrate from the transgenic tissues to non-transgenic ones. Regarding *rolD*, this reduces the growth and promotes early blossom in tobacco (Trovato *et al.* 1997; Mauro *et al.* 1996). This gene may be a candidate for fruit tree transformation.

3.5.2 Light perception modification

Plant growth and reproduction can be modified by changing light perception. By over- or down-expressing light receptors, it is possible to modify some characters specifically regulated by phytochromes, such as plant development, circadian rhythms, apical dominance, blossom, growth and fruit ripening, photosynthesis products partitioning, development of photosynthetic systems, transpiration control and hormone synthesis (Vince-Prue and Canham 1983; Tucker 1976; Muleo and Thomas 1993; Muleo and Thomas 1997). Several phytochromes are present in the plants, e.g. in *Arabidopsis*, five phytochrome-like coding regions (A–E) have been identified (Sharrock and Quail 1989; Clark *et al.* 1994), and in tomato there is evidence that more than five are present (Hauser *et al.* 1995). Since phytochrome genes share considerable sequence homology, the isolation of a large number of gene fragments and cDNAs is rather easy (Robson and Smith 1997). Research using reporter genes with region promoters of *phyA* and *phyB* revealed that both promoters are expressed in most tissues except in pollen in which only *phyB* is expressed. In addition both endogenous and transgenic phytochromes are produced and are exposed to many of the same degradative and signalling mechanisms (Robson and Smith 1997). *Phys* from several herbaceous plants have been isolated (Robson and Smith, 1997) and recently also from fruit crops (Muleo, pers. comm.) and the expression of transgenes of both *phyA* and *phyB* affects a number of responses in both monocotyledonous and dicotyledonous species. The major function of *phys* in mature plants is the regulation of the 'shade avoidance syndrome'. The consequence of this phenomenon is that the resources are channelled towards extension growth of stems and petioles to the detriment of storage and reproductive organs. Reduction of shade avoidance syndrome could be a big advantage particularly in a monoculture, including modern orchards in which the plants are placed very close and susceptible to shade each other with high competition for light. *PhyA* seems to be a major candidate for reducing the response to shade by constitutive expression in plants. Transgenic herbaceous plants over-expressing *phyA* show short internode, resulting in decrease of stem elongation, reduction of petiole length, increased chlorophyll content, delayed leaf senescence and decrease of apical dominance (Cherry *et al.* 1991b).

Studies on fruit crops such as cherry rootstock 'Colt' and *Citrus* sp. (Gentile, pers. com.) over-expressing rice *phyA* are under way. *In vitro* growing shoots of the cherry rootstock 'Colt' over-expressing *phy*A of rice have demonstrated a reduction of apical dominance with red and far-red light treatments (Muleo and Iacona 1998). This indicates that the excess of red and far-red light, generated in orchards with high-density planting, could modify the distribution of the photosynthesis assimilates among the vegetative growing organs.

3.5.3 Root system and rooting ability modification

Horticulturally valuable cultivars or rootstocks often show very poor rooting ability. Rooting can be improved by inoculating *A. rhizogenes* by wounding the basal part of *in vitro* microcuttings. These methods induce rooting in recalcitrant species such as almond (Rugini 1984; Strobel and Nachmias 1985, Damiano *et al.* 1995; Archilletti *et al.* 1995), walnut (Caboni *et al.* 1996)) and also in other woody species such as olive, grape, apple (Rugini 1986; Tepfer and Casse-Delbart 1987; Patena *et al.* 1988; Scorza 1991; Owens 1995; Gribaudo and Schubert 1990). According to our results in olive and cherry, few roots became transgenic, it seems that these results support the hypothesis that the partial integration of T-DNA has a possible inductive role on the non-transgenic neighbour cells or perhaps some unknown substances are present in the bacterial secretions (Rugini *et al.* 2000a). Rooting ability, number of roots and mass of roots increases when *rol* genes are overexpressed in plants, such as in kiwi fruit expressing *rolABC* genes (Rugini *et al.* 1991, 1997, 2000b), in apple 'M.26' rootstock and grape both expressing RiT-DNA (Lambert and Tepfer 1992; Nakano *et al.* 1994) or *rolB* (Zhu 2001; Welander *et al.* 1998).

In species which naturally make suckers from roots and are recalcitrant to regenerate shoots *in vitro*, a simple infection with *A. rhizogenes* induces root formation and makes transgenic shoots easily selectable to show different morphology, i.e. rootstock MRS/5 (Rugini and Gutierrez-Pesce 1999).

3.5.4 Juvenility modification

Some fruit crops show a long juvenility period which delays reproductive development, making traditional cross-breeding difficult. Much research has been devoted to accelerating the flowering process and some genes controlling flower initiation in *Arabidopsis* have been identified (Yanotsky 1995; Simpson *et al.* 1999). Two of them, LEAFY (LFY) and APETALAI (API), were successfully used to transform Citrus seedlings (Pena *et al.* 2001), inducing flower initiation in one-year-old plantlets. Constant leafy and SPL3 were transferred in banana (Sagi *et al.* 1998).

3.6 Fruit quality

In the past, traditional breeding paid particular attention to such issues as improving crop yield. However, consumers are paying more and more attention to final product quality and composition. As an example, consumers require fruits with good nutritional properties (vitamins, sugars, proteins, minerals, antioxidants and others) and are increasingly interested in functional ingredients which may help reduce the risk of certain cancer or cardiovascular disorders (e.g. resveratrol, licopene, flavonoids, oils with proper saturated/unsaturated fatty acid ratio, antioxidants, etc.). The increasing demand for freshness together with increasingly complex supply chains increases the need for fruits with a longer shelf life (Table 3.2).

New varieties are needed for organic agriculture, since the demand for organic fruits increases yearly but suitable genotypes, possibly resistant to major pests and diseases, with their toxins may be more dangerous than chemical residues. In addition, fruits should be harvested ripened for optimal quality but at the moment this is not possible as they would degrade rapidly.

3.6.1 Oil composition

At present, two molecular strategies can be used to modify oil composition and content:

1. alteration of the major fatty acid level by suppressing or over-expressing a specific key enzyme in lipid biosynthesis
2. creation of an unusual fatty acid.

By anti-sense suppression or co-suppression of oleate desaturase it is possible to increase oleic acid (C18:1) by more than three-fold (from 24% to 80%) in the oil of transgenic soybean. The same strategy was adopted to increase stearate (C18:0) by up to 30% both in canola and soybean oils. Unusual fatty acids can be produced in a plant by transferring a gene encoding the specific biosynthetic enzyme. An example can be seen in canola which naturally does not produce laurate (C12:0), while a new transgenic genotype does contain laurate. The oil content of some nut crops used for cosmetics, such as almond, could be increased or their composition could be modified by these techniques.

3.6.2 Protein modification

Research in this area has a number of objectives including:

- improving the functionality of a target crop protein
- increasing the essential amino acid content of the crop
- expressing the storage protein gene in parts of the plant other than seeds
- reducing the content of those proteins with specific allergenic properties.

Work has been done, for example, on the transfer of genes encoding proteins rich in desirable amino acids (usually methionine and lisine) from other species.

Table 3.2 Genetic modification of fruit crops for increasing fruit quality

Fruit crop	Technique	Alien Gene(s)	System/Plasmid or selective agents	Origin of plant material	Property in *planta*	Authors
Apple, Pear, Strawberry	T	*thaumatin*	*A.t.*	Leaf pieces	Sweeter (not determined)	Dolgov *et al.* 1999b
Kiwi, strawberry, grape	T	*defh9-iaaM*	*A.t.*	Leaf	Partenocarpy	Mezzetti *et al.* (in preparation)
Kiwi fruit (A. *Chinensis*)	T	*hEGF*	*A.t.*	Leaf	Not determined	Kobayashi *et al.* 1996
Peach (*P. persica*) (cv Redhaven)	T	*EGases-encoding cDNA*	PB	Cells	Not determined	Trainotti *et al.* 1997
Strawberry (*Fragaria X ananassa*)	T	*SAMase*	*A.t.*	Leaf	Not determined	Mathews *et al.* 1995b
Walnut (*J. nigra X J. regia*)	T	*Chs* (*antisense*)	*A.t.*	Somatic embryos	Low chalcone synthase, no quercitin	El-Euch *et al.* 1998

The maize gene encoding the protein zein increased methionine content by over 80% in transgenic soybean seeds. Allergenic protein has been reduced by the use of an antisense gene in rice (Tada *et al.* 1996) and the same strategies could be applied in fruit crops. It has also been possible to produce the β-casein protein in plants, avoiding the gastric and intestinal disorders in some children when fed with bovine milk (Arakawa *et al.* 1998).

Sweeter fruits, but with low calories, can now be produced by expressing the super-sweet protein thaumatin, isolated first in *Thaumatococcus daniellii* Benth. The gene has been isolated and sequenced (Edens *et al.* 1982) and introduced in potato and cucumber (Witty and Harvey, 1990; Szwacka *et al.* 1996) inducing sweet tasting phenotype. Thaumatin-like protein was isolated also from maize (Malehorn *et al.* 1994) which is similar to zeamatin and to α-amylase trypsin inhibitor. Overexpression in insect cells and in plants showed antifungal activity. Similar proteins were noted in cherry (Fils-Lycaon *et al.* 1996) and grapefruit (Tattersall *et al.* 1997) during ripening, demonstrating a probable antifungal role in ripened fruits and, at the same time, conferring a sweet taste.

3.6.3 Carbohydrate modification

Two main approaches have been successfully attempted in herbaceous plants:

1. qualitative or quantitative change of an existing compound (usually sucrose or starch)
2. introduction of a novel high value product or products (usually non-calorific carbohydrates such as fructans, bacterial cyclodestrins).

Increasing starch content in plants is possible by modifying the enzymes responsible for its synthesis such as the ADP glucose pyrophosphorylase (ADPGPP), the starch synthase, and branching enzymes. An increase of 20–30% in starch content was obtained in transgenic potato through regulation of the ADPGPP bacterial enzyme. The resulting fried potatoes had better flavour, reduced calories, improved texture and a less greasy taste (Stark *et al.* 1996). Fructan-encoding gene of onion has been transferred to chicory (Vijn *et al.* 1997) and bacterial cyclodestrins to potato (Oakes *et al.* 1991). The same strategy increased the solids in tomato.

3.6.4 Nutrients, antibodies, secondary metabolites and vaccines

A gene encoding the enzyme phytoene synthase, which condenses two molecules of geranyl geranyl diphosphate to get β-carotene (provitamin A) synthesis, was expressed in rice endosperm (Burkhardt *et al.* 1997). The biosynthetic pathway for regulating Vitamin C content has been investigated by Wheeler *et al.* (1998). Plants could also produce antibodies (Ma and Hein 1995) and vaccines (Arntzen 1998). Antibodies against bacteria associated with dental caries (Ma *et al.* 1998), antigens for certain forms of diabete (Ma *et al.* 1997), and some new vaccines (Arakawa *et al.* 1998) can be produced in fruits particularly banana.

At present there is active research on flavonoid biosynthesis which is thought to possess health-promoting properties such as antioxidant, vasodilatory actions which may protect against cardiovascular diseases, particularly for one group such as flavonols (e.g., quercitin and kaempferol) and pigmentary flavonoids known as anthocyanins which play an important role in flower and leaf colour in plant. Attention is directed toward calchone isomerase, an enzyme involved in flavonol biosynthesis. Tomato fruits over-expressing constitutively *Petunia* calchone isomerase gene contained high levels of quercitin glycosides and moderate levels of kaempferol (Muir *et al.* 2001). This gene may be used to transform many fruit crops, reinforcing their natural red colour e.g., plum, grape, strawberry, and some *Citrus* sp. (e.g., orange, grapefruit).

3.6.5 Ripening and the control of flavour, texture and shelf-life

By regulating the activity of enzymes involved in fruit ripening, such as cell wall-degrading enzyme polygalacturonase, or ethylene biosynthesis, it is possible to control or delay fruit softening allowing the fruit to stay longer on the plant for greater flavour and texture development, and improving its shelf-life. Both antisense technology and over-expression of metabolising enzymes have been used, for example, in tomato (Smith *et al.* 1988a). Several strategies have been developed to control ethylene production in plant tissues, including:

- antisense 1–amino-cyclopropane-1–carboxylic acid (ACC, precursor of ethylene) synthase (Oeller *et al.* 1991)
- ACC oxidase (Hamilton *et al.* 1990)
- reduction of ACC by over-expressing in plants the alien gene of *Pseudomonas* sp, ACC deaminase, which converts it to α-ketobutyrate (Klee *et al.* 1991)
- lowering the substrate for ACC synthase by over-expressing the S-adenosyl methionine hydrolase (SAMase) gene (Mathews *et al.* 1995a)

Among fruit crops, ripening-controlled strawberry, banana, and pineapple are expected to be on the market in a few years. Work has been done on gene expression during fruit development and ripening for fruit crops such as peach (Trainotti *et al.* 1997), avocado (McGravey *et al.* 1990), grape (Robinson *et al.* 1997; Tattersal *et al.* 1997), sweet orange (Alonso *et al.* 1995) and strawberry (Wilkinson *et al.* 1995). In addition there have been studies of gene transfer in apple by using inhibition of ethylene and polygalacturonase biosynthesis (Table 3.2).

3.6.6 Fruit size

In some self-sterile species such as kiwi or strawberry and also in other seed-fruit species, like table grape or *Citrus* spp. where the seeds are disliked by consumers, the introduction of the parthenocarpy trait may allow control of fruit development even under environmentally prohibitive conditions for pollination and may be used in fruit crops to standardise and increase the fruit size. However, it is well known that in some fruits the quality of the parthenocarpy fruits are inferior to fruit containing seeds. Parthenocarpy trait is often polygenic and therefore more difficult to deal with breeding programmes. Methods for achieving parthenocarpy include spraying of growth regulators, genetic mutation, or altering plant ploidy level. Transgenic parthenocarpic tobacco, eggplant and tomato have been successfully obtained (Rotino *et al.* 1997, 1999). These plants contain the coding region of the *iaaM* gene for the enzyme tryptophanmonoxigenase in their genome which converts tryptophan to indolacetamide, a precursor of IAA, under the control of the placental-ovule-specific *defh9* gene regulator sequence. The expression of chimeric *defh9-iaaM* starts during early flower development producing marketable fruit. These genes mimic the hormonal effects of pollination and embryo development by increasing the content and/or the activity of auxin specifically in the ovule. Among fruit crops, kiwi fruit, grape and strawberry are already transformed and awaiting field evaluation (Mezzetti, pers. com.).

3.7 Biotic stress

There is an increasing demand by consumers for fruits free of pesticide and other residues, but cultivation without their use is only partially possible by using suitable resistant genotypes in a suitable environment. Plants have developed several natural defence strategies to protect themselves against attack of pathogen and pest diseases (Hammond-Kosack and Jones 1996). Concerning pathogen infection, strategies fall mainly into two groups:

1. specific mechanisms responsible for pathogen recognition and control by a specific resistance against a specific pathogen, with a hypersensitive disease resistance response (HR), hampering the diffusion of pathogen to healthy tissues by formation of necrotic lesion;
2. general mechanisms that confer resistance to a broad range of pathogens, occurring either in resistant or susceptible plants, but are able to control the pathogen. The synthesis of antimicrobial metabolites, litic enzymes, pathogenesis-related proteins and other compounds strengthening the cell wall, are involved. The resistance normally depends on the early response of the plant to pathogen attack, which lead to a rapid accumulation of reactive oxygen species (ROS) namely oxidative burst (Lamb and Dixon 1997), with an accumulation of H_2O_2 which functions as a diffusable signal for the induction of cellular protectant genes (Delledonne *et al.* 1999); nitric oxide co-operates in the induction of hypersensitive cell death.

In some cases the plants react to pathogens by accumulating high levels of specific proteins which are toxic or inhibitory against both pathogens and pests (Broekaert *et al.* 1995) such as RIP proteins, effective against insects and fungi; while other proteins seem to be more specific. Overexpressing the genes by genetic engineering or induced mutation (Barbieri *et al.* 1997; Maddaloni *et al.* 1999) in plant cells under toxin or culture filtrate pressure are the two main strategies currently used to produce resistant plants. More research is needed to discover new molecular signals and the efficacy of the promoters of some genes involved in the defence, maybe the reinforcement of the promoters is sufficient to enhance plant resistance.

3.7.1 Virus resistance

Plant viruses reduce both the quantity and quality of crop yields by direct damage to plants, increasing sensitivity to adverse climatic conditions and to the direct pathogens. They cause billions of dollars of losses every year to fruit crops worldwide, second only to the impact of fungal diseases (Waterworth and Hadidi 1998). In several fruit crops virus diseases represent a particular problem, for example in grape with GCMV and GFLV, in *Prunus* spp. with Sharka and in some tropical species, such as papaya, with PRSV (Gonsalves 1998).

At present viral diseases are controlled in a number of ways including: planting virus-free plants, maintaining plant health, controlling plant pathogens which can be virus vectors, and by crossprotection (Alrefai and Korban 1995). However, these techniques provide only limited protection from viral attack. Whilst, in the case of fungi, chemical defences are available, such remedies are either not effective in the case of viruses or can make the impact of the virus even worse. The preventive use of resistant genotypes is thus essential (Khetarpal *et al.* 1998). Two types of fitovirus transgenic resistance are available:

1. pathogen-derived resistance (PDR) (used most at present)
2. resistance induced by sequences of alien DNA.

PDR is conferred to the plants by genes from the virus itself, cloned and transferred to the host genome (Sanford and Johnston 1985). PDR is developed when the viral gene products or virus-related sequences in the plant genome interferes with the virus infection cycle (Table 3.3). The mechanisms which confer PDR are not yet well understood, varying with the nature of the gene used (Carr and Zaitlin 1993; Fitchen and Beachy 1993; Baulcombe 1994; Kaniewski and Lawson 1998; Yie and Tien 1998; Martelli *et al.* 1999; Smyth 1999). Transgenic plants for the virus coat protein gene provide the most common strategy for gene transfer. The other strategies include antisense nucleic acids, satellite sequences, defective interfering molecules and non-structural genes (replicase, protease, movement proteins), antibodies, and interferon-related proteins (Gadani *et al.* 1990; Baulcombe 1994; Grumet 1994; Kaniewski and Lawson 1998; Wilson 1993).

Although a large number of crop plants has been successfully engineered using such strategies, for fruit crops only the coat protein strategy has been applied to confer PDR to potyvirus, nepovirus and closterovirus groups. Studies demonstrate that this strategy is very promising, although in papaya Tennant *et*

Table 3.3 Genetic modification of fruit crops for virus resistance

Fruit crop	Alien gene(s)	System/Plasmid	Origin of plant material	Resistance in *planta*	Authors
Apricot (*P. armeniaca*)	*CP-PPV**	*A.t.* (pBinPPVm)	Cotyledons of immature embryos	Confirmed *vs* PPV	Laimer da Camara Machado *et al.* 1992
'Carrizo' cytranger (*C. sinensis X P. trifoliata)* and Sour Orange (*Citrus aurantium*)	*CP-CTV*	*A.t.*	Stem segment	Not determined	Moore *et al.* 1993
Grape (*Vitis* spp) several rootstocks	*CP-GFLV*	*A.t.*	Embryogenic calli	Not confirmed	Xue *et al.* 1999
Grape (*Vitis berlandieri X V. riparia*) (110 Richter rootstock)	*CP-GCMV*	*A.t.*	Somatic embryos	Not confirmed	Gall *et al.* 1994
Grape (*Vitis vinifera*) L cv Thompson seedless	*TomRSV-CP*	PB followed by *A.t.*	Somatic embryos	Not determined	Scorza *et al.* 1996
Grape (*Vitis vinifera*) L cv Chardonnay and *Vitis* rootstocks 41B, SO4	*CP-GFLV*	*A.t.* LBA4404	Anther derived embryos	Not determined	Mauro *et al.*, 1995
Grape (*Vitis berlandieri X V. riparia*) (110 Richter rootstock)	*CP-GFLV*	*A.t.*	Somatic embryos	Not determined	Ktastanova *et al.* 1995
Grapefruit (*Citrus paradisi* Macf.)	*uncp*	*A.t.*	Epicotyls	Not determined	Yang *et al.* 2000
Mexican lime (*Citrus aurantifolia* Swing) seedlings	*CP-CTV*	*A.t. EHA 105*	Stem segment	Not determined	Dominquez *et al.* 2000

Papaya (*Carica papaya* L.)	*CP-PRSV*	PB	Thin layer embryogenic tissues	Not determined	Cai *et al.* 1999
Papaya (*Carica papaya L.*)	*CP-PRV-4*	PB (pGA482GC/ cpPR-V-4)	Zygotic embryos, hypocotils	Increased resistance for some PRV strains	Fitch *et al.* 1992; Cheng *et* al. 1996
Papaya (*Carica papaya*)	*CP-PRV-4*	*A.t.*	Zygotic embryos, hypocotils	Not determined in *planta*	
Papaya (*Carica papaya*)	*CP-PRSV*	*A.t.*	Zygotic embryos,	Not detérmined	Yeh *et al.* 1998
Papaya (*Carica papaya*) cv Kamiya	*CP-PRSV*	PB	Embryogenic calli	Confirmed	Fitch *et al.* 1998
Papaya (*Carica papaya*) L	*CP-PRSV*		Embryos	Not determined	Cheng *et al.* 1996
Plum (*P. domestica*)	*CP-PPV*	*A.t.* (pGA482gg/ PPV-CP-33)	Hypocotyl slices	Confirmed *vs* PPV	Scorza *et al.* 1994
Plum (*P. domestica*) cv Stanley	*CP-PRV*	*A.t.* (pGA482GG/ cpPRV-4)	Hypocotyl slices	Confirmed *vs* PPV	Scorza 1991; Scorza *et al.* 1995; Ravelonandro *et al.* 1997
Plum (*P. domestica*) L cv Bluefree	*CP-PPV*	*A.t.*	Leaf discs	Confirmed	Machado *et al.* 1994
Strawberry (*Fragaria X Ananassa*)	CP-SMYEL	*A.t.*	Leaves	Not determined	Finstad *et al.* 1995
Troyer cytranger (*C. sinensis X P. trifoliata*)	*CP-CTV*	*A.t.*	—	Not determined	Gutierrz *et al.* 1992

CP = Coat Protein; *CTV* = Citrus tristeza virus (Clostero virus group); *PPV* = Plum pox virus (Potyvirus Group); *GCMV* = Grapevine chrome mosaic virus (Nepovirus Group); *GFLV* = Grapevine fan leaf virus (Nepovirus Group); *PRV or PRSV* = Papaya ring spot virus +(Potyvirus group); *Cp-SMYEL* = Strawberry mild yellow edge luteovirus *al.* (1994) reported that CP-PRV was effective in protecting from some virus isolates but not from others. Recent studies by Singh *et al.* (1997) demonstrated that in tobacco, as a model plant, transgenic plants expressing a defective replicase gene of cucumber mosaic virus (CMV-FNY), acquired resistance to various banana isolates of CMV, suggesting this approach is worth further development. In most cases resistance has been successfully tested *in vivo* or indirectly by testing the accumulation of coat protein by ELISA or Western blot analysis or *gus* gene expression in the transgenic tissues. Examples of the resistance induced by sequences of alien DNA are not yet available but it should be possible to obtain them since in some species, such as *Citrus* spp., resistance to CTV is present in *Poncirus trifoliata* and is known to be controlled by a dominant gene at the Ctr locus.

Developing transgenic fruits for virus resistance may lead to possible risks. These include:

- transcapsidation, when nucleic acids of a virus are covered by the coat protein belonging to another virus expressed by the transgenic plant (Farinelli *et al.* 1992; Greene and Allison 1994; Robinson *et al.* 1999; Buzkan *et al.* 2000). This problem is, however, already frequent in nature, with virus multiple infection (Creamer and Falk 1990; Hobbs and McLaughlin 1990; Bourdin and Lecoq 1991; Buzkan *et al.* 2000)
- recombination of nucleic acid expressed by the transgenic plants with nucleic acids of the virus occurring in transgenic plants, producing new more virulent viruses (Rybicki 1994; Dolja *et al.* 1994; Miller *et al.* 1997; Aziz and Tepfer 1999; Smith *et al.* 2000). This problem is also very common in nature and, together with mutations, is responsible for much viral evolution (Roossinck 1997). According to the studies of Miller *et al.* (1997), Jacquemond and Tepfer (1998), and other scientists, transgenic plants expressing viral sequences do not represent a source of risk greater than those already present in nature;
- genetic depletion caused by abandoning susceptible varieties in favour of transgenic ones. This is a false problem since resistance can be conferred to susceptible varieties by biotechnologies;
- compatible wild species which could become resistant following pollination with transgenic pollen produced by the transgenic crops. This is not usually a problem in areas where fruit crops are cultivated because there are no wild relatives, except for the area of origin of the crop in question.

Several fruit crops have been transformed with virus coat proteins; some of them showed resistance in field conditions, others have not been tested yet (Table 3.3). In Mexican lime expression analysis showed no correlation between coat protein expression and transgene copy number or integration pattern (Dominguez *et al.* 2000). An indirect strategy to fight viruses is to make plants resistant to their vectors. Yang *et al.* (2000) for example, have tried to make plants resistant to aphids, which are the vectors of grapefruit tristeza virus.

3.7.2 Fungal resistance

Among diseases, fungi are the main cause of yield loss in fruit crops. They are controlled by several traditional techniques including quarantine, sanitation, breeding and clonal selection of resistant varieties and application of fungicides. However, resistant cultivars, with the onset of new strains of virulent pathogens, tend to become susceptible over time. In addition, the unrestrained use of fungicides, as well as increasing production costs and degrading the environment, induce new forms of resistance within pathogens, forcing the development of new pesticides. These problems have encouraged the search for biotechnological solutions to combating fungal disease (Table 3.4).

At present research is focused on identifying the genes involved in resistance, both those encoding for enzymes involved in the biosynthesis of toxic compounds for fungi and those encoding toxin proteins which directly inhibit fungal growth (Cornelissen and Melchers 1993; Terras *et al.* 1998) with the aim of introducing them in susceptible plants or substituting their inefficient antifungal gene promoters with more efficient ones. Several proteins have been reported with antifungal activity; they were classified into at least 11 classes named pathogenesis-related proteins (PRs). Some of them also showed antiviral and antibacterial activities.

Some defence-related genes encode enzymes involved in:

1. phenylpropanoid metabolism;
2. hydrolytic enzymes, such as chitinases and β-1,3–glucanases;
3. hydroxyproline-rich glycoproteins (cell wall proteins);
4. inhibitors of fungal enzymes, such as PGIP.

Plant β-1,3–Glucanases (PR-2) and chitinases (PR-3) represent potential antifungal hydrolases which act synergistically to inhibit fungal growth *in vitro* (Mauch *et al.* 1988). In addition, β-1, 3–Glucanases release glycosidic fragments from both the pathogen and the host cell walls which could act as signals in the elicitation of host defences (Keen and Yoshikawa 1983; Hahn *et al.* 1989; Takeuchi *et al.* 1990).

Many of the genes induced by plant disease-resistance responses encode proteins with direct antifungal activity (AFPs) *in vitro* (Lamb *et al.* 1992; Terras *et al.* 1998). Identification of such anti-fungal proteins were isolated from plants and from the fungus itself such as *Tricoderma harsianum* (Neuhaus *et al.* 1991; Mikkelsen *et al.* 1992; Melchers *et al.* 1993) and from humans. They include: Defensins, small cysteine-rich peptides, 2S albumins, chitin-binding proteins, lipid-transfer proteins, hydrogen-peroxide-generating enzymes (Terras *et al.* 1993; Garcia-Olmedo *et al.* 1998), stilbene synthase (Hain *et al.* 1990), ribosome inactivating proteins (Stripe *et al.* 1992; Longemann *et al.* 1992), lysozyme from humans, osmotin (PR-5) and osmotin-like protein (Liu *et al.* 1994; Zhu *et al.* 1996), polygalacturonase-inhibiting protein, thaumatin and several others.

Several herbaceous plants have been engineered with some success by using single genes (*chitinase*, *defensin*, *osmotin*, etc.) or multiple genes (*osmotin* +

Table 3.4 Genetic modification of fruit crops for fungal disease resistance

Fruit crops	Technique	Alien gene(s)	System/Plasmid or selective agents	Origin of plant material	Resistance in *planta*	Authors
Apple (*Malus X domestica*)	T	*amp*	*A.t.*	Leaf	Confirmed	Broothaarts *et al.* 1998
Apple (*Malus X domestica*)	T	endochitinase	*A.t.*	Leaf	Confirmed for apple scab	Norelli *et al.* 2000
Apple (*Malus X domestica*)	S.V.	–	Culture *filtrate* of *P. cactorum*	Shoot regenerants	*Phytophthora cactorum*	Rosati *et al.* 1990
Apple (*Malus X domestica*) Borkh McIntosh	T	*ThEn-42*	*A.t.*	Maternal	*Venturia inequalis* Apple scab	Bolar *et al.* 2000
Apple rootstock N545	T	*RS-AFP-2* (*defensin*)	*A.t.* supervirulent strain CBE 21	Leaf pieces	Not determined	Dolgov *et al.* 1999a
Apple (*Malus X domestica*)	T	*Defensin*	–	–	Not determined	De Bondt *et al.* 1996b
Apple (*Malus X domestica*), cv Melba	T	*Thaumatin II*	*A.t.* CBE21, 35S promoter	Leaf pieces	Not determined	Dolgov *et al.* 1999b
Banana (*Musa* spp) Dwarf Parfitt (AAA-Cavendish)	SV	–	Irradiation	Shoots	*Fusarium* wilt-race 4	Smith *et al.* 1995
Banana (*Musa* AAA group)	S.V.	–	–	Adventitious buds	*Fusarium oxysporum* f. sp. *cubense* race 4	Hwang 1990

Banana (*Musa* spp)	S.V.	–	–		*Fusarium oxysporum* f. sp. *cubense*	Matsumoto *et al.* 1999a b
Banana (*Musa* sp., AAB, Silk) cv. Maçã	S.V.	–	Fusaric acid	Multiple bud clumps	*Fusarium oxysporum* f. sp. *cubense* race 1	Matsumoto *et al.* 1995
Kiwi fruit (*Actinidia deliciosa*) cv Hayward	T	*osmotin* pKYLX71\ -35S	*A.t.*	Maternal	*Botrytis*	Rugini *et al.* 1999
Lemon (*Citrus limon*)	S.V.	–	–		Mal secco disease (*Phoma tracheiphila*)	Nadel and Spiegel-Roy 1987
Lemon (*Citrus limon*) cv Femminello continella	S.V.	–	Toxin of *P. tracheiphila*	Nucellar calli	Mal secco disease (*P. tracheiphila*)	Gentile *et al.* 1992
Lemon (*Citrus limon*) cv Femminello siracusano	S.V.	–	Toxin of *P. tracheiphila*	Protoplasts	*P. tracheiphila*	Deng *et al.* 1995
Mango (*Mangifera indica*)	S.V.	–	–	Embryogenic suspension	*Colletotrichum gloeosporioides*	Liz *et al.* 1991
Mango (*Mangifera indica*)	S.V.	–	Partial purified phytotoxin of C. gloesporides	Seedlings	*Colletotrichum gloeosporioides*	Jayasanker *et al.* 1999
Olive (*Olea europaea*, L.) cv Canino	T	*osmotin* pKYLX71\ -35S	*A.t.*	Maternal somatic embryogenesis	Not determined	Rugini *et al.* 2000a
Pear (*Pirus domestica*)	T	*RS-AFP-2 (defensin)*	*A.t.* supervirulent strain CBE 21	Leaf pieces	Not determined	Dolgov *et al.* 1999a
Pear (*Pyrus communis*) cv Burakovka	T	*Thaumatin II*	*A.t.* CBE21, 35S promoter	Leaf pieces	Not determined	Dolgov *et al.* 1999b

Table 3.4 Continued

Fruit crops	Technique	Alien gene(s)	System (Plasmid or selective agents)	Origin of plant material	Resistance in *planta*	Authors
Strawberry (*Fragaria X Ananassa*)	S.V.	–	–	Shoot regenerants	*Fusarium oxysporum* f. sp *fragariae*	Toyoda *et al.* 1991
Strawberry (*Fragaria* X Ananassa)	S.V.	–	Pectic enzymes of *R. fragariae*	Leaf petioles	*Rhizoctonia fragariae*, *Botrytis cinerea*	Orlando *et al.* 1997
Strawberry (*Fragaria X Ananassa*)	S.V.	–	–	Shoots	*Phytophthora cactorum*	Battistini and Rosati 1991
Strawberry (*Fragaria X Ananassa*)	S.V.	–	–	Receptacles	Calcareos soil and fungal diseases	Carli *et al.* 1993
Strawberry (*Fragaria X Ananassa*)		–	–	–	–	
Strawberry (*Fragaria X ananassa*) cv Feyerverk	T	*Thaumatin II*	*A.t.* CBE21, 35S promoter	Leaf pieces	Not determined	Dolgov *et al.* 1999b
Swingle (*Citrus aurantifolia*)	S.V.	–	–	Zygotic and mature material	Canker	Raman and Dhillon 1990

chitinase + PR1) (Veronese *et al.* 1999a) and work on woody plants such as pear, apple, strawberry, olive is also in progress (Dolgov *et al.* 1999a; Rugini *et al.* 2000a). Correlation between the level of expression of antifungal proteins in the leaves and resistance has been observed in several herbaceous transgenes. In field trials the olive plants expressing the *osmotin* gene of tobacco showed reduction of growth (Rugini *et al.* 2000a; D'Angeli *et al.* 2001) similarly to the apple plants engineered with the endochitinase gene (Bolar *et al.* 2000).

Research aims at isolating pure compounds (toxins) from fungi, i.e., specific pectic enzymes, malseccin, fusicoccin, fusaric acids, and others to be used as selective pressure on plant cell or tissue culture to recover resistant genotypes (Table 3.4), although the resistance acquired by the cells is not always maintained by the derived regenerated plant. However, Orlando *et al.* (1997) demonstrated that pectic enzymes of *Rhizoctonia fragariae* were effective in selecting strawberry plants resistant to some fungi. Unfortunately the fruits appeared pale red in colour and a little sour, since horto-di-hydroxiphenols in the leaves were increased by over 40%. In the absence of pure compounds the crude culture filtrate of the pathogen could be applied (Hammerschlag and Ognianov 1990). *In vitro* mutation by ionising rays combined with toxin-used selection seems a promising strategy for future work for fruit crops also.

3.7.3 Bacterial resistance

Every year bacterial diseases cause loss of yield on both tropical and temperate fruit trees. They have effects varying from death of the entire plant to loss of quality of fruits. Important bacterial diseases of fruit trees are fire blight (apple, pear, quince and other ornamental species of *Rosaceae* caused by *Erwinia amylovora*), bacterial blight and canker of stone fruits (by *Pseudomonas syringae)*, blight of persian walnut (by *Xanthomonas campestris* pv. *Juglandis)* and canker of citruses (by *Xanthomonas citri*). Research on resistance to bacterial diseases has focused on genes producing anti-microbial proteins like lytic peptids (cercopins, attacins and synthetic analogs: shiva-1, SB-37), and lysozymes (egg white, T4 bacteriophage and human lysozyme). Transformation of apple Malling 26 by attacin E (Norelli *et al.* 1994; Borejsza-Wysocka *et al.* 1999), and pear cv. Passe Crassane by attacin E and SB-37 (Reynoird *et al.* 1999a, b; Mourgues *et al.* 1998) for resistance to *E. amylovora*, are examples of this approach (Table 3.5). Recently, relationships between attacin expressed in transgenic apple and disease resistance were detected using immunoblot assays with the fusion attacin polyclonal antibody (Ko *et al.* 1999).

Recent advances in our understanding of *harpin* gene clusters of *P. syringae* and *E. amylovora*, the apoplast conditions for the expression of these genes, their products and secretion systems, and their effects on host plants, have contributed to clarify the interaction between bacteria and host cells. Transformation of apple rootstock Malling 26 by *hrpN* of *E. amylovora* is one of the first examples of enhancement of natural plant defences (Abdul Kader *et al.* 1999). Other strategies against bacteria effective also in fruit crops are represented by the

Table 3.5 Genetic modification of fruit crops for bacteria disease resistance

Fruit crop	Technique	Alien gene(s)	System/Plasmid or selective agents	Origin of plant material	Resistance in *planta*	Authors
Apple (*Malus X domestica*) M26	T	*AttE*	*A.t.*	Leaf segment	*Erwinia amylovora*	Norelli *et al.* 1994
Apple (*Malus X domestica*) cv Galaxy	T	*AtE*; *T4 lysozyme*	*A.t.*	Leaf segment	*Erwinia amylovora*	Ko *et al.* 1999
Apple (*Malus X domestica*) Bork, cv Royal Gala, M7	T	*AttE*; *SB-37*; *Shiva-1*; *lisozyme*	*A.t.*	Leaf segment	*Erwinia amylovora*	Norelli *et al.* 1999
Apple (*Malus X domestica*) rootstock M26	T	AttE	A.t.	Maternal	*Erwinia amylovora*	Norelli *et al.* 1994
Apple (*Malus X domestica*) Borkh	T	*cecropin MB39*	*A.t.*	Maternal	*Erwinia amylovora*	Liu *et al.* 2001
Apple (*Malus X domestica*) Borkh cv Greeleaves	S.V.	–	–	Leaves	*Erwinia amylovora*	Chevreau 1998
Apple (*Malus X domestica*)	S.V.	–	*E. amylovora*	Leaves	*Erwinia amylovora*	Donovan *et al.* 1994
Grape (*Vitis vinifera*), cv Thompson seedless	T	*Shiva-1*	PB followed by *A.t.*	Somatic embryos	–	Scorza *et al.* 1996
Peach (*Prunus persica*)	S.V.	–	Culture filtrate of *X. campestris*	Callus from immature zygotic embryos	*Xanthomonas campestris* pv. pruni and *P. syringae* pv. *syringae*	Hammerschlag 1988, 1990a b; 2000; Hammerschlag and Ognjanov 1990
Pear (*Pyrus communis*)	S.V.	–	–	Stem and root callus	*Erwinia amylovora*	Viseur *et al.* 1987
Pear (*Pyrus communis*) cv Passe Crassane	T	*AttE*	*A.t.*	Leaf segment	*Erwinia amylovora*	Mourgues *et al.* 1998; Reynoird *et al.* 1999

induction of overproduction of H_2O_2 in the plant cells. Hydrogen peroxide triggers local hypersensitive cell death, exerts direct antimicrobial activity (Peng and Kùc 1992) and is involved in the reinforcement of plant cell wall (Bolwell *et al.* 1995). Glucose oxidase (*Gox*) gene from *Aspergillus niger*, which induces the production of H_2O_2, increased the level of resistance to *Erwinia carotovora* and *Phytophthora infestans* in potato (Wu *et al.* 1995) and to *Pseudomonas syringae* and *Xanthomonas campestris* in tomato (Caccia *et al.* 1999).

The resistance or susceptibility to pathogens can be modified by overexpressing hormone genes (Fladung and Gieffers 1993; Storti *et al.* 1994). These authors found an increase of resistance to fungi in tomato transgenic plants overexpressing auxin- and cytokinin-synthesising genes (*iaaH* or *iaaM, ipt*) from *Agrobacterium tumefaciens*, when the equilibrium of phytormone of transgenic plants was modified in favour of auxins. Transgenic kiwifruit plants and their transgenic offspring (resulting from crossing *rolABC* staminate cv GTH X normal pistillate Hayward) artificially infected with *Pseudomonas siryngae* and *P. viridiflava*, became more sensitive to these bacteria than untransformed plants (both cv. 'GTH' and 'T1 offspring' non-carrying *rol* genes) (Rugini *et al.* 1999; Balestra *et al.* 2001). A positive correlation between bacterium susceptibility and high nitrogen content in the leaves characterised the transgenic plants (Magro and Rugini, unpublished). Other characters, modified by foreign genes, are reported in Table 3.5.

3.7.4 Nematode resistance

Many fruit crops are attacked by nematodes of the species *Meloidogyne* spp., *Xiphinema* spp. and *Longidorus* spp. (Brown *et al.* 1993; Ploetz *et al.* 1994; Nyczepir and Halbrendt 1993). Nematodes are difficult to eradicate from infected soils and control is normally via nematocides, resistant cultivars, and appropriate crop husbandry techniques. However, resistant rootstocks are very rare (Roberts 1992) and chemical treatments are expensive and not always effective since egg-containing cysts formed by some nematodes are very resistant to chemicals and can survive for years in the soil. Plants respond to infection with a variety of defence strategies including production of phytoalexins, deposition of lignin-like material, accumulation of hydroxyproline-rich glycoproteins, expression of PR-proteins and with an increase of lytic enzymes. Genes involved in nematode resistance have been identified in *Beta procumbens* and *Solanum tuberosum* (*$Hs1^{pro1}$* and GPA2) and have been cloned (Stiekema *et al.* 1999).

Two strategies of genetic engineering for introducing resistance to nematodes have been suggested (Sijmons *et al.* 1994):

1. introduction of an effector gene whose product is addressed to the parasite or its excretion
2. introduction of an effector gene whose product is addressed to the plant cells which feed the nematodes.

Table 3.6 Genetic modification of fruit crops for nematode resistance

Fruit crop	Technique	System/ Plasmid or selective agents	Origin of plant material	Resistance in *planta*	Authors
Peach (*Prunus persica*)	S.V.	Culture filtrate	Seedlings	*Meloidogyne incognita*	Hashmi *et al.* 1995

Potential anti-nematode genes have been reported and seem to be effective when they are constitutively expressed in plants (see Table 3.6). Usually these also are involved in the control of insects:

1. genes over-expressing collagenases which damage the animal cuticle (Havstad *et al.* 1991);
2. exotoxin of *B. thuringiensis* (Devidas and Rehberger 1992) or other feeding inhibitor such as the cowpea trypsin inhibitor. This approach is based on the very localised expression of a phytotoxin gene responsible for the inhibition of development or maintenance of feeding structures of nematodes within the plant system. Genes encoding lipases, transcription factors, nucleases, proteases, and glucanases have been suggested (Sijmons et al. 1994);
3. anti-nematode monoclonal antibodies (Schots *et al.* 1992).

Molecular information on nematode resistance is limited, the availability of specific nematode-responsive regulatory plant sequences could represent an important goal and the durability of resistance is believed to depend on the combination of different chimeric constructs and strategies (Barthels *et al.* 1999).

3.7.5 Insect and pest resistance

Several strategies to control insects and pests in fruit crops are currently used. Control by insecticides is the most used, followed by biological control, use of resistant plants and other integrated insect pest control techniques. The massive use of pesticides, beyond environmental pollution, induces insecticide resistance among the insect populations, leading to new insecticides and their increased use. Resistance is controlled by partially recessive or co-dominant (additive) genes and involves a small number of loci. Resistant plants in fruit crops are rare, and their development is difficult. Genetic engineering offers new approaches to more rapid deployment of anti-insect strategies in fruit crops (Table 3.6).

Resistance to *Bt* protein has been studied with the aim of retarding the spread of resistance in the insect population and the strategies involve:

- use of high doses of insecticidal protein in order to kill the homozygous resistant insects, driving the insects away from the crop;
- multiple target strategy, aiming to use multiple insecticides;
- the refuge strategy aims to express in plants a dose of insecticide able to kill heterozygous insects and the survival of homozygous ones. The density of the homozygous population can be kept low by planting a mixture of transgenic resistant plants and non-transgenic ones on which susceptible insects could survive and mate with homozygous resistant insects thus creating heterozygous ones (Alstad and Andow 1995).

Several plants have been engineered with the aim of killing phytophagous insects by the following strategies:

1. genes encoding insecticidal crystal protein from *Bacillus thuringiensis*
2. proteinase inhibitors
3. lectins
4. alpha-amylase inhibitors
5. chitinases
6. polyphenol oxidases and peroxidases
7. lipoxygenases
8. ribosome inactivating proteins (RIPs) and promising insecticidal proteins, isolated from microbial culture filtrates, such as cholesterol oxidase, Vip3A and Tca (Escobar and Dandekar 2000).

Genes encoding insecticidal crystal protein from Bacillus thuringiensis

Individual strains of *Bacillus thuringiensis* have been characterised and classified according to their insecticidal activity (Schnepf *et al.* 1998). This activity is due to their insecticidal crystal proteins (ICPs) which are toxic for several important insect pests for fruit and nut crops. There are several individual ICPs which are highly specific to a particular insect order, including Lepidoptera, Diptera and Coleoptera. Recently *B. thuringiensis* (Bt) strains have been isolated with activity against Hymenoptera, Homoptera, Orthoptera, Mallophaga, nematodes, mites and protozoa (Schnepf *et al.* 1998; Escobar and Dandekar 2000). Since the individual ICPs are released from parasporal crystal in response to alkaline conditions (pH 9–10) in the midgut of the target insect, toxicity may depend on solubilisation of some ICPs and the number and type of receptors in the midgut microvillae of insects (Du *et al.* 1994; McGauhey and Whalon 1992). Genes encoding these ICPs have been cloned from Bt and are referred to as *cry* genes (*cryIAa*, *cryIAb*, *cryIAc*). They have been transferred to several plants including fruit crops (Table 3.7). However, in some transgenic plants, such as apple and walnut, the wild type gene sequences encoding the *cryIA(c)* revealed no expression. Subsequently the gene sequences were restructured and these problems eliminated, giving rise to successful transgenic plants (Dandekar *et al.* 1994). It has been demonstrated that damage by such transgenic plants to non-target insect populations is less than that caused by chemical pesticides (Losey *et al.* 1999).

Table 3.7 Genetic modification of fruit crops for insect resistance

Fruit crop	Alien gene(s)	System/Plasmid or selective agents	Origin of plant material	Insect target	Expressed protein	Authors
Apple (*Malus X domestica*) cv Greenleaves	*CryIAa (c)*	*A.t.*	Leaf segments	Lepidoptera	CryIAc	Dandekar 1991; Dandekar 1992
Apple (*Malus X domestica*) cv Greenleaves	*CpTI; CryIAa*	*A.t.*	Leaf segments	Lepidoptera Coleoptera	CpTI	James *et al.* 1992; 1993
Cranberry (*Vaccinium macrocarpon*)	*CryIAa*	*A.t.*		Lepidoptera	CryIAa	Singh and Sansavini 1998
Cranberry (*Vaccinium macrocarpon*) Ait.	*Icp*	PB	Stem section	–	–	Serres and Stang, 1992
Grape (*Vitis vinifera*)	*CryIAa*	*A.t.*		Lepidoptera	CryIAa	Singh and Sansavini 1998
Grape (*Vitis vinifera*)	*gna*	*A.t.*		Lepidoptera	GNA	Coghlan 1997
Grapefruit (*Citrus paradisi* Macf.)	*gna*	*A.t.*		Aphids	GNA	Yang *et al.* 2000
Juneberry (*Amelanchier laevis*)	*Btk-icp*	*A.t.*	Basal cut end of shoots	–	Toxin HD73	Hajela *et al.* 1993
Juneberry (*Amelanchier laevis*)	*cryC*	A.t.		Lepidoptera, Coleoptera	Unspecified Ccry gene	Krattiger 1997
Pear (*Pyrus communis*)	*cryC*	*A.t.*		Lepidoptera, Coleoptera	Unspecified Ccry gene	Krattiger 1997
Persimmon (*Diospiros kaki*)	CryIAc	*A.t.*	Leaf discs	Lepidoptera	CryIAc	Tao *et al.* 1997
Strawberry (*Fragaria X ananassa*)	*cpti*	*A.t.*		–	Expressed CPTI	James *et al.* 1992
Strawberry (*Fragaria X ananassa*) some cultivars	*cpti*	*A.t.*		Vs *Otiorhynchus sulcatus*	Expressed CPTI	Graham *et al.* 1996
Walnut (*Juglans regia*) hybryds & cv Sanland	*CryIAc*	*A.t.*	Somatic embryos	Lepidoptera	Cry1Ac (vs codling moth)	Dandekar *et al.* 1992; Dandekar 1994, 1998

Proteinase inhibitors

Amino acids are essential for herbivorous insect survival and are metabolised by proteinase in the insect gut. The availability of one amino acid rather than another depends on the pH of the insect gut (Wolfson and Murdock, 1990). In plants, proteinase inhibitors that are highly specific for some classes of insect proteinases are induced in response to mechanical or insect damage, reversibly binding the active site of proteinases and forming an inactive complex (Laskowski *et al.* 1987; Boulter 1993) which reduces the availability of amino acids necessary for insect nutrition. A potent natural insecticide was found in cowpea and identified in tripsin inhibitor (CpTI), which confers resistance to cowpea seed weevil. Over-expression of this proteinase gene in plants may have an insecticidal activity by also increasing the hyperproduction of proteinases in the insect gut, depleting the insect's metabolic reserve of sulphur-containing amino-acids (Broadway and Duffey 1986). Although several crops have been transformed and the protein has been expressed, insecticidal effect has been reported in a few cases. Since the insects seem to develop a rapid resistance following a continuous ingestion of proteinase inhibitors (Jongsma *et al.* 1995; Broadway 1995, Girard *et al.* 1998), the effectiveness of this strategy for plant protection still needs to be established.

Lectins

Lectins are proteins with specific carbohydrate binding activity. Many of the over 300 purified lectins from seeds are toxic for animals and seem to have multiple roles in plant physiology. They are targeted to the vacuoles or secreted extracellularly. Some of them are toxic for Coleopteran, Lepidopteran, Dipteran and Homopteran insects (Van Damme *et al.* 1998). Considering that the last group of insects, up to now, is not controlled by any other insecticidal protein, lectins have aroused particular interest (Gatehouse and Gatehouse, 1998). An insect diet containing mannose-specific snowdrop lectin (GNA) is effective in reducing larval growth of Coleoptera and in reducing fecundity of adults of peach and potato aphids (*Myzus persicae*) (Sauvion *et al.* 1996), while N-acetylglucosamine from castor bean and wheat germ (WGA) is toxic for some Lepidoptera. The mode of action is not clear yet, though it seems to be linked to endocytosis in the intestine (Zhu-Salzman *et al.* 1998). Up to now, expression of insecticidal lectins in transgenic plants has provided relatively low protection against Lepidopteran and Homopteran pests, and their potential toxicity to mammals limits their use in the growing of fresh fruits.

Alpha-amylase inhibitors

The presence of these proteins in seeds seems to protect them from insect attack by inhibiting midgut-α-amylase, reducing the ability of the insect to catabolise starch (Baker *et al.* 1991). They were isolated from common bean, named αAI-1 and αAI-2, and are active mainly against Coleopteran α-amylase. Application in fruit crops, by overexpressing them in the tissues, seems at the moment limited, since inhibition of gut-α-amylase may not be an antinutritive deterrent to insects that feed upon leaves or phloem sap and, in addition, could limit the activity of mammalian α-amylase.

Chitinases

Besides their activity against fungal pathogens, chitinases also have potential for insect control, since the exoskeleton and peritrophic gut membrane of insects are constituted by chitin. Up to now, plant chitinases have not shown high insecticidal activity (Kramer and Muthukrishnan, 1997). However, following the studies of Ding *et al.* (1998), who fed tobacco budworm with transgenic-chitinase tobacco foliage and sub-lethal doses of ICP of *B. thuringiensis*, transgenic plants overexpressing both chitinase and ICPs seem to provide a promising strategy in insect control. It seems that the degradation of peritrophic gut membrane, operated by chitinase, increases the accessibility of the ICP to epithelian cell membrane receptors, enhancing the insecticidal property (Ding *et al.* 1998).

Polyphenol oxidases and peroxidases

Both polyphenol oxidases (PPOs) and peroxidases catalyse the oxidation of phenolic compounds to reactive quinones (Steffens *et al.* 1994), perhaps complimenting each other in a generalised antinutritive response (Duffey and Felton 1991). Following cell lysis, the contact of PPOs or peroxidases with the phenolic substrate produces quinones which can irreversibly degrade nucleophilic amino-acids that are essential to insect diet, limiting herbivorous insect growth. Studies with potato, either overexpressing PPO or reducing it by antisense-PPO, increased or reduced insect mortality respectively (Steffens *et al.* 1994). Antisense-peroxidases in tobacco plants, however, did not reduce larval susceptibility. Peroxidases also induce rapid lignification response at plant wound sites, which may play a role in insect defence. However, abnormality due to high lignin content and root system mass reduction was evident in some transgenic tobacco plants. Plants overexpressing anionic peroxidases caused in some cases, but not in others, an increase of mortality of Lepidopteran and Coleopteran larvae, including the woody transgenic sweetgum. Both strategies may have a limited application in fruit crops.

Lipoxygenases

Lipoxygenases catalyse degradation (peroxidation) of free unsaturated fatty acids, which are essential in insect diet. Since the activity of plant lipoxygenases has been associated with tissue wounding, their increase in the wounded tissues could represent an active antinutrient strategy for reducing damage caused by herbivorous insects (Felton *et al.* 1994; Royo *et al.* 1999), reducing also palatability (Duffey and Stout 1996) and regulating the expression of other wound-responsive defence genes (e.g. through jasmonic acid) from pathogen attack.

Ribosome inactivating proteins (RIPs) and other compounds
Studies on these groups of plant proteins are quite recent (Barbieri *et al.* 1997). RIPs were isolated from microbial culture filtrates. They are similar to compounds such as *cholesterol oxidase* (Shen *et al.* 1997), *Vip3A* (Yu *et al.* 1997), and *Tca* (Bowen and Ensign 1998), and show insecticidal properties when added to the insect diet. In addition a new class of molecules named *neuropeptides* is under observation, they include proctolin, schistocerca allatostatin-5, locustanin 2 (Kelly *et al.* 1990). The insects exposed to them experienced dismetabolic effects, growth inhibition and death (Tortiglione *et al.* 1999).

At present only BtICP transgenics maintain acute toxicity in the absence of other insecticidal proteins (Roush 1998; Escobar and Dandekar 2000). Field trial experiments demonstrated a rapid evolution, similar to chemical insecticides, of resistance to insecticidal proteins such as α-amylase, proteinase inhibitors ICP of *B. thuringensis* (Michaud 1997).

To maintain long-term insect resistance in transgenic crops, several strategies have been suggested (Roush 1998; Escobar and Dandekar 2000).

- cultivation of multilines containing different insect resistant genes
- transgene pyramiding (multiple insecticidal proteins expressed in a single crop cultivar)
- cultivation of susceptible plants for insect refugia
- expression of the transgene only in specific plant tissues.

Since insects show different levels of resistance, according to the homozygosity, high resistance in insects acquired after consuming tissues of transgenic crops could be reduced by crossing with the large non-resistant insect population colonising non-transgenic refugia crops (Gould 1998). Few fruit crop genotypes have yet been transformed with generally poor results. However, genes such as the *ipt* gene, under a wound-inducible promoter, have produced insect pest resistance in tobacco plant (Smigocki *et al.* 1993). Pyramiding transgenics, particularly with *BtICP* and *cholesterol oxidase*, seem to be good candidates for future transgene strategies, possibly using inducible or *situ*-specific promoters.

3.7.6 Resistance to herbicides

Fruit crops resistant to herbicides may be useful in some species, whether for nursery plants or field cultivation. In woody plants this characteristic may not be essential, since a mature trunk prevents uptake of the herbicide, though it remains important to avoid spreading herbicide chemicals on the herbaceous organs. In the case of some herbaceous fruit crops such as strawberry, the trait may also not be essential because other techniques are used to control weed growth, such as covering the soil with plastic sheeting or straw mulch. However, some resistant genotypes of fruit species to herbicides were produced, mainly by gene transformation (Table 3.8). Genes for resistance to both commercial and developing herbicides are available. They have been isolated from micro-organisms or plants (*bar*, *aro*A, *als*). The *bar* (phosphinothricin acetyl-transferase) gene, isolated from *Streptomyces hygroscopicus* (Murakami *et al.* 1986), confers resistance to herbicide Basta. The *aro*A gene confers Glyphosate tolerance in plants (Shah *et al.* 1986; Padgette *et al.* 1989), while the mutant *aro*A, isolated from *Salmonella tytyphimurium* (Fillatti *et al.* 1987), confers resistance to Glyphosphate. The *als* (acetolactate synthase) genes, isolated from species such as tobacco (Chaleff and Ray 1984), yeast (Yadav *et al.* 1986) and *Arabidopsis* (Smith *et al.* 1988b), confer tolerance to sulfonylurea. A number of transformed fruit plants have shown strong resistance to herbicides, tolerating in some cases 3–5 times the recommended field concentration.

The possibility that transgenic crops could become weeds themselves, or that the resistance gene, by hybridisation, could integrate into the population of wild species, thus creating a new herbicide-resistant weed, is not considered a significant problem. The main fruit crops have never represented weeds themselves and they are not compatible with common weed species. It is also well known that conventional weed control produces selection of resistant weeds with the long-term use of the same herbicide.

3.8 Abiotic stress resistance

In response to stress plants are able to adjust their morphology, phenology, physiology and biochemistry. Adaptations to water and salt stresses, for example, are reported by Jain and Selvaraj (1997). Stress resistance is an efficient approach in reducing losses of yield due to adverse climatic conditions. Since such changes are regulated by genes, research has concentrated on characterising and isolating genes induced by stress. Stress not only induces the expression of genes but can also inactivate them, including the foreign genes in transgenic plants. Advances in understanding the effectiveness of stress responses, and the distinction between pathology and adaptive advantages, are based on the analysis of transgenic and mutant plants, in particular the analysis of mutant defective *Arabidopsis* (Hasegawa *et al.* 2000).

At present work has been carried out almost exclusively on model herbaceous plants and yeasts, even though preliminary work is in progress on some fruit crops. With available technology it is possible to improve crops for drought and salt tolerance by both gene transformation and somaclonal variation (Table 3.9). The improvement in drought-salt tolerances can also improve cold resistance. The physiological and genetic mechanisms involved in the main abiotic stresses are synthetically explained below in order to understand the ongoing work to improve these traits.

Table 3.8 Genetic modification of fruit crops for herbicide tolerance

Fruit crop	Technique*	Gene(s)	System/ Marker gene	Explant(s)	Resistance to	Authors
Apple (*Malus X domestica*) cv Royal Gala	T	*als*	*A.t.*	Leaf pieces	Sulfonylurea	Yao *et al.* (1995)
Damil GM61/1 rootstock (*P. dawyckensis*) Inmil GM9 rootstock (*P. incisa X serrula*) Cherry (*P. avium*) cv Summit	T	*bar*	*A.t.*	Meristems, Shoots, Embryogenic calli	BASTA	Druart *et al.* 1998
Orange (*Citrus sinensis*)	S.V.		2,4-D	Nucellar callus	2,4-D	Spiegel-Roy *et al.* 1983
Papaya (*Carica papaya* L.)	T	*bar*	PB	Zygotic embryos	–	Cabrera-Ponce *et al.* 1995
Pepino (*Solanum uricatum*) Ait.	T	*als*	*A.t.* (*nptII; gus*)	Leaf	Sulfonylurea	Atkinson and Gardner (1991)
Tamarillo (*Cyphomandra betaceae*)	T	*als*	*A.t.*	Leaf	Sulfonylurea	Atkinson and Gardner (1993)

*Technique = gene transformation; S.V. = somaclonal variation

Table 3.9 Genetic modification of fruit crops for various stress tolerance

Fruit crop	Technique	Alien gene(s)	System/Plasmid or selective agents	Origin of plant material	Resistance in *planta*	Authors
Kiwi fruit (*Actinidia deliciosa*) cv Hayward	S.V.	–	Synthetic carbohydrates or NaCl	Stem	Drought resistance	Muleo *et al.* 1996
Kiwi fruit (*Actinidia deliciosa*) cv Hayward and GTH	T	*rolABC*	*A.t.*	Leaf discs	Drought resistance	Rugini *et al.* 2000b
Kiwi fruit (*Actinidia deliciosa*) cv Hayward and Tomuri	S.V.	–	High pH (7.5)	Leaf	Lime tolerant	Marino and Battistini 1990; Marino *et al.* 1998
Quince A (*Cydonia oblonga*)	S.V.	–	Low Fe	Leaf discs	Fe-deficient	Dolcet-Sanjuan *et al.* 1992
Orange (*Citrus sinensis*)	S.V.	–	–	Nucellar and ovular callus	Salt tolerance	Kochba *et al.* 1982; Spiegel-Roy and Ben-Hayyim 1985; Ben-Hayyim and Goffer 1989
Colt rootstock (*Prunus avium X pseudocerasus*)	S.V.	–	–	Protoplast callus (root, stem)	Salt and drought tolerance	Ochatt and Power 1989
Sour Cherry (*Prunus. cerasus)*	T	*AFP antifreeze*	*A.t.*	Maternal leaf	Cold tolerance	Dolgov 1998
Strawberry (*Fragariaxananassa*)	T	*antifreeze*		Leaf	Cold tolerance	Hightower *et al.* 1991
Kiwi (*A. chinensis*)	T	*hEGF*	*A.t.*	Leaf	–	Kobayashi *et al.* 1996
Citrus spp	T	*HAL2 of yeast*	*A.t.*	–	Salt tolerance	Cervera 2000

3.8.1 Salt stress

One-quarter of the cultivated land in the world is salt-affected, mostly by sodium chloride (NaCl) which is the most prevalent salt in saline soils, although Ca, Na and Mg sulphate, K chloride and Na carbonate are also common. Salt stress causes various types of damage including growth reduction, photosynthesis inhibition, reactive oxygen species production, attenuated nutrient uptake, membrane disorganisation, toxic metabolite production and ionic steady states modification. After salt stress cytosolic Ca^{2+} increases, transported from apoplast and intracellular compartments, leading to activation of stress response pathway. Recently attention has been paid to ATPase in regulating calciums in the cells. However, many genes are involved in the control of salt stress, but ion transport and organic osmolytes seem to be particularly promising targets, maybe combined in a pyramidal approach (Jain and Selvaraj 1997). Somaclonal variation induced by natural or better synthetic osmolites, such as glucose-like synthetic compounds: 3-O-Methylglucose or N-Methyl-D-glucamine seems to be a promising approach (Boscherini *et al.* 1999; Muleo *et al.* 1996) regenerated kiwi fruit plants resistant to water stress from callus undertaken under osmotic stresses with both these osmolites.

3.8.2 High pH stress

The high pH level in the soil reduces the iron uptake and, inside the plant, it reduces the iron reductase activity, the enzyme which converts the Fe^{3+} transported in the sap to Fe^{2+}, the sole form utilised by the cells. At present, the most suitable strategy to reduce high pH stress seems to be somaclonal variation. An actively limestone tolerant somaclone kiwi fruit has been reported by Marino and Bertazza (1998) and Marino *et al.* (1998) by *in vitro* selection at high pH during regeneration (Marino and Battistini 1990). Muleo *et al.* (1996) produced similar kiwi plants from cultured calli in high concentrations of glucose-like synthetic compounds: 3-O-Methylglucose and N-Methyl-D-glucamine. Both compounds induced a similar type of stable genetic variation in regenerated plants, ascertained by molecular analysis.

3.8.3 Cold stress

In temperate climates, woody plants are exposed to freezing stress involving extremes of low temperature during winter, including possible ice encasement, and unseasonal episodes of frost at other times of the year. Trees have evolved an ability to acclimatise to this stress before and during winter dormancy. Different tissues respond in different ways when exposed to freezing temperatures (Sakai and Larcher 1987; Wisniewski and Arora 1993). Acclimatisation changes involve carbohydrate metabolism, composition of plasma membranes and accumulation of unique classes of proteins with putative cryoprotective function (Chen *et al.* 1995). To avoid cold stress plants adopt two main strategies:

1. Freezing tolerance: (formation of extracellular ice and consequent loss of cellular water which increases solute concentration with lowering of the freezing point of the cells).
2. Freezing avoidance: by deep supercooling, that is the isolation of cellular water from the dehydratative and nucleating effects of extracellular ice. This ability is species-dependent and is influenced by the ice nucleating agent which may have either plant or bacterial origin (see review of Wisniewski and Arora 2000). Normally in fruit trees supercooling is around −2 to −4°C (Ashworth and Kieft 1995).

In ice-nucleating-active (INA) compounds, the genes responsible for the Ice^{+} phenotype, have been cloned from different species (Warren 1995). Other than specific proteins (antifreeze), such as dehydrins (a subgroup of LEA proteins), directly involved in cold acclimatisation, proteins such as bark storage proteins (BSPs), some enzyme systems (plasma membrane ATPase; gluthatione reductase) and phytochrome are also associated with adaptation to cold stress. Antifreeze proteins reduce the freezing temperature of water and modify the morphology of ice crystals. They were reported in fish, where they are synthesised in the liver and secreted in the blood (Gong *et al.* 1996), but they are also common in insects, fungi, bacteria and plants (Duman and Olsen 1993; Griffith and Ewart 1995). Antifreeze proteins isolated in plants have different structures and probably different modes of action. They are expressed during cold acclimatisation and after ABA treatment. Protein isolated from winter bark tissue of peach (PCA60) is an amphypathic α-elical (Wisniewski *et al.* 1999), that in winter rye (AFP) is similar to a pathogenesis-related protein (Hon *et al.* 1994, 1995). The peach dehydrin gene (*ppdhn 1*) has been cloned and sequenced. In cold-acclimatised blackcurrant plants dehydrin-like genes (*dhns*) have also been isolated (Lanham *et al.* 2001). Finally, many cold-responsive genes, expressed in response to ABA (Pearce 1999), are currently being studied.

Several proteins, identified as up-regulated in response to cold, are also up-regulated in response to other types of stress. Osmotin has been identified in response to salt stress and, subsequently, to cold and other abiotic stresses. β-1, 3 endoglucanase-like, chitinase-like and thaumatin-like proteins have antifreeze properties and antimicrobial functions *in vivo* (Pearce 1999). Their corresponding genes have been isolated and transferred to plants included fruit crops.

Three promoter elements, responsible for abiotic stress induction, have been identified. They include ABA-responsive element (ABRE), drought-responsive element (DRE) and heat-shock-responsive element (HSE) (Zhang *et al.* 2000). Promoters involved in the cold stress have not yet been identified. Several groups of genes have been targeted in improving abiotic stress resistance in plants (Zhang *et al.* 2000), encoding for:

1. enzymes for the biosynthesis of compatible compounds
2. enzymes for scavenging active oxygen species
3. heat shock proteins (HSPs)

4. late embryogenesis-abundant (LEA) proteins
5. enzymes modifying membrane lipid saturation
6. transcription factors
7. proteins required for ion homeostasis.

These general strategies are discussed below, followed by a discussion of the techniques used to combat particular types of abiotic stress.

3.8.4 Genes encoding enzymes for the biosynthesis of compatible compounds

Under stress, plants accumulate some osmolytes, such as sugars (mainly sucrose and fructose), sugar alcohols like glycerol, methylated inositols (D-ononitol), mannitol, complex sugars (fructans, threalose, raffinose), charged metabolites (proline, glycine, betaine, ectoine, dimethyl sulfonic propionate, polyamines) and ions (K^+) (see review by Hasegawa *et al.* 2000). It is believed that their accumulation facilitates osmotic adjustment, contributing to stress tolerance. Transgenic plants (see review by Bajaj *et al.* 1999), overexpressing one or more osmolytes did not always result, as expected, in resistance to stresses such as cold or high salt concentration. In other cases the transgenes became resistant only with low concentration of osmolytes and were unable to play a role in osmotic adjustments (Zhang *et al.* 1999). This evidence suggests a different function for osmolytes, or their efficacy should depend on the presence around the organs (i.e. thylakoids and plasma membrane) to protect them. Transgenic rice, which overexpresses mannitol, after transformation with a bacterial gene *mtL* that encodes mannitol 1–phosphate dehydrogenase, tolerates high salinity. Tolerance to high temperature was induced in *Arabidopsis thaliana*, transgenic for *codA* gene for choline oxidase (overexpressing glycine betaine), both during seed germination and plant growth (Alia *et al.* 1998). Although plants tolerating high degree of salt contain high levels of polyamines, transgenic rice plants overexpressing oat arginine decarboxilase gene, under a constitutive promoter, did not develop properly (Capell *et al.* 1998). Since plant response varies according to the phenological phase, more research is needed on the influence of various stress intensities and development stages in plants. Most transgenic plants overexpressing osmoprotectant genes under constitutive promoters showed severe developmental disorders, necessitating the use of stress-inducible promoters.

3.8.5 Enzymes for scavenging active oxygen species

When a plant is suffering from stress, such as extreme temperature, salinity, drought or flooding, it overproduces active oxygen species, a process which is dangerous to cells. To reduce the damaging effects of active oxygen, plants have evolved enzymatic and nonenzymatic mechanisms to reduce oxidative stresses by producing antioxidants (ascorbate peroxidase, superoxide dismutase, gluthatione reductase, etc.) (Zhang *et al.* 2000). Transgenic herbaceous plants overexpressing genes encoding enzymes with a capacity to reduce oxidative stresses resulted, in some cases, in a higher tolerance to abiotic stresses (Bajaj *et al.* 1999). Plants more resistant to water and freezing stress, or tolerant to chilling, have been reported in alfalfa and tobacco respectively by McKersie *et al.* (1999) and Sen Gupta *et al.* (1993) by an overproduction of SOD. Heat stress tolerance can be also increased by an overproduction of SOD (mitochondrial Fe/Mn-SOD; chloroplastic Cu/Zn-SOD) and bacterial catalase (Sairam *et al.* 1997; review of Hasegawa *et al.* 2000). Cold and salt stresses can also be alleviated by overproduction of gluthatione S-transferase or gluthatione peroxidase (Roxas *et al.* 1997).

3.8.6 Heat shock proteins (HSPs)

In response to water stress, heat shock proteins (HSP) have been identified in some plants. Modification of the cell content of the HSP family via genetic engineering has been done in herbaceous plants, both over-expressing them (Schoffi *et al.* 1987) or inhibiting them by antisense gene (Lee and Schoffi 1996). Encouraging results were obtained, indicating a possible use of this strategy for increasing thermal tolerance in crop species, possibly by the use of appropriate promoters.

3.8.7 Late embryogenesis-abundant (LEA) proteins

LEAs are low molecular weight proteins that are highly accumulated in the embryos at the late stage of seed development. The transcription genes encoding them are activated by osmotic stress, and it has been hypothesised that they play a role in desiccation tolerance during seed development and in response to dehydration, salinity and cold stresses (Close 1997), however their physiological role is not known, but only a correlation between gene expression and LEA accumulation and dehydration is certain. They are classified into groups according to the stress type. Rice transgenic plants and their offspring, constitutively overexpressing HVA1 protein, demonstrated improved tolerance to water deficit and salt stress conditions (Xu *et al.* 1996). Improved salinity and freezing resistance has been observed in yeast transformed with LE25 (Imai *et al.* 1996).

3.8.8 Enzymes modifying membrane lipid saturation

Since plant membranes are very sensitive to all kind of stresses, the increase in levels of unsaturated membrane lipids, which increase membrane fluidity, may increase plant chilling tolerance and, as a consequence, photosynthetic activity. This is confirmed in cyanobacteria in which the level of unsaturation of fatty acids in membrane lipids was directly proportional to the chilling tolerance and to photosynthesis (Gombos *et al.* 1992). Transgenic tobacco plants demonstrated greater chill tolerance than wild-type tobacco if acyl-ACP-glycerol-3-phosphate acyltransferase (GPAT) from chilling-tolerant *Arabidopsis* was used. In contrast, no chilling tolerance was obtained when GPAT from chilling-sensitive squash was used (Murata *et al.* 1992).

3.8.9 Transcription factors

Specific regulatory genes can induce the whole cascade of cellular changes necessary for rendering the plants tolerant to stresses. They include dehydration response element (DRE) and two gene families which have domains that bind ethylene-responsive elements (Hasegawa *et al.* 2000), *DREB1* (which includes the previously identified *CBF1*) and *DREB2*. Overexpressing in plants a single stress-inducible transcription factor (DRE1A) enhanced plant tolerance to salinity, freezing and drought was observed (Kasuga *et al.* 1999). Freezing tolerance in transgenic plants of *Arabidopsis* was obtained by overexpressing a transcriptional activator *CBF1,* which induced the expression of four *COR* genes (Jaglo-Ottosen *et al.* 1998). These results are encouraging, but more research is needed on identification and isolation of more stress-responsive transcription factor genes.

3.8.10 Proteins required for ion homeostasis

To tolerate high salt levels in the cytoplasm the plant cells use ions for osmotic adjustment and sequester them away from the salt-sensitive metabolic reactions. Ion homeostasis in saline environments is dependent on transmembrane transport protein that mediate ion fluxes, including H^+ translocating ATPases and pyrophosphatases, Ca^{2+}-ATPases, secondary active transporters and channels (Hasegawa *et al.* 2000). The genes related to vacuolar Na^+/H^+ antiports, and some proteins signaling stress (Apse *et al.* 1999; Liu and Zhu 1998), maintain the ion homeostasis in the cells, facilitating the compartmentalisation of excess Na^+ into vacuoles. Transgenic plants of *Arabidopsis* for a single gene *AtNHX1,* encoding a vacuolar Na^+/H^+ antiports protein, resulted in higher tolerance to salt than untransformed varieties. Future work should include a strategy that reduces both ion toxicity and water loss due to osmotic stress combining genes able to regulate homeostasis, accumulate compatible solutes, antioxidants and expressing transcription factors in order to produce plants resistant to different abiotic stresses (Zhang *et al.* 2000).

3.9 Plant breeding: the use of molecular markers

Most biotechnological research has been concerned with identifying the genetic basis of particular cultivars and particular traits within cultivars. Such work is both a necessary foundation for the targeted modification of such traits but also assists conventional breeding programmes. Molecular markers based on PCR technology, such as RAPDs, AFLPs and SSRs, have started to replace classical markers such as RFLPs and isozymes in characterising varieties. For some species, such as grapevine, apple and citrus, the richness of the germplasm and difficulties in cultivar identification make DNA markers particularly suitable in distinguishing genotypes.

Traditional breeding of fruit trees is made particularly difficult by long generation times, the space occupied by plants under selection and the slow changes in plant characteristics obtainable within any one generation, due to the largely diffused heterozygosity at most of the loci. For this reason genome mapping, aimed at identifying molecular markers tightly linked to the traits under selection, is particularly important in fruit tree species in permitting early selection of the most interesting genotypes. Genetic linkage maps are now available for most species and, as a consequence, numerous markers can be used in Marker-Assisted Selection (MAS). Some of these markers are discussed for particular species in the following sections.

3.9.1 Apple

Cultivar identification and phylogeny

The need for rapid and reliable identification of apple cultivars has driven the need to produce consistent molecular markers for identification purposes. Microsatellites have been identified which distinguish almost all the cultivars (Guilford *et al.* 1997; Gianfranceschi *et al.* 1998). RAPD markers have been used to establish phylogenetic relationships among closely related *Malus* species (Zhou and Li 2000).

Genome mapping

Linkage maps have been constructed for apple (Hemmat *et al.* 1994; Conner *et al.* 1997). Recently, it has been possible to identify molecular markers able to combine the genetic maps built from separate progenies (Chevrau *et al.* 1999). and linkage maps of the apple cultivars 'Prima' and 'Fiesta' were aligned using multi-allelic markers (Maliepaard *et al.* 1998). RAPD markers have been identified to estimate the position and effects of quantitative trait loci (QTL) for traits influencing juvenile tree growth and development in two apple cultivars (Conner *et al.* 1998) or the columnar growth habit (Hemmat *et al.* 1997).

Most of the efforts have been devoted to the identification of molecular markers tightly linked to the genes conferring resistance to apple scab. There are multiple resistance loci but the *Vf* gene is considered the major gene controlling the disease. Numerous markers have been identified which are linked to *Vf* gene (Yang *et al.* 1997; Hemmat *et al.* 1998; Tartarini *et al.* 1999). Conversion of these markers to PCR based markers has made it feasible to use MAS in breeding programmes. A detailed linkage map of the scab resistance region has been constructed (King *et al.* 1998; Xu and Korban, 2000) and numerous genetic markers identified, representing an important prerequisite for map-based cloning of genes (Patocchi *et al.* 1999). In this respect a BAC library spanning the genomic region containing the apple scab resistance gene *Vf* has been constructed (Patocchi *et al.* 1999). Some sequence-characterised amplified regions (SCARs), tightly linked to the *Vf* gene, have been developed (Xu *et al.* 2001). Markers linked to other sources of resistance, such as the *Vm* gene have also been identified (Cheng *et al.* 1998).

Genes related to reproduction and fruit ripening
Numerous genes and cDNA clones involved in the ripening process have been recently identified and characterised, such as:

- the 1-aminocyclopropane-1-carboxylic acid (ACC) synthase gene (Md-ACS1) (Sunako *et al.* 1999; Harada *et al.* 2000)
- the *Mdh* genes, differentially expressed in the early stage of fruit development (Dong *et al.* 1999, 2000)
- the polygalacturonase inhibiting protein (PGIP) (Yao *et al.* 1999)
- genes expressed in flower development, such as the MADS-box genes (Sung *et al.* 1999, 2000).

A comprehensive study has been conducted for cloning and identifying the S-alleles controlling self-incompatibility (Broothaerts *et al.* 1995; Janssens *et al.* 1995; Kitahara *et al.* 2000; Matsumoto and Kitahara, 2000).

Other genes
Genes controlling other traits, such as programmed cell death (Dong *et al.* 1998) or root formation have been investigated as well as the polymorphisms of the superoxide dismutase (SOD) genes (*Sod-1*, *Sod-3*, *Sod-4*). PCR-based molecular markers have been identified to detect the presence of alternaria blotch of apple (Johnson *et al.* 2000).

3.9.2 Pear

Cultivar identification and phylogeny
Pear polymorphisms and genetic diversity have been assessed by the use of AFLP and RAPD markers (Oliveira *et al.* 1999; Monte-Corvo *et al.* 2000) and SSRs previously selected in apple (Yamamoto *et al.* 2001). The use of cpDNA-RFLPs has led to a better understanding of the relationships between oriental and occidental pear species.

Genes related to reproduction and fruit ripening
The S-RNase-alleles associated with self-incompatibility of the Japanese pear, *Pyrus pyrifolia* Nakai, have been cloned and identified (Ishimizu *et al.* 1996; Norioka *et al.* 1995, 1996; Sassa *et al.* 1997; Ishimizu *et al.* 1999). First attempts to elucidate the molecular basis of fruit ripening in Japanese pear (*Pyrus pyrifolia*) have been made and some genes involved in ripening identified, such as the beta-D-xylosidase-like gene, a possible senescence-related gene (Itai *et al.* 1999b). Other genes involved in ethylene signal transduction have been located

(Itai *et al.* 2000), as well as the 1-aminocyclopropane-1-carboxylic acid (ACC) synthase gene (Itai *et al.* 1999a) controlling ethylene levels, and the beta-galactosidase (Tateishi *et al.* 2001).

3.9.3 Peach

Cultivar identification and phylogeny

The molecular fingerprinting of nectarine and peach varieties has been performed by the use of AFLPs (Manubens *et al.* 1999) and microsatellites (Cipriani *et al.* 1999; Sosinski *et al.* 2000; Testolin *et al.* 2000). Peach rootstocks have been identified by RAPD markers (Lu *et al.* 1996).

Genome mapping

Linkage maps have been constructed on an interspecific cross between peach and almond (Foolad *et al.* 1995), on an almond x peach F_2 progeny (Joobeur *et al.* 1998) and on a F_2 population derived from a cross peach x nectarine (Dirlewanger *et al.* 1998). QTLs controlling fruit quality have been mapped by Dirlewanger et al. (1999). A reciprocal translocation between 'Garfi' almond and 'Nemared' peach chromosomes was detected by developing a map between these cultivars (Jauregui *et al.* 2001). AFLP (Lu *et al.* 1998) and a codominant marker (Lu *et al.* 1999) linked to the root-knot nematode resistance trait have been identified in peach rootstocks.

Genes related to reproduction and fruit ripening

The genes encoding ethylene biosynthetic enzymes have been identified and their regulation has been studied (Mathooko *et al.* 2001). Other genes have been characterised, encoding proteins of photosystem II (Bassett *et al.* 1998; Chung *et al.* 1998), or the ABP (auxin-binding protein) genes (Ohmiya *et al.* 1998) and their expression in leaves at various developmental stages has been observed (Sakanishi *et al.* 1998). Numerous molecular probes have been identified for the detection of virus (Heuss *et al.* 1999) and phytoplasma (Green *et al.* 1999) infections on peach.

3.9.4 Apricot

Cultivar identification and phylogeny

Variability within the apricot species has been assessed by RFLP markers (de Vicente *et al.* 1998).

Genome mapping

Molecular markers linked to self-compatibility (Tao *et al.* 2000) and to male sterility (Badenes *et al.* 2000) have been identified. The gene encoding for the polyphenol oxidase (PPO) has been characterised and its regulation defined (Chevalier *et al.* 1999).

3.9.5 Cherry

Cultivar identification and phylogeny

Numerous microsatellite (SSR) markers have been used to screen the germplasm of sweet cherry, sour cherry, black cherry (Downey and Iezzoni 2000) and tetraploid cherry (Cantini *et al.* 2001).

Genome mapping

Inheritance and linkage relationships of isozymes were established in two interspecific cherry progenies (Boskovic *et al.* 1997; Boskovic and Tobutt 1998). A genetic linkage map has been constructed in sour cherry on a cross progeny of tetraploid cultivars using RFLP markers (Wang *et al.* 1998) and QTL were located on the same map for some flower and fruit traits (Wang *et al.* 2000). RAPD and SCAR markers were identified in Myrabolan plum for a major dominant gene (*Ma1*), controlling root-knot nematode resistance (Lecouls *et al.* 1999). To avoid confusion in the assignment of sweet cherry cultivars to cross-compatibility groups, significant research has been devoted to the identification of markers linked to the S alleles or to the direct identification of S-RNase sequences (Boskovic *et al.* 2000).

Genes related to reproduction and fruit ripening

The S-alleles were identified, characterised and cDNAs were cloned (Tao *et al.* 1999). Twenty-five genomic DNA fragments, representing the six most common alleles, were cloned and sequenced and four new S-alleles were characterised (Wiersma *et al.* 2001). The gene of a thaumatin-like protein, abundantly expressed in ripe cherry fruits, has been identified (Fils-Lycaon *et al.* 1996).

3.9.6 Citrus

Cultivar identification and phylogeny

The diversity within the *Citrus* genus and the identification of putative parents was determined using DNA amplified fingerprinting (Luro *et al.* 1995), inter-simple sequence repeat (ISSR) (Fang and Roose 1997), isozymes and RFLPs (Fang *et al.* 1997). The phylogeny of the genus and the genetic origin of important species were investigated by RAPD, SCAR and cpDNA markers (Nicolosi *et al.* 2000). The distribution of *copia*-like retrotransposons throughout the Citrus genome has also been investigated (Asins *et al.* 1999) demonstrating a higher abundance in their genome in comparison to the genome of some *Prunus* species.

Genome mapping

Genetic linkage maps of citrus have been constructed (Cai *et al.* 1994; Kijas *et al.* 1997; Sankar and Moore, 2001) and some genes related to virus (Gmitter *et al.* 1996; Deng *et al.* 1997; Mestre *et al.* 1997; Fang *et al.* 1998; Cristofani *et al.* 1999) and nematode (Ling *et al.* 2000) resistance were mapped. Molecular markers linked to QTLs governing apomixis (Garcia *et al.* 1999) or yield and seed number (Garcia *et al.* 2000) were identified. The *Ctv* gene, controlling citrus tristeza virus resistance, was localised within a genomic region by a map-based cloning strategy and through chromosome walking (Deng *et al.* 2001a, 2001b).

3.9.7 Grape

Cultivar identification and phylogeny

A large set of markers has been produced to characterise grape cultivars, with particular reference to RFLPs (Bourquin *et al.* 1993), AFLPs (Cervera *et al.* 1998), RAPDs (Stravrakakis and Biniari 1998), SCARs (Xu and Bakalinski 1996), and microsatellites (Botta *et al.* 1995; Bowers *et al.* 1996, 1999a; Cipriani *et al.* 1994; Lamboy and Alpha 1998; Sefc *et al.* 1997, 1999; Thomas and Scott 1993; Thomas *et al.* 1994). The origin of the classic European wine grapes has been the subject of much speculation and the parental relationships were analysed by means of microsatellite loci in more than 300 cultivars (Bowers and Meredith 1997; Bowers *et al.* 1999b). For identification purposes cultivar-specific SCAR primers from single bands have been obtained for PCR fingerprinting (Vidal *et al.* 2000) and the optimum combination of RAPD/microsatellites has been established (Tessier *et al.* 1999). The microsatellite conservation across 15 different *Vitis* species was studied, demonstrating the possibility of extending the use of microsatellite markers to wild germplasm and inter-specific hybrids (Di Gaspero *et al.* 2000). The geographic origin of grape cultivars has also been investigated by microsatellite markers, detecting a significant genetic differentiation among cultivars sampled from seven European vine-growing regions (Sefc *et al.* 2000). To decipher homonyms and synonyms in grapevine within the varietal group of 'Schiave', AFLP and SSR markers were used (Fossati *et al.* 2001).

Genome mapping

Molecular marker-based linkage maps have been constructed for *Vitis* on interspecific cross populations using RAPD and RFLP markers (Lodhi *et al.* 1995) or RAPDs, AFLPs, microsatellites and CAPs where one microsatellite was linked to a single locus controlling sex in grapes (Dalbo *et al.* 2000).

Genes related to reproduction and fruit ripening

The global gene expression pattern has been studied in leaf and grape berries (Ablett *et al.* 2000) as well as SSRs derived from the ESTs in order to use them for mapping and genotyping (Scott *et al.* 2000). The expression of anthocyanin pathway genes (Boss *et al.* 1996), putative vacuolar invertase cDNAs related to sugar accumulation (Davies and Robinson 1996), have been studied in developing berries. cDNA clones encoding osmotin-like protein or alcohol dehydrogenase enzyme (Sarni-Manchado *et al.* 1997; Tesniere and Verries 2000), or putative cell wall and stress response proteins (Davies and Robinson

2000), have been cloned and characterised from ripening grape berries, as well as the thaumatin-like protein that accumulates at very high levels in conjunction with the onset of sugar accumulation and berry softening (Tattersall *et al.* 1997). Agamous and Shatterproof homologues (*Vvmads1*), isolated by differential display, are expressed both in flowers and developing berries (Boss *et al.* 2001). cDNAs induced by powdery mildew infection have been analysed (Jacobs *et al.* 1999).

3.10 Future perspectives

Different biotechnology methodologies, including gene transfer, somaclonal variation under selective pressure and protoplast hybridisation, have been available for genetic improvement in fruit crops by using explants both from zygotic and maternal origin. Although much effort has been focused on developing efficient protocols to regenerate transgenic plants by using reporter or marker, transgenic plants with some desirable agronomic traits have already been produced in several fruit species by using genes from both plant and non-plant origin. Efforts focused on resistance to biotic stress and fruit ripening have been the major areas of research in past years, while less work has been done on altering growth rates and providing cold stress resistance. Some obstacles still exist for some species in fundamental methodology, including gene transfer, genetic selection and efficient protocols for regeneration from cell and tissue cultures with maternal origin. However, it seems possible to overcome these last limitations following the recent contribution of a 'double regeneration system' which allows one to obtain and maintain morphogenesis in callusin for a long time in several fruit species including apple, cherry and olive. Up to now *Agrobacterium*-mediated genes have been widely used; in the future maybe microprojectile bombardment, which allows transformation of organelles, avoiding the spread of transgenic pollen, seems to be a more promising technique of transformation. Selection is often a laborious, time- and space-consuming process. It is advisable to use new selectable markers (*gfp, lecI, ipt, rolA, rolB, rolC, pmi, xyla*) in place of the old ones (*nptII, hpt, pat/bar*) to improve public acceptance and reduce the percentage of loss which can be quite high (more than 40%) in fruit crops such as apple, pear, banana, grapevine, citrange and sweet orange. Particularly interesting seems the suggested use of *ipt* gene from *A. tumefaciens* as a selectable gene, since it drastically induces regeneration from transformed cells and the identification of regenerants can be made visually since they change in morphology without the use of antibiotics or herbicides.

The abnormality of the plants, due to the continued expression of the marker gene, can be eliminated by using a MAT (Multi-Auto-Transformation) vector. This approach combines the use of genes that stimulate growth and morphogenesis for positive selection of transformed cells with an excision mechanism to remove the markers and allow recovery of plants with normal phenotypes. The vector, which includes *ipt* and maize transposable element *Ac* for removing the *ipt* gene, seems to be promising (Ebinuma *et al.* 1997). In addition *xyla* (xylose isomerase) (Haldrup *et al.* 1998) and *pmi* (Phosphomannose isomerase) genes, both confer the capacity for utilisation of normally non-metabolisable substrate. While *lecI, ipt, rolA, B, C* genes promoting growth and/or morphogenesis should be used because plants can be visually screenable. Finally, for the explants with a very high regeneration capacity it is advisable to eliminate any marker gene and select early or late by *in vitro* screening or in the greenhouse on physiology and morphology parameters according to the metabolism that has been modified (to combat toxins, culture filtrate of pathogens, salt drought resistance, etc.). Many genes have already been isolated from several species, which may be introduced in fruit crops singly or associated among them. Transformation experiments with multiple genes are in progress in several plants including olive and strawberry. The critical areas for future research will be mainly two: (i) the identification and evaluation of genes for useful traits with their specific promoters, and (ii) regulation of foreign gene expression in transgenic fruit trees.

3.11 Abbreviations used in this chapter

amp = antimicrobe peptide
A.rh = *Agrobacterium rhizogenes*
A.t. = *Agrobacterium tumefaciens*
ALS = acetolactate synthase (herbicide resistance)
APH = hygromycin phosphotransferase (hygromycin resistance)
attE or *attacinE* = lytic protein attacin E
BAR = phosphinorthiricin acetyltransferase (herbicide resistance)
btk-ICP = insecticidal crystal protein (*B. thuringensis* var. Kurstak)
cec = cecropin
CpTI = cowpea trypsin inhibitor
*cry*IAa, *cry*IAc = insecticidal crystal protein genes (ICPs) from *B. thuringensis*)
EP = electroporation
GCMV-CP = grapevine chrome mosaic virus-coat protein gene
GFLV-CP = grapevine fan leaf virus coat protein gene
GNA = *Galanthus nivalis* agglutinin
gus = β-glucuronidase gene
hEGF = human epidermal growth factor
HPT = hygromycin phosphotransferase (hygromycin resistance)
ipt = isopentyl transferase (cytokinin synthesis)
NOS = nopaline synthase
NPTII = neomycin phosphotransferase (kanamycin resistance)
OSH1 = homebox rice
PB = particle bombardment
PG = polyethylene glycol mediated

PGIP = polygalacturonase inhibitor protein
PPV-CP = plum pox virus-coat protein gene
PRV-CP = papaya ringspot virus-coat protein gene
RiT-DNA = root inducing T-DNA of *A. rhizogenes*
SAMase = S-adenosyl methionine hydrolase
S.V. = somaclonal variation
T = gene transformation
ThEn-42 = endochitinase from *Trichoderma harsianum*
TomRSV-CP = tomato ringspot virus coat protein gene
uncp = untranslatable coat protein gene of the *Citrus* tristeza virus

3.12 References and further reading

The reference section is divided into a general section followed by references specific to particular fruits.

General

ABDUL KADER A.M., NORELLI J.L., ALDWINCKLE H.S., BAUER D.W., BEER S.V., 1999. 'Evaluation of the *hrp N* gene for increasing resistance to fire blight in transgenic apple'. *Acta Hort.* 489: 247–250.

ALIA H.W., SAKAMOTO A., MURATA N., 1998. 'Enhancement of the tolerance of *Arabidopsis* to high temperatures by genetic engineering of the synthesis of glycine betaine'. *Plant J.* 16: 155–161.

ALONSO J.M., CHAMARRO J., GRANELL A., 1995. 'Evidence of the involvement of ethylene in the expression of specific RNAs during maturation of sweet orange, a non climateric fruit'. *Plant Mol. Biol.* 29: 385–390.

ALREFAI R.H., KORBAN S.S., 1995. 'Cross protection against virus diseases in fruit trees'. *Fruit. Var. J.* 49: 21–30.

ALSTAD D.N., ANDOW D.A., 1995. 'Managing the evolution of insects resistant to transgenic plants'. *Science.* 268: 1894–1896.

APSE M.P., AHARON G.S., SNEDDEN W.A., BLUMWALD E., 1999. 'Salt tolerance conferred by overexpression of a vacuolar W/W antiport in *Arabidopsis*'. *Science*, 285: 1256–1258.

ARAKAWA T., CHONG D.K., LANGRIDGE W.H., 1998. 'Efficacy of a food plant-based oral cholera toxin B subunit vaccine'. *Nature Bio/Technol.* 16: 292–297.

ARCHILLETTI T., LAURI P., DAMIANO C., 1995. '*Agrobacterium*-mediated transformation of almond leaf pieces'. *Plant Cell Rep.* 14: 267–272.

ARNTZEN C.J., 1998. 'Pharmaceutical foodstuffs – oral immunization with transgenic plants'. *Nature Medicine*, 4: 502–503.

ASHWORTH E.N., KIEFT T.L., 1995. 'Ice nucleation activity associated with plants and fungi'. In R.E. Lee, C.J. Warren, and L.V. Gusta (eds). *Biological Ice Nucleation and Its Applications*, APS Press, St, Paul, MN, pp. 137–162.

ASINS M.J., MESTRE P.F., NAVARRO L., CARBONELL E.A., 1999. 'Strategies to search for new Citrus Tristeza Virus resistant genotypes in a germplasm bank'.

In: G.T. Scarascia Mugnozza, E. Porceddu and M.A. Pagnotta (eds) *Genetics and Breeding for Crop Quality and Resistance.* Kluwer Academic Publishers, The Netherlands, pp. 251–256.

ATKINSON R.G., GARDNER R.C., 1991. 'Agrobacterium-mediated transformation of pepino and regeneration of transgenic plants'. *Plant Cell Rep.* 10: 208–212.

ATKINSON R.G., GARDNER R.C., 1993. 'Regeneration of transgenic tamarillo plants'. *Plant Cell Rep.* 12: 347–351.

AZIZ R., TEPFER M., 1999. 'Recombination in RNA viruses and in virus-resistant transgenic plants'. *J. General Virology* 8: 1339–1346.

BAJAJ S., TARGOLLI L., LIU L.R., HO T.H.D., WU R., 1999. 'Transgenic approaches to increase dehydration-stress tolerance in plants'. *Mol. Breeding* 5: 493–503.

BAKER J.E., WOO S.M., THRONE LE., FINNEY P.L., 1991. 'Correlation of α-amylase inhibitor content in eastern soft wheats with development parameters of the rice weevil (Coleoptera: Curculionidae).' *Environ. Entomol.* 20: 53–60.

BALESTRA G.M., RUGINI E., VARVARO L., 2001. 'Increasing of susceptibility to *Pseudomonas syringae* pv. *syringae* and *Pseudomonas viridiflava* of transgenic *rolABC* genes kiwi plants and its inheritance in T1 offspring'. *J. Phytopathol.* 149: 189–194.

BARBIER M., DULIEU H.L., 1980. 'Effets genetiques observés sur des plantes de tabac regenerees a partir de cotyledons par culture in vitro'. *Annales Amelioration Plantes* 30, 321–344.

BARBIERI L., VALBONESI P., BONORA E., GORINI P., BOLOGNESI A., STIRPE F., 1997. 'Polynucleotide: adenosine glycosidase activity of ribosome inactivating proteins: Effect on DNA, RNA, and poly-(A)'. *Nucl. Ac. Res.* 25: 518–522.

BARTHELS N., KARIMI M., VERCAUTEREN I., VAN MONTAGU M., GHEYSEN G., 1999. 'Integration of nematode-responsive regulatory sequences from *Arabidopsis thaliana* into nematode control strategies'. In: G.T. Scarascia Mugnozza, E. Porceddu and M.A. Pagnotta (eds.) *Genetics and Breeding for Crop Quality and Resistance.* Kluwer Academic Publishers, The Netherlands, pp. 203–210.

BATTISTINI C., ROSATI P., 1991. 'In vitro evaluation of somaclonal strawberry (*Fragaria x Ananassa* "Brighton") variants for susceptibility to *Phytophthora cactorum*'. In: A. Dale and W.W. Lubby (eds.) *The strawberry into the 21st Century.* 121–123. Timber Press, Portland, Oregon.

BAULCOMBE D., 1994. 'Replicase-mediated resistance. A novel type of transgenic resistance in transgenic plants'. *Trends in Microbiology* 2: 60–63.

BEHRINGER F.J., DAVIES P.J., 1992. 'Indole-3-acetic levels after phytochrome-mediated changes in stem-elongation rate dark- and light-grown *Pisum* seedlings'. *Planta* 188: 85–92.

BEN-HAYYIM G., GOFFER Y., 1989. 'Plantlet regeneration from NaCl-selected salt-tolerant callus culture of Shamouti orange (*Citrus sinensis* L. Osbeck)'. *Plant Cell Rep.* 7: 680–683.

BOERJAN W., VAN MONTAGU M., 1999. 'Molecular marker and genetic engineering strategies to improve wood quality in poplar'. In: G.T. Scarascia Mugnozza, E. Porceddu and M.A. Pagnotta (eds) *Genetics and Breeding for Crop Quality and Resistance.* Kluwer Academic Publishers, The Netherlands, pp. 271–282.

BOLAR J.P., NORELLI J.L., WONG K.W., HAYES C.K., HARMAN G.E., ALDWINCKLE H.S., 2000. 'Expression of endochitinase from *Tricoderma harzianum* in transgenic apple increases resistance to apple scab and reduces vigour'. *Phytopathology*, 90: 72–77.

BOLWELL G.P., BUTT V.S., DAVIES D.R. ZIMMERLIN A., 1995. 'The origin of the oxidative burst in plants'. *Free Rad. Res. Comm.* 23: 517–532.

BOREJSZA-WYSOCKA E.E., NORELLI J.L., KO K., 1999. 'Transformation of authentic M. 26 apple rootstock for enhanced resistance to fire blight'. *Acta Hort.* 489: 259–266.

BOSCHERINI G., MULEO R., MONTAGNI G., CINELLI F., PELLEGRINI M.G., BERNARDINI M., BURATTI M., 1999. 'Characterization of salt tolerant plants derived from a *Lycopersicon esculentum* Mill. somaclone'. *J. Plant Physiol.* 155: 613–619.

BOULTER D., 1993. 'Insect pest control by copying nature using genetically engineered crops'. *Phytochem.* 34: 1453–1466.

BOURDIN D., LECOQ H., 1991. 'Evidence that heteroencapsidation between two polyviruses is involved in aphid transmision of a non-aphid-transmissibile isolate from mixed infections'. *Phytopathology*, 81: 1459–1464.

BOWEN D.J., ENSIGN J.C., 1998. 'Purification and characterization of a high-molecular weight insecticidal protein complex produced by the entomopathogenic bacterium *Photorhabdus luminescens*'. *Appl. Environ. Microbiol.* 64: 3029–3035.

BOWLER C., SLOOTEN L., VANDENBRANDEN S., DE RYCKE R., BOTTERMAN J., 1991. 'Manganese superoxide dismutase can reduce cellular damage mediated by oxygen radicals in transgenic plants'. *EMBO J.* 10: 1723–1732.

BROADWAY R.M., 1995. 'Are insects resistant to plant protease inhibitors?' *J. Insect Physiol.* 41: 107–116.

BROADWAY R.M., DUFFEY S.S., 1986. 'Plant proteinase inhibitors: mechanism of action and effect on the growth and digestive physiology of lama, *Heliothis zea* and *Spodoptera exiqua*'. *J. Insect Physiol.* 32: 827–833.

BROEKAERT W.F., TERRAS F.R.G., CAMMUE B.P.A., OSBORN R.W., 1995. 'Plant defensins: novel antimicrobial peptides as components of the host defense system.' *Plant Physiol.* 108: 1353–1358.

BROOTHAERTS W., BONDT A., DE WITTE K., PAUWELS E., KEULEMANS J., 1998. 'Apple transformation'. *Med. Facul. Landb. Toeq. Biol. Weten. Univ. Gent.* 62: 1413–1419.

BROWN D.J.F., DALMASSO A., TRUDGILL D.L., 1993. 'Nematode pests of soft fruits and vines'. In: K. Evans, D.L. Trudgill, and J.M. Webster (eds). *Plant parasitic nematodes in temperate agriculture.* CAB Int., Wallingford, UK, pp. 427–462.

BURK L.G., MATZINGER D.F., 1976. 'Variation among anther-derived haploids from an inbred line of tobacco'. *J. Heredity* 67: 381–384.

BURKHARDT P.K, BEYER P., WUN J., KLOTI A., ARMSTRONG G., SCHLEDZ M., VON LINTING J., POTRYKUS I., 1997. 'Transgenic rice (*Oryza sativa*) endosperm expressing daffodil (*Narcissus pseudonarcissus*) Phytoene synthase accumulates phytoene, a key intermediate of provitamin A biosynthesis'. *Plant J.* 11: 1071–1078.

BUZKAN N., MINAFRA K., SAIDARELLI P., CASTELLANO M.K., MARTELLI G.P., 2000. 'Heteroencapsidation in transgenic and non-transgenic *Nicotiana* plants infected by Grapevine virus A and B'. *Extended Abstracts*, 13th Meeting 1CVG, Adelaide 2000, 38.

CABONI E., TONELLI M., FALASCA G., DAMIANO C., 1996. 'Factors affecting adventitious shoot regeneration in vitro in the apple rootstock "Jork 9"'. *Adv. Hort. Sci.* 10: 146–150

CABRERA-PONCE J.L., VEGAS-GARCIA A., HERRERA-ESTRELLA L., 1995. 'Herbicide resistant transgenic papaya plants produced by an efficient particle bombardment transformation method'. *Plant Cell Rep.* 15: 1–7.

CABRERA-PONCE J.L., VEGAS-GARCIA A., HERRERA-ESTRELLA L., 1996. 'Regeneration of transgenic papaya plants via somatic embryogenesis induced by *A. rhizogenes*'. *In Vitro Cellular Dev. Biol. Plant* 32: 86–90.

CACCIA R., DELLEDONNE M., BALESTRA G.M., VARVARO L., SORESSI G.P., 1999. 'Plant-bacterial pathogen interaction modified in transgenic tomato plants expressing the Gox gene encoding glucose oxidase'. In: G.T. Scarascia Mugnozza, E. Porceddu and M.A. Pagnotta (eds) *Genetics and Breeding for Crop Quality and Resistance*. Kluwer Academic Publishers, The Netherlands, pp. 119–125.

CAI W.Q., GONSALVES C., TENNANT P., FERMIN G., SONZA M., SARINDU N., JAN F.J., ZHU H.Y., GONZALVES O.D., 1999. 'A protocol for efficient transformation and regeneration of *Carica papaya* L'. *In Vitro Cell. Develop. Biol. Plant* 35: 61–69.

CAPELL T., ESCOBAR C., LIU H., BURTIN D., LEPRI O., CHRISTOU P., 1998. 'Overexpression of the oat arginine decarboxylase cDNA in transgenic rice (*Oryza sativa* L.) affects normal development patterns in vitro and results in putrescine accumulation in transgenic plants.' *Theor. Appl. Genet.* 97: 246–254.

CAPONE I., CARDARELLI M., TROVATO M., COSTANTINO P., 1989. 'Upstream non-cooling region which confers polar expression to Ri plasmid root inducing gene *rolB*'. *Mol. Gen. Genet.* 216: 239–244.

CARDARELLI M., MARIOTTI D., POMPONI M., SPANÓ L., CAPONE I., COSTANTINO P., 1987. '*Agrobacterium rhizogenes* T-DNA gene capable of inducing hairy root phenotype'. *Mol. Gen. Genet.* 209: 475–480.

CARLI A., PELLICANÒ D., MEZZETTI B., ROSATI P., 1993. '*Phytophthora cactorum* resistance evaluation with ion leakage method of petal and receptacle regenerants in the susceptible strawberry cv. "Pajaro"'. *Acta Hort.* 384: 422–426.

CARR P., ZAITLIN M., 1993. 'Replicase-mediate resistance'. *Seminars in Virology* 4: 330–347.

CERVERA M., 2000. 'Generation of transgenic citrus plants with the tolerance-to-salinity gene HAL2 from yeast'. *J. Hort. Sci.* 75: 26–30.

CERVERA M., JUAREZ J., NAVARRO A., PENA J.A., DURAN-VILA N., NAVARRO L., PENA L., 1998. 'Genetic transformation and regeneration of mature tissues of woody fruit plants bypassing the juvenile stage'. *Transgenic Res.* 7: 51–59.

CHALEFF R.S., RAY T.B., 1984. 'Herbicide resistant mutants from tobacco cell cultures'. *Science* 223: 1148–1151.

CHEN T.H.H., BURKE M.J., GUSTA L.V., 1995. Freezing tolerance in plants. In: R.E. Lee C. J. Warren, L.V. Gusta (eds) *Biological Ice Nucleation and Its Applications*, APS Press, St. Paul, MN,. 115–136.

CHENG Y.H., YANG J.S., YEH S. D., 1996. 'Efficient transformation of papaya by coat protein gene of papaya ringspot virus mediated by *Agrobacterium* following liquid phase wounding of embryogenic tissues with carborundum'. *Plant Cell Rep.* 16: 127–132.

CHERRY J.R., HERSHEY H.P., VIERSTRA R.D., 1991a. 'Characterization of tobacco expressing functional oat phytochrome: domains responsible for the rapid degradation of Pfr are conserved between monocots and dicots'. *Plant Physiol.* 96: 775–785.

CHERRY J.R., HONDRED D., KELLER J.M., HERSHEY H.P., VIERSTRA R.D., 1991b. 'The use of transgenic plants to study phytochrome'. In: B. Thomas and B.C. Johnson (eds) *Phytochrome properties and biological action*. Springer-Verlag.

CHEVREAU E., 1998. 'Fire blight resistance and genetic trueness-to-type of four somaclonal variants from the apple cultivar'. *Euphytica* 104: 199–205.

CHILDS K.L., LU J.L., MULLET J.E., MORGAN P.W., 1995. 'Genetic regulation of development in *Sorghum bicolor*. 10. Greatly attenuated photoperiod sensitivity in a phytochrome-deficient sorghum possessing a biological clock but lacking a red light-high irradiance response'. *Plant Physiol.* 108: 345–351.

CLARK T., MATHEW S., SCERROCK R.A., 1994. 'The phytochrome apoprotein family in Arabidopsis is encoded by five genes: the sequences and expression of phyD and phyE'. *Plant Mol. Biol.* 25: 413–427.

CLOSE T., 1997. 'Dehydrins: A commonality in the response of plants to dehydration and low temperature'. *Physiol. Plant* 100: 291–296.

COGHLAN A., 1997. 'Altered genes turn the worm'. *New Sci.* 153: 19.

CORNELISSEN B.J.C., MELCHERS L.S., 1993. 'Strategies for control of fungal diseases with transgenic plants'. *Plant Physiol.* 101: 709–712.

CREAMER R., FALK B.W., 1990. 'Direct detection of transcapsidated barley yellow dwarf luteovirus in double infected plants'. *J. Gen. Virology* 71: 211–217.

CZAPLA T.H., LANG B.A., 1990. 'Effect of plant lectins on the larval development of European corn borer and Southern corn rootworm'. *J. Econ. Entomol.* 83: 2480–2485.

D'AMATO F., 1975. 'The problem of genetic stability in plant tissue and cell cultures'. In: O.H. Frankel and J.G. Hawkes (eds) *Crop Genetic Resources for Today and Tomorrow*. Cambridge University Press, Cambridge, pp. 333–348.

D'ANGELI S., GUTIERREZ-PESCE P., ALTAMURA M.M., BIASI R., RUGGIERO B., MUGANU M., BRESSAN R., RUGINI E., 2001. 'Genetic transformation of olive tree (*Olea europaea* L.) with *osmotin* gene and *in situ* protein localisation in the transgenic tissues'. Italian Society of Agriculture Genetics, XLV Annual Congress, Salsomaggiore Terme (PR) Italy.

DAGERLIND A., FRIBERG K., BEAN A.J., HOKFELD T., 1992. 'Sensitive mRNA detection using unfixed tissue: combined radioactive and non radioactive in situ hybridisation'. *Histochemistry* 98: 39–49.

DAMIANO C., ARCHILLETTI T., CABONI E., LAURI P., FALASCA G., MARIOTTI D., FERRAIOLO G., NISHIO T., DORE C., 1995. '*Agrobacterium* mediated transformation of almond: in vitro rooting through localized infection of *A. rhizogenes* w.t.' *Acta Hort.* 392: 161–169.

DANDEKAR A.M., 1991. 'To conduct a planned release of genetically engineered apple plants'. *Biotechnology Permit* 91: 218–223.

DANDEKAR A.M., 1992. 'Transformation'. In: F.A.Q. Hammerschlag and R.E. Litz (eds), *Biotechnology of fruit crops.* CAB Int. Wallingford, UK, pp. 141–168.

DANDEKAR A.M., MARTIN L.A., MCGRANAHAN G.H., 1988. 'Genetic transformation and foreign gene expression in walnut tissue'. *J. Am. Soc. Hort Sci.* 113: 945–949.

DANDEKAR A.M., MCGRANAHAN G.H., VAIL P.V., URATSU S.L., LESLIE C.A., TEBBET J.S., 1994. 'Low levels of expression of wild type *Bacillus thuringiensis* var kurstaki cry 1(c)sequences in transgenic walnut somatic embryos'. *Plant Sci.* 96: 151–162.

DE BONDT A., EGGERMONT K., PENNINCKX L., GODERIS L., BROEKAERT W.F., 1996a. '*Agrobacterium*-mediated transformation of apple (*Malus x domestica* Borkh.): an assessment of factors affecting regeneration of transgenic plants.' *Plant Cell Rep.* 15: 549–554.

DE BONDT A., ZAMAN S., BROEKAERT W., CAMMUE B., KEULEMANS J., 1996b. 'Genetic transformation of apple for increased fungal resistance'. *EUCARPIA Symp. on Fruit Breeding and Genetics*. Oxford, 1–6/9/96 p. 5.

DELLEDONNE M., XIA Y., DIXON R., LAMB C., 1998. 'Nitricoxide functions as a signal in plant disease resistance'. *Nature* 394: 585–588.

DELLEDONNE M., XIA Y., DIXON R.A., LORENZONI C., LAMB C., 1999. 'Nitric oxide signalling in the plant hypersensitive disease resistance response'. In: G.T. Scarascia Mugnozza, E. Porceddu and M.A. Pagnotta (eds) *Genetics and Breeding for Crop Quality and Resistance.* Kluwer Academic Publishers, The Netherlands, pp. 127–133.

DENG X.X., GROSSER J.W., GMITTER F.G.J., 1992. 'Intergeneric somatic hybryd plants from protoplasts fusion of *Fortunella crassifolia* "Meiwa" with

Citrus sinensis cultivar "Valencia"'. *Sci Hort.* 49: 55–62.

DENG Z.N., GENTILE A., DOMINA F., NICOLOSI E., TRIBULATO E., 1995. 'Selecting lemon protoplast for insensitivity to *Phoma tracheiphila* toxin and regenerating tolerant plants'. *J. Am. Soc. Hort. Sci.* 120(6): 902–905.

DEVIDAS P., REHBERGER L.A., 1992. 'The effect of exotoxin from *Bacillus thuringiensis* on *Meloidogyne incognita* and *Caenorhabditis elegans*'. *Plant & Soil* 145: 115–120.

DING X., GOPALAKRISHNAN B., JOHNSON L.B., WHITE F.F., WANG X., MORGAN T.D., KNUMER K.J., MUTHUKRISHNAN S., 1998. 'Insect resistance of transgenic tobacco expressing an insect chitinase gene'. *Transgenic Research.* 7: 77–84.

DIX, P.J., STREET, H.E. 1976. 'Selection of plant cell lines with enhanced chilling resistance'. *Annals of Botany* 40: 903–910.

DOLCET-SANJUAN R., MOK D.W.S., MOK M.C., 1992. 'Characterization and in vitro selection for iron efficiency in *Pyrus* and *Cydonia*'. *In Vitro Cellular and Developmental Biology-Plant* 28: 25–9.

DOLGOV S.V., 1998. 'Genetic transformation of sour cherry (*Cerasus vulgaris* Mill.) leaf disks'. Biotechnology. In: Y.P.S. Bajaj (ed.) *Agriculture and Forestry*, vol. 44, 29–39. Springer-Verlag, Berlin.

DOLGOV S. V., LEBEDEV V.G., ANISIMOVA S.S., LAVROVA N., SERDOBINSKIY L.A., TJUKAVIN G.B., SHADENKOV S.A., LUNIN V.G., 1999a. 'Phytopathogen resistance improvement of horticultural crops by plant-defensin gene introduction'. In: G.T. Scarascia Mugnozza, E. Porceddu and M.A. Pagnotta (eds) *Genetics and Breeding for Crop Quality and Resistance.* Kluwer Academic Publishers, The Netherlands, pp. 111–118.

DOLGOV, S.V., LEBEDEV V.G., FIRSOV A.P., TARAN S.A., TJUKAVIN G.B., 1999b. 'Expression of thaumatin II gene in horticultural crops'. In: G.T. Scarascia Mugnozza, E. Porceddu and M.A. Pagnotta (eds) *Genetics and Breeding for Crop Quality and Resistance.* Kluwer Academic Publishers, The Netherlands, pp. 165–172.

DOLJA V.V., HALDERMAN R., ROBERTSON N.L., DOUGHERTY W.G., CARRINGTON J.C. 1994. 'Distinct functions of capsid protein in assembly and movement of tobacco etch potyvirus in plants'. *EMBO J.* 13: 1482–1491.

DOMINGUEZ A., GUERRI J., CAMBRA M., NAVARRO L., MORENO P., PEGNA L., 2000. 'Efficient production of transgenic citrus plants expressing the coat protein gene of citrus tristeza virus'. *Plant Cell Rep.* 19: 427–433.

DONOVAN A.M., MORGAN R., VALOBRA-PIAGNANI C., RIDOUT M.S., JAMES D.J., GARRET C.M.E., 1994. 'Assessment of somaclonal variation in apple. I. Resistance to the fire blight pathogen', *Erwinia amylovora. J. Hort. Sci.*, 69: 105–113.

DOWD P.F., LAGRIMINI L.M., HERMS D.A., 1998a. 'Differential leaf resistance to insects of transgenic sweetgum (*Liquidambar styraciflua*) expressing tobacco anionic peroxidase'. *Cell. Mol. Life Sci.* 54: 712–720.

DRUART P.H., DELPORTE F., BRAZDA M., UGARTE-BALLON C., DA CAMARA MACHADO A., LAIMER DA CAMARA MACHADO M., JACQUEMIN J., WATILLON

B., 1998. 'Genetic Transformation of Cherry trees'. *Acta Hort.* 468: 71–76.

DU C., MARTIN, A.W., NIKERSON, K.W., 1994. 'Comparison of disulfide contents and solubility at alkaline pH of insecticidal and noninsecticidal *Bacillus thuringiensis* protein crystals'. *Appl. Environ. Microbiol.* 60: 3847–3853.

DUFFEY S.D., STOUT M.J., 1996. 'Antinutritive and toxic components of plant defense against insects'. *Arch. Insect Biochem. Phys.* 32: 3–37.

DUFFEY S.S., FELTON G.W., 1991. 'Enzymatic antinutritive defenses of the tomato plant against insects'. In: P. Hedin, (ed.) *Naturally Occurring Pm Bioregulators*, American Chemical Society, Washington, DC, pp. 166–197.

DUMAN J.G., OLSEN T.M., 1993. 'Thermal hysteresis protein activity in bacteria, fungi, and phylogenetically diverse plants'. *Criobiology* 30: 322–328.

EBINUMA H., SUGITA K., MATSUNAGA E., YMAKADO M., 1997. 'Selection of marker free transgenic plants using the isopentenyl transferase gene'. *Proc. Natl. Acad. Sci. USA*. 94: 2117–2121.

EDENS L., HESLINGA L., LEDEBOER A.M., MAAT J., TOONEN M.Y., VISSER C., VERRIPS C.T., 1982. 'Cloning of cDNA encoding the sweet-testing plant protein thaumatin and its expression in *Escherichia coli*'. *Gene* 18: 1–12.

EL-EUCH C., HAY-ALLEMAND C., PASTUGLIA MDUMAS P., CHARENTIER J.P., APELLI P., JOUANIN L., 1998. 'Expression of antisense chalcone synthase RNA in transgenic hybryd walnut microcutting, effect on flavonoid content and rooting ability'. *Plant Mol. Biol.* 38: 467–479.

ESCOBAR M., DANDEKAR A.M., 2000. 'Development of insect resistance in fruit and nut tree crops'. In: S.M. Jain and S.C. Minocha (eds) *Molecular Biology of Woody Plants*, Vol. 2 Kluwer Academic Publishers, Dordrecht, pp. 395–417.

ESTRUCH J.J., CHRIQUI D., GROSSMANN K., SCHELL J., SPENA A., 1991a. 'The plant oncogene *rolC* is responsible for the release of cytokinins from glucoside conjugates'. *EMBO J.* 10: 2889–2895.

ESTRUCH J.J., SCHELL J., SPENA A., 1991b. 'The protein encoded by *rol B* plant oncogene hydrolyses indole glucosides'. *Embo J.* 10: 3125–3128.

FALASCA G., REVERBERI M., LAURI P., CABONI E., DE STRADIS A., ALTAMURA M.M., 2000. 'How *Agrobacterium rhizogenes* triggers de novo root formation in a recalcitrant woody plant: an integrated histological, ultrastructural and molecular analysis'. *New Phytol.* 145: 77–93.

FARINELLI L., MALNOE P., COLLET G.F., 1992. 'Heterologous encapsidation of potato virus Y strain 0 (PVYO) with the transgenic coat protein of PVY strain N (PVYN) in *Solanum tuberosum* cv. Bintje'. *Biotechnol.* 10: 1020–1025.

FELTON G.W., BI J.L., SUMMERS, C.B., MUELLER A.J., DUFFEY S.S., 1994. 'Potential role of lipoxygenase in defense against insect herbivory'. *J. Chem. Ecol.* 20: 651–666.

FILIPPINI F., ROSSI V., MARIN O., TROVATO M., COSTANTINO P., DOWNEY P.M., LOSCHIAVO F., TERZI M., 1996. 'A plant oncogene as a phosphatase'. *Nature* 379: 499–500.

FILLATTI J.J., HAISSI G., MCCOWN B., COMAI L., RIEMENSCHNEIDER D., 1987.

'Development of glyphosate tolerant *Populus* plants through expression of a mutant *avrA* gene from *Salmonella typhimurium*'. In: J.W. Hanover and D.E. Keathley (eds), *Genetic manipulation of woody plants*. Plenum Press, New York, pp. 243–249.

FILS-LYCAON B.R., WIERSMA P.A., EASTWELL K.C., SAUTIERE P., 1996. 'A cherry protein and its gene, abundantly expressed in ripening fruit has been identified as thaumathin-like'. *Plant Physiol.* 111: 269–273.

FINSTAD K., MARTIN R.R., RAMSDELL D.C., BARBA M., 1995. 'Transformation of strawberry for virus resistance'. *Acta Hort.* 385: 86–90.

FIRSOV A.P., DOLGOV S.V., 1997. 'Agrobacterial transformation of *Actinidia kolomikta*'. *Acta Hort.* 447, 323–327.

FITCH M.M.M., MANSHARDT R.M., GONSALVES D., SLIGHTOM J.M., 1992. 'Virus resistant papaya plants derived from tissues bombarded with the coat protein gene of papaya ringspot virus'. *Biotechnol.* 10: 1466–1472.

FITCH M.M.M., MOORE P., LIONG T., DREW R.A., 1998. 'Progress in transgenic papaya (*Carica papaya*) research: transformation for broader resistance among cultivars and micropropagating selected hybrid transgenic plants'. *Acta Hort.* 461: 315–319.

FITCHEN J.H., BEACHY R.N., 1993. 'Genetically engineered protection against viruses in transgenic plants'. *Annu. Rev. Microbiol.* 47: 739–763.

FLADUNG M., GIEFFERS W., 1993. 'Resistance reactions of leaves and tubers of *rol* C transgenic tetraploid potato to bacterial and fungal pathogens. Correlation with sugar, starch and chlorophyll content'. *Phys and Mol. Plant Path.* 42, 123–132.

GADANI F., MANSKY L.M., MEDICI R., MILLE W.A., HULL J.N., 1990. 'Genetic engineering of plants for virus resistance'. *Arch. Viral.* 115: 1–21.

GALL O.L., TORREGROZA L., DANGLOT Y., CANDRESSE T., BOUQUET A., 1994. 'Agrobacterium mediated genetic transformation of grapevine somatic embryos and regenerations of transgenic plants expressing the coat protein of grapevine chrome mosaic nepovirus (GCMV)'. *Plant Sci.* 102: 161–170.

GARCIA-OLMEDO F., MOLINA A., ALAMILLO J.M., RODRIGUEZ-PALENZUELA P., 1998. 'Plant defense peptides'. *Biopolymers* 47: 479–491.

GATEHOUSE A.M.R., GATEHOUSE L.A., 1998. 'Identifying proteins with insecticidal activity: Use of encoding gem to produce insect-resistant transgenic crops'. *Pestic. Sci.* 52: 165–175.

GENTILE A., TRIBULATO E., DENG Z.N., GALUN E., FLUHR R., VARDI A., 1993. 'Nucellar callus of "Femminello" lemon, selected for tolerance to *Phoma tracheiphila* toxin, shows enhanced release of chitinase and glucanase into the culture medium'. *Theor. Appl. Genet.* 86: 527–532.

GENTILE A., LA MALFA S., DENG Z.N., DOMINA F., NICOLOSI E., TRIBULATO E. 1999. 'Agrami transgenici: prime esperienze con i geni *rol*'. *Riv. Frutticoltura*, 1: 59–61

GIRARD C., LE METAYER M., BONADE-BOTTINO M., PHAM-DELEGUE M.H., JOUANIN L., 1998. 'High level of resistance to proteinase inhibitors may be conferred by proteolytic cleavage in beetle larvae'. *Insect Biochem. Mol. Biol.* 28: 229–237.

GMITTER G.F., GROSSER J.W., MOORE G.A., 1992. 'Citrus'. In: F.A. Hammerschlag and R.E. Litz (eds), *Biotechnology of perennial fruit crops.* CAB Int., Wallingford, UK, pp. 335–370.

GOMBOS Z., WADAL H., MURATA N., 1992. 'Unsaturation of fatty acids in membrane lipids enhances tolerance of the cyanobacterium *Synechocystis* PCC6803 to low-temperature photoinhibition'. *Proc. Natl. Acad. Sci. USA* 89: 9959–9963.

GONG Z., EWART K.V., HU Z., FLETCHER G.L., HEW C.L., 1996. 'Skin antifreeze protein genes of the winter flounder, *Pleuronectes americanum,* encode distinct and active polypeptides without the secretory signal and prosequences'. *J. Biol. Chem.* 271: 4106–4112.

GONSALVES D., 1998. 'Control of papaya ringspot virus in papaya: a case study'. *Annual Review of Phytopathology* 415–437.

GOULD F., 1988. 'Evolutionary biology and genetically engineered crops'. *BioScience* 38: 26–33.

GOULD F., 1998. 'Sustaining the efficacy of Bt toxins'. In: R.W.F Hardy and J.B. Segelken (eds) *National Agricultural Biotechnology Report 10*, pp. 77–86.

GRAHAM J., GORDON S.C., WILLIMSON B., 1996a. *Proc. Brighton Crop Protection Conf., Pests and Diseases.* Brighton, UK, pp. 777–782.

GRAHAM J., GREIG K., MCNICOL R.J., 1996b. 'Transformation of blueberry without antibiotic selection'. *Ann. App. Biol.* 128: 557–564.

GREENE A.E., ALLISON R.F., 1994. 'Recombination between viral RNA and transgenic plant transcripts'. *Science* 263: 1423–1425.

GRIBAUDO L., SCHUBERT A., 1990. 'Grapevine root transformation with *Agrobacterium rhizogenes*'. *Vitis* (special issue): 412–418.

GRIFFITH M., EWART K.V., 1995. 'Antifreeze proteins and their potential use in frozen foods'. *Biotechnol.* 13: 375–402.

GROSSER J.W., GMITTER Jr F.G., TUSA N., RECUPERO R.G., CUCINOTTA P., 1996. 'Further evidence of a cybridization requirement for plant regeneration from citrus leaf protoplasts following somatic fusion'. *Plant Cell Rep.* 15(9): 672–676.

GROSSER J.W., JIANG J., MOURAO-FO F.A.A., LOUZADA E.S., BAERGEN K., CHANDLER J.L., GMITTER Jr F.G., 1998. 'Somatic hybridization, an integral component of citrus cultivar improvement'. *Scion Improvement. HortScience*, 33: 1057–1059.

GRUMET R., 1994. 'Development of virus resistance plants via genetic engineering'. *Plant Breed. Rev.* 12: 47–79.

GUIS M., BEN AMOR M., BOTONDI R., AYUB R., LATCHÈ A., BOUZAYEN M., PECH J.C., 1999. 'Control of melon ripening by genetic engineering'. In: G.T. Scarascia Mugnozza, E. Porceddu and M.A. Pagnotta (eds) *Genetics and Breeding for Crop Quality and Resistance.* Kluwer Academic Publishers, The Netherlands, pp. 307–312.

GUO W.W., DENG X.X., 2001. 'Wide somatic hybrid of *Citrus* with its related genera and their potential in genetic improvement'. *Euphytica* 118: 175–183.

GUPTA A.S., HEINEN J.L., HOLADAY A.S., BURKE J.J., ALLEN R.D., 1993a. 'Increased resistance to oxidative transgenic plants that overexpress chloroplastic Cu/Zn-superoxide dismutase'. *Proc. Natl. Acad. Sci. USA* 90: 1629–1633.

GUPTA A.S., WEBB R.P., HOLADAY A.S., ALLEN R.D., 1993b. 'Overexpression of superoxide dismutase protects plants from oxidative stress: induction of ascorbate peroxidase in superoxide dismutase-overexpressing plants'. *Plant Physiol.* 103: 1067–1073.

GUTIERREZ E.A., MOORE G.A., JACONO C., MCCAFFERY M., CLINE K., 1992. 'Production of transgenic Citrus plants expressing the Citrus tristeza virus coat protein gene'. *Phytopathology* 82: 1148 (abstract).

GUTIERREZ-PESCE P., TAYLOR K., MULEO R., RUGINI E., 1998. 'Somatic embryogenesis and shoot regeneration from transgenic roots of cherry rootstock "Colt" (*Prunus avium x Prunus pseudocerasus*) mediated by pRi 1855 T-DNA of *Agrobacterium rhizogenes*'. *Plant Cell Rep.* 17: 574–578.

HAHN M.G., BUCHELI P., CERVONE F., DOARES S.H., O'NEILL R.A., DARVILL A., ALBERSHEIM P., 1989. 'Roles of cell wall constituents in plant-pathogen interactions'. *Plant-Microbe Interactions.* 3: 131–181.

HAIN R., BIESELER B., KINDEL H., SCHRODER G., STOCKER R., 1990. 'Expression of a stilbene sythase gene in *Nicotiana tabacum* results in synthesis of the phytoalexin resveratron'. *Plant Mol. Biol.* 15: 325–335.

HAJELA R.K., HAJELA N., BOLYARD M.G., BARNES W.M., STICKLEN M.B., 1993. 'A simple transformation system using adventitious shoot multiplication of Juneberry'. *Hort. Sci.* 28: 330–332.

HALDRUP A., PETERSEN S.G., OKKELS F.T., 1998. 'Positive selection: a plant selection principle based on xylose isomerase, an enzyme used in the food industry'. *Plant Cell Rep.* 18: 76–81.

HALL H.K., SKIRVIN R.M., BRAAN W.F., 1986. 'Germplasm release of "Lincoln logan" a tissue culture-derived genetic thornless "loganberry"'. *Fruit Var. J.* 40: 134–135.

HAMILTON A.J., LYCETT G.W., GRIERSON D., 1990. 'Antisense gene that inhibits synthesis of the hormone ethylene in transgenic plants'. *Nature* 346: 284–287.

HAMMERSCHLAG F.A., 1988. 'Selection of peach cells for insensitivity to culture filtrates of *Xanthomonas campestris* pv. *pruni* and regeneration of resistant plants'. *Theor. Appl. Genet.* 76: 865–869.

HAMMERSCHLAG F.A., 1990a. 'Resistant responses of plants regenerated from peach callus to *Xanthomonas campestris* pv. *pruni*'. *J. Am. Soc. Hort. Sci.* 115, 1034–1037.

HAMMERSCHLAG F.A., 1990b. 'Phenotypic stability of bacterial leaf spot and bacterial canker resistance in peach regenerants'. *VII Int. Cong. Plant Tissue and Cell Culture*, Amsterdam, p. 155.

HAMMERSCHLAG F.A., 1992. 'Somaclonal variation'. In: F.A. Hammerschlag and

R.E. Litz (eds) *Biotechnology of Perennial Fruit Crops.* CAB Int., Wallingford, UK, pp. 35–55.

HAMMERSCHLAG F.A., 2000. 'Resistant responses of peach somaclone 122-1 to *Xanthomonas campestris* pv. *pruni* and to *Pseudomonas syringae*' pv. *syringae*. *Hort. Sci.* 35: 141–143.

HAMMERSCHLAG F.A., OGNIANOV V., 1990. 'Somaclonal variation in peach: Screening for resistance to *Xanthomonas campestris* pv. *pruni* and *Pseudomonas syringae* pv *syringae*'. *Acta Hort.* 280: 403–408.

HAMMERSCHLAG F.A., SMIGOCKI A.C., 1998. 'Growth and in vitro propagation of peach plants transformed with the shooty mutant strain of *Agrobacterium tumefaciens*'. *Hort. Sci.* 33, 897–899.

HAMMOND-KOSACK K.E., JONES J.D.G., 1996. Plant disease resistance genes. *Annu. Rev. Plant Physiol. Plant Mol. Biol.* 48: 575–607.

HARTMAN J.B., VUYLSTEKE D., 1999. 'Breeding for fungal resistance in *Musa*'. In: G.T. Scarascia Mugnozza, E. Porceddu, and M.A. Pagnotta (eds) *Genetics and Breeding for Crop Quality and Resistance.* Kluwer Academic Publishers, The Netherlands, pp. 83–92.

HASEGAWA P.M., RAY A., BRESSAN R.A., 2000. 'Plant cellular and molecular responsis to high salinity'. *Annu. Rev. Plant. Physiol. Plant Mol. Biol.* 51: 463–499.

HASHMI G.P., HAMMERSCHLAG F.A., HUETTEL R.N., 1995. 'Growth, development, and response of peach someclones to the root-knot nematode, *Meloidogyne incognita*'. *J. Am. Soc. Hort. Sci.* 120: 932–937.

HAUSER B.A., CORDONNIER-PRATT M.M., DANIEL-VEDELE F., PRATT L.H., 1995. 'The phytochrome gene family in tomato includes a novel subfamily'. *Plant Mol. Biol.* 29: 1143–1155.

HAVSTAD P., SUTTON D., THOMAS S., SENGUPTA-GOPALAN C., KEMP J., 1991. 'Collagenase expression in transgenic plants: an alternative to nematicides'. *Mol. Biol. Plant Growth Develop.* p. 354 (abstract).

HIGHTOWER R., BADEN C., PENZES E., LUND P., DUNSMUIR P., 1991. 'Expression of antifreeze proteins in transgenic plants'. *Plant Mol. Biol.* 17: 1013–1021.

HOBBS H.A., MCLAUGHLIN M.R., 1990. 'A non aphid-transmissible isolate of bean yellow mosaic virus-Scott that is transmissibile from mixed infections with pea mosaic virus-204-1'. *Phytopathology* 80: 268–272.

HOLEFORS A., ZHONGTIAN X., WELANDER M., 1998. 'Transformation of the apple rootstock M26 with the *rolA* gene and its influence on growth'. *Plant Sci.* 136, 69–78.

HON W.C., GRIFFITH M., CHONG P., YANG D.S.C., 1994. 'Extraction and isolation of antifreeze proteins from winter rye leaves'. *Plant Physiol.* 104: 971–980.

HON W.C., GRIFFITH M., MLYNARZ A., KWOK Y.C., YANG D.S.C., 1995. 'Antifreeze proteins in winter rye are similar to pathogenesis-related proteins'. *Plant Physiol.* 109: 879–889.

HWANG S.C., 1985. 'Ecology and control of fusarial wilt of banana'. *Plant Protection Bulletin* (*Taiwan*), 27: 233–245.

HWANG S.C., 1990. 'Somaclonal resistance in Cavendish banana to Fusarium wilt'. In: R.C. Ploetz (ed.), *Fusarium Wilt of Banana.* APS Press, St. Paul, pp. 121–125.

HWANG S.C., KO W.H., 1987. Somaclonal variation of bananas and screeening for resistance to Fusarium wilt. In: G.J. Persley and E.A. De Langhe (eds), *Banana and Plantain Breeding Strategies.* ACIAR Proceedings, No. 21, Canberra, pp. 151–156.

IMAI R., CHANG L., OHTA A., BRAY E.A., TAKAGI M., 1996. 'A lea-class gene of tomato confers salt and freezing tolerance when over-expressed in *Saccharomyces cerevisiae*'. *Gene* 170: 243–248.

JACKSON S.D., 1999. 'Multiple signalling pathways control tuber induction in potato'. *Plant Physiol.* 119: 1–8.

JACQUEMOND M., TEPFER M., 1998. 'Satellite RNA-mediated resistance to plant viruses: are the ecological risks well assessed?' In: A. Hadidi, R.H. Khetarpal and H. Koganezawa (eds) *Plant Virus Disease Control* American Phytopathological Society Press, St. Paul, pp. 94–120.

JAGLO-OTTOSEN K.R., GILMOUR S.J., ZARKA D.G., SCHABENBERGER O., THOMASHOW M.F., 1998. 'Arabidopsis CBF1 over-expression induces COR genes and enhances freezing tolerance'. *Science* 280: 104–106.

JAIN R.K., SELVARAJ G., 1997. 'Molecular genetic improvement of salt tolerance in plants'. *Biotechnol. Ann. Rev.* 3.

JAIN S.M., 2001. 'Tissue culture-derived variation in crop improvement'. *Euphytica*, 118: 153–166.

JAMES D.J., PASSEY A.J., BAKER S.A., 1994. 'Stable gene expression in transgenic apple tree tissues and segregation of transgenes in the progeny: preliminary evidence'. *Euphytica* 77: 119–121.

JAMES, D.J., PASSEY A.J., BAKER S.A., WILSON F.M., 1996. 'Transgene display stable patterns of expression in apple fruit and Mendelian segregation in the progeny'. *Bio/Technol.* 14: 56–60.

JAMES D.J., PASSEY A.J., ESTERBROOK M.A., SOLOMON M.G., BARBARA D.J., 1992. *Phytoparasitica*, 20: 83–87.

JAMES D.J., URASTU S., CHENG J.S., NEGRI P., VISS P., DANDEKAR A.M., 1993. 'Acetosyringone and osmoprotectants like betain or prolin synergistically enhance Agrobacterium-mediated transformation of apple'. *Plant. Cell. Rep.* 12: 559–563.

JAYASHANKAR S., LITZ R.E., GRAY D.J., MOON P.A., 1999. 'Responses of embryogenic mango cultures and seedling bioassays to a partially purified phytotoxin produced by a mango leaf isolate of *Colletotrichum gloeosporioides* Penz'. *In Vitro Cell and Develop. Bio. (Plant)* 35: 475–479.

JAYNES J.M., 1993. 'Use of genes encoding novel lytic peptides and proteins that enhance microbial disease resistance in plants'. *Acta Hort.* 336: 33–39.

JONGSMA M.A., BAKKER P.L, PETERS J., BOSCH D., STIEKEMA W.J., 1995. 'Adaptation of *Spodoptera exigua* larvae to plant procinase inhibitors by induction of gut proteinase activity insensitive to inhibition'. *Proc. Natl. Acad. Sci.*

USA 92: 8041–8045.

JOUANIN L., VILAINE F., TOURNEUR J., TOURNEUR C., PAUTOT V., MULLER J.F., CABOCHE M., 1987. 'Transfer of a 4.3-kb fragment of the TL-DNA to regenerated tobacco plants'. *Plant Sci.* 53: 53–63.

KANEYOSHI J., KOBAYASHI S., 1999. 'Characteristics of transgenic trifoliate orange (*Poncirus trifoliata* Raf.) possessing the *rolC* gene of *Agrobacterium rhizogenes* Ri plasmid'. *J. Jap. Soc. Hort. Sci.* 68(4): 734–738.

KANIEWSKI W.K., LAWSON E.C., 1998. 'Coat protein and replicase-mediated resistance to plant viruses'. In: A. Hadidi, R.H. Khetarpal and H. Koganezawa (eds) *Plant Virus Disease Control.* American Phytopathological Society Press, St. Paul, pp. 65–78.

KASUGA M., LITI Q., MIURA S., YAMAGUEHI-SHINOZAKI K., SHINOZAKI K., 1999. 'Improving plant drought, salt, and freezing tolerance by gene transfer of a single stress-inducible transcription factor'. *Nature Biotechnol.* 17: 287–291.

KEEN N.T., YOSHIKAWA M., 1983. 'β-1, 3 endoglucanase from soybean releases elicitor-active carbohydrates from fungus cell walls'. *Plant Physiol.* 71: 460–465.

KELLY T.J., MASLER E.P., MENN J.J., 1990. 'Insect neuropeptides: new strategies for insect control'. *Pesticides and Alternatives*, Elsevier Science, pp. 283–297.

KHETARPAL R.K., MAISONNEUVE B., MAURY Y., CHALHOUB B., DINANT S., LECOQ H., VARMA A., 1998. 'Breeding for resistance to plant viruses'. In: A. Hadidi, R.H. Khetarpal and H. Koganezawa (eds) *Plant Virus Disease Control.* American Phytopathological Society Press, St. Paul, pp. 1–32.

KLEE H.J., HAYFORD M.B., KRETZMER K.A., BARRY G.F., KISHMORE G.M., 1991. 'Control of ethylene synthesis by expression of a bacterial enzyme in transgenic tobacco plants'. *Plant Cell* 3: 1187–1193.

KO K., NORELLI J.L., BROWN S.K., ALDWINCKLE H.S., DURING K., 1999. 'Galaxy line transgenic for attacin E and T4 lysozyme genes have increased resistance to fire blight'. In: A. Altman, M. Ziv and S. Izhar (eds) *Plant Biotechnology and In Vitro Biology in the 21st Century*, vol. 36. Kluver Academic Publishers, Dordrecht, pp. 507–511.

KOBAYASHI S., UCHIMIYA H., 1989. 'Expression and integration of a foreign gene in orange (*Citrus sinensis* Osb.) protoplast by direct DNA transfer'. *Japan. J. Genet.* 64: 91–97.

KOBAYASHI S., NAKAMURA Y., KANEYOSHI J., HIGO H., HIGO K., 1996. 'Transformation of kiwifruit (*Actinidia chinensis*) and trifoliate orange (*Poncirus trifoliata*) with a synthetic gene encoding the human epidermal growth factor (HEGF)'. *J. Japan. Soc. Hort. Sci.* 64: 763–769.

KOCHBA J., BEN-HAYYIM G., SPIEGEL-ROY P., SAAD S., NEUMANN H., 1982. 'Selection of stable salt-tolerant callus cell lines and embryos in *Citrus sinensis* and *C. aurantium*'. *Zeitschrift für Pflanzenphysiologie* 106, 111–118.

KRAMER K.J., MUTHUKRISHNAN S., 1997. 'Insect chitinases: Molecular biology and potential use as biopesticides'. *Insect Biochem. Molec. Biol.* 27: 887–900.

KRASTANOVA S., PERRIN M., BARBIER P., DEMANGEAT G., CORNUET P., BARDONNET N., OTTEN L., PINCK L., WALTER B., 1995. 'Transformation of grapevine rootstocks with coat protein gene of grapevine fanleaf nepovirus'. *Plant Cell. Rep.* 14: 550–554.

KRATTIGER A.F., 1997. Insect Resistance in Crops: A case study of *Bacillus thuringiensis* (Bt) and its transfer to developing countries. International Service for the Acquisition of Agri-Biotech Applications, AmeriCenter at Cornell University, USA.

KUKSOVA V.B., PIVEN N.M., GLEBA Y.Y., 1997. 'Plant cell variation and in vitro induced mutagenesis in grapevine'. *Plant Cell Tissue and Organ Culture* 49: 17–27.

KUSABA S., KANO-MURAKAMI Y., MATSUOKA M., FUKUMOTO M., NISHIO T., DORE C., 1995. 'A rice homeobox containing gene altered morphology of tobacco and kiwifruit. Genetic improvement of horticultural crops by biotechnology'. XXIVth Int. Hort. Cong., 21–27 August 1994, Kyoto, Japan. *Acta Hort.* 392: 203–208.

LAMB C., DIXON R.A., 1997. 'The oxidative burst in plant disease resistance'. *Annu. Rev. Plant Physiol. Plant. Mol. Biol.* 48: 251–275.

LAMB C.J., RYALS J.A., WARD E.R., DIXON R.A., 1992. 'Emerging strategies for enhancing crop resistance to microbial pathogens'. *Biotechnol.* 11: 1436–1445.

LAMBERT C., TEPFER D., 1992. 'Use of *Agrobacterium rhizogenes* to create transgenic apple trees having an altered organogenic response to hormones'. *Theor. Appl. Genet.* 85: 105–109.

LANHAM P.G., KEMP R.J., JONES H., BRENNAN R.M., 2001. 'Expression of dehydrin-like genes in response to chilling in leaves of blackcurrant, *Ribes nigrum* L'. *J. Hort. Sci. Biotechnol.*, 76: 201–207.

LARKIN P.J., SCOWCROFT W.R., 1981. 'Somaclonal variation – a novel source of variability from cell cultures for plant improvement'. *Theor. Appl. Genet.* 60, 197–214.

LASKOWSKI W., KATO I., KOHR W.J., PARK S.J., TAHIRO M., WHATLEY H.F., 1987. 'Positive Darwinian selection in evolution of protein inhibitors of serine proteinases'. *Cold Spring Harbor Symp. Quant. Biol.* 545–550.

LEE J.W., SCHOFFI F., 1996. 'An Hsp70 antisense gene affects the expression of HSP70ffiSC70, the regulation of HSF, and the acquisition of thermo-tolerance in transgenic *Arabidopsis thaliana*'. *Mol. Gen. Genet.* 252: 11–19.

LEVINE A., TENHAKEN R., DIXON R., LAMB C., 1994. 'H_2O_2 from the oxidative burst orchestrates the plant hypersensitive disease resistance response'. *Cell* 79: 583–593.

LI J., NAGPAL P., VITART V., MCMORRIS T., CHORY J., 1996. 'A role for brassinosteroids in light-dependent development of Arabidopsis'. *Science* 272: 398–401.

LITZ R.E., MATHEWS V.H., HENDRIX R.C., YURGALEVITCH C., 1991. 'Mango somatic cell genetics'. *Acta Hort.* 291, 133–140.

LIU D., RAGHOTHAMA K.G., HASEGAWA P.M., BRESSAN R.A., 1994. 'Osmotic over expression in potato delays development of disease symptoms'. *Proc. Natl. Acad. Sci. USA* 91: 1888–1892.

LIU J., ZHU J.K., 1998. 'A calcium sensor homologue required for plant salt tolerance'. *Science* 280: 1943–1945.

LIU Q., INGERSOLL J., OWENS L., SALIH S., MENG R., HAMMERSCHLAG F., 2001. 'Response of transgenic Royal Gala apple (*Malus X domestica* Borkh.) shoots carrying a modified cecropin MB39 gene to *Erwinia amylovora*'. *Plant Cell Rep.* 20: 306–312.

LIU Q., KASUGA M., SAKUMA Y., ABE A., MIURA S., 1998. 'Two transcription factors, DREB1 & DREB2, with an EREBP/AP2 DNA binding domain separate two cellular signal transduction pathways in drought- and low-temperature responsive gene expression, respectively, in *Arabidopsis*'. *Plant Cell.* 10: 1391–1406.

LOEW R., ROCKEL B., KIRSCH M., RATAJCZAK R., HORTENSTEINER S., 1996. 'Early salt stress effects on the differential expression of vacuolar H^+-ATPase genes in roots and leaves of *Mesembryanthemum crystallinum*'. *Plant Physiol.* 110: 259–65.

LOLLETTI D., 1999. Miglioramento genetico della fragola e valutazione agronomica. Thesis of University of Tuscia, Faculty of Agriculture Viterbo (Italy).

LONGEMANN J., JACK G., TOMMERUP H., MUNDAY J., SCHELL J., 1992. 'Expression of barley ribosome-inactivating protein lead to increased fungal protection in transgenic tobacco plants'. *Biotechnol.* 10: 305–308.

LOSEY J.E., RAYOR L.S., CARTER M.E., 1999. 'Transgenic pollen harms monarch larvae'. *Nature* 399–214.

LOUZADA E.S., GROSSER J.W., GMITTER F.G., NIELSEN B., CHANDLER J.L., DENG X.X., TUSA N., 1992. 'Eight new somatic hybrid citrus rootstocks with potential for improved disease resistance'. *Hort. Sci.* 27: 1033–1036.

MA J.K., HEIN M.B., 1995. 'Immunotherapeutic potential of antibodies produced in plants'. *Trends in Biotechnology.* 13: 522–527.

MA J.K., HIKMAT B.Y., WYCOFF K., 1998. 'Characterization of a recombinant plant monoclonal secretory antibody and preventative immunotherapy in humans'. *Nature Medicine* 4: 601–606.

MA S.W., ZHAO D.L., YIN Z.Q., 1997. 'Transgenic plants expressing autoantigens fed to mice to induce oral immune tolerance'. *Nature Medicine* 3: 793–796.

MACHADO A.C., KATINGER H., MACHADO M.L.C., 1994. 'Coat protein-mediated protection against plum pox virus in herbaceous model plants and transformation of apricot and plum'. *Euphytica* 77: 129–134.

MACHADO M.L.C., MACHADO A.C., HANZER V., WEISS H., REGNER F., STEINKELLNER H., MATTANOVICH D., PLAIL R., KNAPP E., KALTHAFF B., KATINGER H., 1992. 'Regeneration of transgenic plants of *Prunus armeniaca* containing the coat protein gene of plum pox virus'. *Plant Cell Rep.* 11: 25–29.

MADDALONI M., FORLANI F., BALMAS V., DONINI G., CORAZZA L., FANG H., PINCUS S., MOTTO M., 1999. 'The role of b-32 protein in protecting plants against pathogens'. In: G.T. Scarascia Mugnozza, E. Porceddu and M.A. Pagnotta (eds) *Genetics and Breeding for Crop Quality and Resistance.* Kluwer Academic Publishers, The Netherlands, pp. 77–82.

MALEHORN D.E., BORGMEYER J.R., SMITH C.E., SHAH D.M., 1994. 'Characterization and expression of an antifungal zeamatin-like protein (Zlp) gene from *Zea mays*'. *Plant Physiol.* 106: 1471–1481.

MARIANI C., DE BEUCKELEER M., TRUETTNER J., LEEMANS J., GOLDBERG R.B., 1990. 'Induction of male sterility in plants by a chimaeric ribonuclease gene'. *Nature* 347: 737–741.

MARIANI C., GOSSELE V., DE BEUCKELEER M., DE BLOCK M., GOLDBERG R.B., DE GREEF W., LEEMANS J., 1992. 'A chimaeric ribonuclease-inhibitor gene restores fertiliy to male sterile plants'. *Nature* 357: 384–387.

MARINO G., BATTISTINI S., 1990. 'Leaf-callus growth, shoot regeneration and somaclonal variation'. In: *Effect of medium pH. Acta Hort.* 280: 37–44.

MARINO G., BERTAZZA G., 1998. 'Selection pressure effects of medium pH during regeneration on successive performance of leaf derived cv Hayward kiwifruit *(Actinidia deliciosa)* somaclones cultured on proliferation media with variable pH'. *J. Hort. & Biotech.* 73: 664–669.

MARINO G., BERTAZZA G., BUSCAROLI C., 1998. 'In vivo growth and tolerance to lime-induced iron chlorosis of leaf-derived cvs Tomuri and Hayward kiwifruit *(Actinidia deliciosa)* somaclones'. *J. Hort. & Biotech.* 73: 670–675.

MARTELLI G.P., GALLITELLI D., RUSSO M., 1999. 'An appraisal of pathogen-derived resistance for the control of virus diseases'. In: G.T. Scarascia Mungozza, E. Porceddu, M.A. Pagnotta (eds) *Genetics and Breeding for Crop Quality and Resistance*. Kluwer Academic Publishers, Dordrecht, pp. 223–232.

MARTINELLI L., MANDOLINO G., 1994. 'Genetic transformation and regeneration of transgenic plants in grapevine (*Vitis rupestris* S.)'. *Theor. Appl. Genet.* 88: 621–628.

MATHEWS H., WAGONER W., COHEN C., KELLOGG J., BESTWICK R., 1995a. 'Efficient genetic transformation of red raspberry, *Rubus ideaus* L'. *Plant Cell Rep.* 14: 471–476.

MATHEWS H., WAGONER W., KELLOGG J., BESTWICK R., 1995b. 'Genetic transformation of strawberry: stable integration of a gene to control biosynthesis of ethylene'. *In Vitro Cell Develop. Biology Plant.* 31: 36–43.

MATSUMOTO K., BARBOSA M.L., SOUZA L.A.C., TEIXEIRA J.B., 1995. 'Race 1 Fusarium wilt tolerance on banana plants selected by fusaric acid'. *Euphytica* 84: 67–71.

MATSUMOTO K., BARBOSA M.L., SOUZA L.A.C., TEIXEIRA J.B., 1999a. 'In vitro selection for Fusarium wilt resistance in banana. II. Resistance to culture filtrate of race 1 *Fusarium oxysporum* f.sp. *cubens*'. *Fruits* 54: 151–157.

MATSUMOTO K., SOUZA L.A.C., BARBOSA M.L., 1999b. 'In vitro selection for Fusarium wilt resistance in banana. 1. Co-cultivation techinique to produce culture filtrate of race 1 *Fusarium oxysporum* f.sp. *cubens*'. *Fruits* 54: 97–102.

MAUCH F., HADWIGER L.A., BOLLER T., 1988. 'Antifungal hydrolases in pea tissues. I. Purification and characterisation of two chitinases and two β-1,3-glucanases differentially regulated during development and in response to fungal infection'. *Plant Physiol.* 87: 325–333

MAURO M.C., TOUTAIN S., WALTER B., PINCK L., OTTEN L., COUTOS-THEVENOT P., DELOIRE A., BARBIER P., 1995. 'High efficiency regeneration of grapevine plants transformed with the GFLV coat protein gene'. *Plant Sci.* 112: 97–160.

MAURO M.L., TROVATO M., GALLELLI A., COSTANTINO P., ALTAMURA M.M., 1996. 'The plant oncogene *rolD* stimulates flowering in transgenic tobacco plants'. *Developmental Biology*, 180: 693–700.

MAY G.D., AFZA R., MASON H.S., WIECKO A., NOVAK F.J., ARNTZEN C.J., 1995. 'Generation of transgenic banana (*Musa acuminata*) plants via *Agrobacterium*-mediated transformation'. *Biotechnol.* 13: 486–492.

MAZZARA M., MEZZETTI B., JAMES J.D. NEGRI P., 1998. 'Il gene *rolC* in fragola'. *L'informatore Agrario*, 29: 46–49.

MCGAUHEY, W.H., AND WHALON M.E., 1992. 'Managing insect resistance to *Bacillus thuringiensis* toxins'. *Science* 258: 1451–1455.

MCGRANAHAN G.H., LESLIE C.A., URATSU S., MARTIN L.A., DANDEKAR A.M., 1988. '*Agrobacterium*-mediated transformation of walnut somatic embryos and regeneration of transgenic plants'. *Biotechnol.* 6: 800–804.

MCGRAVEY D.J., YU H., CHRISTOFFERSEN R.E., 1990. 'Nucleotide sequence of a ripening related complementary DNA from avocado fruit'. *Plant Mol. Biol.* 15: 165–168.

MCKERSIE B.D., BOWLEY S.R., JONES K.S., 1999. 'Winter survival of transgenic alfalfa overexpressing superoxide dismutase'. *Plant Physiol.* 119: 839–847.

MCNICOL R.J., GRAHAM J., 1990. 'Genetic manipulation in *Rubus* and *Ribes*'. *Acta Hort.* 262: 41–46.

MCOAUHEY W.H., WHALON M.E., 1992. 'Managing insect resistance to *Bacillus thuringiensis* toxins'. *Science* 258: 1451–1455.

MCPHEETERS K., SKIRVIN R.M., 1983. 'Histogenic layer manipulation in chimeral "Thornless Evergreen" trailing blackberry'. *Euphytica* 32: 351–360.

MEHLENBACHER S.A., 1995. 'Classical and molecular approaches to breeding fruit and nut crops for disease resistance'. *Hort. Sci.* 30: 466–477.

MEINS F.JR., BINNS A.N., 1977. 'Epigenetic variation of cultured somatic cells: Evidence for gradual changes in requirement for factors promoting cell division'. *Proc. Natl. Acad. Sci. USA* 74, 2928–2932.

MELCHERS L.S., PONSTEIN A.S., SELA-BUURLAGE M.B., VLOEMANS S.A., CORNELISSEN B.J.C., 1993. 'In vitro anti-microbial activities of defence proteins and biotechnology'. In: B. Fritig and M. Legrand (eds) *Mechanisms of plant defense responses.* Kluwer Academic Publishers, Dordrecht, The Netherlands, pp. 401–410.

MICHAUD D., 1997. 'Avoiding protease-mediated resistance in herbivorous pests'. *Trends Biotechnol.* 15(1): 4–6.

MIKKELSEN J.D., BERGLUND L., NIELSON K.K., CHRISTIANSEN H., BOJSEN K., 1992. 'Structure of endochitinase genes from sugar beets'. In: C.J. Brine, P.A.

Sanford, and J.P. Zikakis (eds), *Advances in chitin and chitosan.* Elsevier Science Publ., London, UK, pp. 344–35.

MILLER W.A., KOEV G., MOHAN B.R., 1997. 'Are risks associated with transgenic resistance to luteoviruses?' *Plant Disease* 81: 700–710.

MOORE G.A., CUTIERREZ E.A., JACONO A., JACONO C., MCCAFFERY M., CLINE K., 1993. 'Production of transgenic citrus plants expressing the citrus tristeza virus coat protein gene'. *Hort. Sci.* 28: 512.

MOREIRA C.D., CHASE C.D., GMITTER F.G. JR., GROSSER J.W., 2000. 'Inheritance of organelle genomes in citrus'. *Molecular Breeding* 6: 401–405.

MORIGUCHI T., MOTOMURA T., HIDAKA T., AKIHAMA T., OMURA M., 1997. 'Analysis of mitochondrial genomes among *Citrus* plants produced by the interspecific somatic fusion of "Seminole" tangelo with rough lemon'. *Plant Cell Rep* 16: 397–400.

MOTOMURA T., HIDAKA T., MORIGUCHI T., AKIHAMA T., OMURA M., 1995. 'Intergeneric somatic hybrids between *Citrus* and *Atalantia* or *Severinia* by electrofusion, and recombination of mitochondrial genomes'. *Breed Sci.* 45: 309–314.

MOURGUES F., BRISSET M.N., CHEVREAU E., 1998. 'Strategies to improve plant resistance to bacterial diseases through genetic engineering'. *Trends Biotechnology*, 16: 203–210.

MUIR S.R., COLLINS G.J., ROBINSON S., HUGHES S., BOVY A., RIC DE VOS C.H., VAN TUNEN A.J., VERHOEYEN M.E., 2001. 'Overexpression of petunia chalcone isomerare in tomato results in fruit containing increased levels of flavonols'. *Plant Research* 19: 470–474.

MULEO R., IACONA C. 1998. 'Regolazione dell'habitus vegetativo da parte del fitocromo, evidenze fenotipiche in piante di ciliegio sovraesprimenti il fitocromo A di riso'. *Agro. bio. fruit* pp. 14–15.

MULEO R., THOMAS B., 1993. 'The role of light quality in the control of branching in *Prunus cerasifera* shoot tip cultured in vitro'. *European Symp. Photomorphogenesis in Plants* (abstract).

MULEO R., THOMAS B., 1997. 'Effect of light quality on shoot proliferation of *Prunus cerasifera* in vitro are the result of differential effects on bud induction and apical dominance'. *J. Hort. Sci.* 72, 483–491.

MULEO R., BOSCHERINI G., BIGOZZI L., LEVA A.R., BUIATTI M., 1996. 'Differential appearence of specific RAPD polymorphic fragments in Actinidia plants regenerated in high osmotic and saline media'. *Plant Physiol. Biochem.*, special issue, p. 251.

MURAKAMI T., ANZAI H., IMAI S., SATO A., NAGAOKA K., THOMPSON C.J., 1986. 'The bialaphos biosynthetic genes of *Streptomyces hygroscopicus*: molecular cloning and characterisation of the gene cluster'. *Mol. Gen. Genet.* 205: 42–50.

MURASHIGE T., 1974. 'Plant propagation through tissue culture'. *Annual Review of Plant Physiology* 25, 135–166.

MURATA N.W., ISHIZAKI-NISHIZAWA S., HIGASHI S., HIGASH H., TASAKA Y., NISHIDA I., 1992. 'Genetically engineered alterations in chilling sensitivity of plants'. *Nature* 356: 710–713.

NADEL B., SPIEGEL-ROY P., 1987. 'Selection of Citrus lemon culture variants resistant to the mal secco toxin'. *Plant Sci.* 53, 177–182.

NAKANO M., HOSHINO Y., MII M., 1994. 'Regeneration of transgenic plants of grapevine (*Vitis vinifera* L.) via *Agrobacterium rhizogenes*-mediated transformation of embryogenic calli'. *J. Expt. Bot.* 45: 649–656.

NAVARRO L., ORTIZ J.M., JUAREZ J., 1985. 'Aberrant Citrus plant obtained by somatic embryogenesis of nucelli cultured in vitro'. *Hort. Sci.* 20: 214–215.

NEGRI P., MAGNANINI E., CANTONI L., BERARDI G., SANSAVINI S., 1998. 'Gene transfer in woody fruit species: initial trials in growth habit control'. *Proc. Agro. Bio. Frut*, Cesena (Italy) pp. 12–13.

NEUHAUS J.M., AHL-GOY P., HINZ U., FLORES S., MEINS F., 1991. 'High level expression of a tobacco chitinase gene in *Nicotiana sylvestris* susceptibility of transgenic plants to *Cercospora nicotianae* infection'. *Plant Mol. Biol.* 16: 141–151.

NORELLI J.L., ALDWINCKLE H.S., DESTEFANO-BELTRAN L., JAYNES J.M., 1994. 'Transgenic "Malling 26" ["M. 7"] apple expressing the *attacin E* gene has increased resistance to *Erwinia amylovora*'. In: H. Schmidt and M. Kellerhals (eds), *Progress in temperate fruit breeding.* Kluwer Academic Publishers, London, UK, pp. 333–338.

NORELLI J.L., WONG K.W., HAYES C.K., HARMAN G.E., ALDWINCKLE H.S., 2000. Expression of endochitinase from *Tricoderma harzianum* in transgenic apple increases resistance to apple scab and reduces vigour. *Phytopathology*, 90: 72–77.

NORELLI J., MILLS J., JENSEN L.A., MOMOL M.T., ALDUINCHLE H.S., TOBUTT K.R., ALSTON F.H., 1999. *Acta Hort.* 484: 541–546.

NOTHORN M.A., SKIRVIN R.M., 1997. 'Somaclonal variation among ex vitro "Thornless Evergreen" trailing blackberries: the morphological status of selected clones after seven years of field growth'. *J. Am. Soc. Hort. Sci.*, 122(2): 152–157.

NYCZEPIR A.P., HALBRENDT J. M., 1993. 'Nematode pests of deciduous fruit and nut trees'. In: K. Evans, D. L. Trudgill, and J. M. Webster (eds), *Plant parasitic nematodes in temperate agriculture.* CAB Int., Wallingford, UK, pp. 381–425.

NYMAN M., WALLIN A., 1992. 'Transient gene expression strawberry (*Fragaria x ananassa* Duch.) protoplast and the recovery of transgenic plants'. *Plant Cell Rep.* 11: 105–108.

OAKES J.V., SHEWMAKER C.K., STALKER D.M., 1991. 'Production of cyclodextrins, a novel carbohydrate, in the tubers of transgenic potato plants'. *Biotechnol.* 9: 982–986.

OCHATT S.J., 1987. 'Coltura di protoplasti come metodo per il miglioramento genetico nelle piante da frutto'. *Frutticoltura* 49, 58–60.

OCHATT S.J., POWER J.P., 1989. 'Selection for salt-drought tolerance using protoplast- and explant-derived tissue cultures of Colt cherry (*Prunus avium X pseudocerasus*)'. *Tree Physiol.* 5, 259–266.

OELLER P.W., MING-WONG T., TAYLOR L.P., PIKE D.A., THEOLOGIS A., 1991. 'Reversible inhibition of tomato fruit senescence by antisense RNA'. *Science* 254: 437–439.

OLSEN J.E., JUNTTILA O., NILSEN J., ERIKSSON M.E., MARTILUSSEN I., OLSSON O., SANDBERG G., MORITZ T., 1997. 'Ectopic expression of oat phytochrome A in hybrid aspen changes critical daylength for growth and prevents cold acclimatisation'. *Plant J.* 12: 1339–1350.

OLSON P.D., VARNER J.E., 1993. 'Hydrogen peroxide and lignification'. *Plant J.* 4: 887–892.

ORLANDO R., MAGRO P., RUGINI E., 1997. 'Pectic enzyme as a selective press, for in vitro recovering of strawberry plants with a multiple fungal disease resistance'. *Plant Cell Rep.* 16: 272–276.

OWENS L.D., 1995. 'Overview of gene availability, identification and regulation'. *Hort. Sci.* 30: 957–961.

PADGETTE S.R., DELTA-CIOPPA G., SHAH D.M., FRALEY R.T., KISHORE G.M., 1989. 'Selective herbicide tolerance through protein engineering'. In: J. Schell and I.K. Vasil (eds), *Cell culture and somatic cell genetics*, vol. 6, Academic Press, New York, pp. 441–475.

PATENA L., SUTTER E.G., DANDEKAR A.M., 1988. 'Root induction by *Agrobacterium rhizogenes* in difficult-to-root woody species'. *Acta Hort.* 227: 324–329.

PEARCE R.S., 1999. 'Molecular analysis of acclimation to cold'. *Plant Growth Regulation*, 29: 47–76.

PENA L., MARTIN-TRILLO M., JUAREZ J., PINA J.A., NAVARRO L., MARTINEZ-ZAPATER J.M., 2001. 'Constitutive expression of Arabidopsis LEAFY or APETALA1 genes in citrus reduces their generation time'. *Nature Biotech.*, 19: 263–267.

PENG, M., KÙC J., 1992. 'Peroxidase-generated hydrogen peroxide as source of antifungal activity in vitro and on tobacco leaf disks'. *Phytopathology* 82: 696–699.

PÈREZ-MOLPHE-BLALCH E., OCHOA-ALEJO N., 1998. 'Regeneration of transgenic plants of Mexican lime from *Agrobacterium rhizogenes*-transformed tissues'. *Plant Cell Rep.* 17: 591–596.

PLOETZ R.C., ZENTMYER G.A., NISHIJIA W.T., ROHRBACH K.G., OHR H.D., 1994. *Compendium of tropical fruit diseases.* APS Press, St. Paul, MN. pp. 1–84.

POTTER L.R., 1980. 'The effects of barley yellow dwarf virus and powdery mildew in oats and barley with single and dual infections'. *Ann. Applied Biology* 94, 11–17.

RAMAN H., DHILLON B.S., 1990. 'Somaclonal variability for canker resistance in *Citrus aurantifolia* cv. Kagzi lime'. *VII Int. Cong. on Plant Tissue and Cell Culture*, Amsterdam, p. 164 (abstract).

RAMCHARAN C., GONZALEZ A., KNAUSENBERGER W.I., 1985. 'Performance of plantains produced from tissue culture plantlets in St. Croix, U.S. Virgin Islands'. *Proc. of the 3rd Meeting of the International Association for*

Research on Plantain and Banana, Abidjan, pp. 36–39.

RAMI V., RAINA V.R., 2000. 'Genetic fidelity of organized meristem-derived micropropagated plant: a critical reappraisal'. *In Vitro Cell Dev. Biol Plant.* 36: 319–330.

RAVELONANDRO M., SCORZA R., BACHELIER J.C., LABONNE G., LEVY L., DAMSTEEGT V., CALLAHAN A.M., DUNEZ J., 1997. 'Resistance of transgenic *Prunus domestica* to plum pox Virus infection'. *Plant Disease* 81: 1231–1235.

REMOTTI P.C., 1998. 'Somaclonal variation and in vitro selection for crop improvement'. In: S.M. Jain, D.S. Brar and B.S. Ahloowalia (eds), *Somaclonal Variation and Induced Mutations in Crop improvement.* Kluwer Academic Publishers, Dordrecht, pp. 169–70.

REUVENI O., ISRAELI L., DEGANI H., ESHDAT Y., 1985. 'Genetic variability in banana plants multiplied by in vitro techniques'. *Research Report AGPG: IBPGR18SI216.* International Board for Plant Genetic Resources, Rome.

REYNOIRD J.P., MOURGUES F., CHEVREAU E., ALDWINCKLE H.S., 1999a. 'Expression of SB-37 gene in transgenic pears enhanced resistance to fire blight'. *Acta Hort.* 489: 243–244.

REYNOIRD J.P., MOURGUES F., CHEVREAU E., BRISSET M.N., 1999b. 'First evidence of fire blight resistance among transgenic pear clones expressing *attacin E* gene'. *Acta Hort.* 489: 245–246.

ROBERTS P.A., 1992. 'Current status of the availability, development and use of host plant resistance to nematodes'. *J. Nematol.* 24: 213–223.

ROBINSON D., TALIANSKY M., RYABOV E., 1999. 'Expression of luteovirus coat protein in transgenic plants revealed by infection with an umbravirus'. *XI Int. Cong. Virology*, Sydney, 1999: 49.

ROBINSON S.P., JACOB A.K., DRY I.B., 1997. 'A class IV chitinase is highly expressed in grape berries during ripening'. *Plant Physiol.* 114: 771–778.

ROBSON P.R.H., SMITH H., 1997. 'Fundamental and biotechnological applications of phytochrome transgenes'. *Plant, Cell and Environment* 20: 831–839.

ROOSSINCK M.J., 1997. 'Mechanisms of plant virus evolution'. *Annual Review of Phytopathology* 35: 191–209.

ROSATI P., MEZZETTI B., ANCHENARI M., FOSCOLO S., PREDIERI S., FOSCOLO F., 1990. 'In vitro selection of apple rootstock somaclones with *Phytophthora cactorum* culture filtrate'. *Acta Hort.* 280: 409–416.

ROTINO G.L., PERRI E., ZOTTINI M., SOMMER H., SPENA A., 1997. 'Genetic engineering of parthenocarpic plants'. *Nature Biotech.* 15: 1398–1401.

ROTINO G.L., DONZELLA G., BOTTINI M., SOMMER H., FICCADENTI N., CIRILLO C., SESTILI S., PERRI E., PANDOLFINI T., SPENA A., 1999. 'Genetic engineering of parthenocarpic vegetable crops'. In: G.T. Scarascia Mugnozza, E. Porceddu and M.A. Pagnotta (eds) *Genetics and Breeding for Crop Quality and Resistance.* Kluwer Academic Publishers. Netherlands, pp. 301–306.

ROUSH R.T., 1998. 'Two-toxin strategies for management of insecticidal transgenic crops: can pyramiding succeed, where pesticide mixtures have not?' *Phil. Trans. R. Soc. Lord. B.* 353: 1777–1786.

ROXAS, V.P., SMITH R.H.Jr, ALLEN E.R., ALLEN R.D., 1997. 'Overexpression of glutathione S-transferase/glutathione peroxidase enhances the growth of transgenic tobacco seedlings during stress'. *Nat. Biotechnol.* 15: 988–991.

ROYO L., LEON J., VANCANNEYT G., ALBAR J.P., ROSAHL S., ORTEGO F., CASTANERA P., SANCHEZ-SERRANO J.J., 1999. 'Antisense-mediated depletion of a potato lipoxygenase reduces wound induction of proteinase inhibitors and increases weight gain of insect pests'. *Proc. Natl. Acad. Sci. USA* 96: 1146–1151.

RUGINI E., 1984. 'Progress in studies on in vitro culture of Almonds'. *Proc. 41st Conf. on Plant Tissue Culture and its Agricultural Applications.* Nottingham. 17–21 Sept. p. 73.

RUGINI E., 1986. 'Olive (*Olea europaea* L.)'. In: Y.P.S. Bajaj (ed.) *Biotechnology in Agriculture and Forestry 1, Trees I*, Springer-Verlag, Berlin, pp. 253–267.

RUGINI E., GUTIERREZ-PESCE P., 1999. 'Transformation in *Prunus* species'. In: Y.P.S. Bajaj (ed.) *Biotecnology in Agriculture and Forestry*, vol. 44, Springer-Verlag, Berlin, pp. 245–262.

RUGINI E., MARIOTTI D., 1992. '*Agrobacterium rhizogenes* T-DNA genes and rooting in woody species'. *Acta Hort.* 300: 301–308.

RUGINI E., PELLEGRINESCHI A., MENCUCCINI M., MARIOTTI D., 1991. 'Increase of rooting ability in the woody species kiwi (*Actinidia deliciosa* A. Chev.) by transformation with *Agrobacterium rhizogenes rol* genes'. *Plant Cell Rep.* 10: 291–295.

RUGINI E., PELLEGRINESCHI A., MUGANU M., 1994. 'Transformation of kiwi, cherry and papaya with *rol* genes'. In: *V Congress on University and Biotechnology Innovation Brescia*, June 20–21 pp. 68–69.

RUGINI E., CARICATO G., MUGANU M., TARATUFOLO C., CAMILLI M., CAMILLI C., 1997. 'Genetic stability and agronomic evaluation of six-year-old transgenic kiwi plants for *rolABC* and *rolB* genes'. *Acta Hort.* 447: 609–610.

RUGINI E., MUGANU M., PILOTTI M., BALESTRA G.M., VARVARO L., MAGRO P., BRESSAN R., TARATUFOLO C., 1999. 'Genetic stability, transgene hereditability and agronomic evaluation of transgenic kiwi (*Actinidia deliciosa* A. Chev.) plants for *rolABC*, *rolB* and Osmotin genes'. *Fourth International Symposium on Kiwifruit*, Santiago del Chile 11–14 Febr. p. 26.

RUGINI E., BIASI R., MULEO R., 2000a. 'Olive (*Olea europaea* var. *sativa*) Transformation'. In: S.M. Jain and S.C. Minocha (eds), *Molecular Biology of Woody Plants, vol. 2* Kluwer Academic Publishers, pp. 245–279.

RUGINI E., MUGANU M. GUTIERREZ-PESCE P., 2000b. 'Transformation of actinidia spp'. In: S.M. Jain and S.C. Minocha, (eds), *Molecular Biology of Woody Plants, vol. 2* Kluwer Academic Publishers, pp. 191–225.

RYBICKI E.P., 1994. 'A phylogenetic and evolutionary justification for the three genera of Geminiviridae'. *Archives of Virology* 139: 49–77.

SAGI L., REMY S., SWENNEN R., 1998. 'INIBAP', *Ann. Rep. Montpellier* (FRA), pp.

33–36.

SAIRAM R.K., DESHMUKH P.S., SHULCLA D.S., 1997. 'Tolerance of drought and temperature stress in relation to increased antioxidant enzyme activity in wheat'. *J. Agron. Crop Sci.* 178: 171–178.

SAITO W., OHGAWARA T., SHIMIZU J., ISHII S., KOBAYASHI S., 1993. 'Citrus cybrid regeneration following cell fusion between nucellar cells and mesophyll cells'. *Plant Sci.* 88: 195–201.

SAKAI A., LARCHER W., 1987. 'Frost Survival of Plants: responses to freezing stress'. *Ecol. Stud. Anal. Synth.* 62.

SALIBA-COLOMBANI V., CAUSSE M., PHILOUZE J., BURET M., ISSANCHOU S., LESSCHAEVE I., 1999. 'QTLs for organoleptic quality in fresh market tomato'. In: G.T. Scarascia Mugnozza, E. Porceddu and M.A. Pagnotta (eds) *Genetics and Breeding for Crop Quality and Resistance.* Kluwer Academic Publishers, The Netherlands, pp. 291–299.

SANFORD J.C., JOHNSTON S.A., 1985. 'The concept of parasite-derived resistance-deriving resistance genes from the parasite's own genome'. *J. Theor. Biol.* 113: 395–405.

SANSAVINI S., 1989. 'Biotechnology and fruit growing'. *Fruit Var. J.* 43: 75–84.

SAUVION N., RALIBE Y., PEUMANS W.J., VAN DAMME E.J.M., GATEHOUSE J.A., GATEHOUSE A.M.R., 1996. 'Effects of GNA and other mannose binding lectins on development and fecundity of the peach-pot aphid *Myzus persicae*'. *Entomol. Exp. Appl.* 79: 285–293.

SCHMULLING T., SCHELL J., SPENA A., 1988. 'Single genes from *Agrobacterium rhizogenes* influence plant development'. *EMBO J.* 7: 2621–2629.

SCHNEPF E., CRICKMORE N., VAN RIE J., LERECUS D., BAUM J., FEITELSON J., ZEIGLER D.R., DEAN D.H., 1998. '*Bacillus thuringiensis* and its pesticide crystal proteins.' *Microbiol. Mol. Biol. Rev.* 62: 775–806.

SCHOFFI F., RIEPING M., BAUNIANN G., 1987. 'Constitutive transcription of a soybean heat-shock gene by a cauliflower mosaic virus promoter in transgenic tobacco'. *Dev. Genet.* 8: 365–374.

SCHOTS A., DEBOER J., SCHOUTEN A., ROOSEIN J., ZILVERENTANT J.F., 1992. 'Plantibodies: a flexible approach to design resistance against pathogens'. *Neth. J. Plant Pathol.* 98: 183–191.

SCHUERMAN P.L., DANDEKAR A.M., 1993. 'Transformation of temperate woody crops: Progress and potentials'. *Scientia Hort.* 55: 101–124.

SCORZA R., 1991. 'Gene transfer for the genetic improvement of perennial fruit and nut crops'. *Hort. Sci.* 26: 1033–1035.

SCORZA R., ZIMMERMAN T.W., CORDTS J.M., FOOTEN K.L., 1994. 'Horticultural characteristics of transgenic tobacco expressing the *rolC*'. *J. Am. Soc. Hort. Sci.* 119: 1091–1098.

SCORZA R., CORDTS J.M., RAMMING D.W., EMERSHAD R. L., 1995. 'Transformation of grape (*Vitis vinifera* L.) zygotic-derived somatic embryos and regeneration of transgenic plants'. *Plant Cell Rep.* 14: 589–592.

SCORZA R., CORDTS J.M., CRAY D.J., GONSALVES D., EMERSHAD R.W., RAMMING D.W., 1996. 'Producing transgenic "Thompson Seedless" grape (*Vitis vinifera*

L.) plants'. *J. Am. Soc. Hort. Sci.* 121: 616–619.

SEN GUPTA A., HEINEN J.L., HOLADAY A.S., BURKE J.L., ALIEN R.D., 1993. 'Increased resistance to oxidative stress in transgenic plants that overexpress chloroplastic Cu/Zn superoxide dismutase'. *Proc. Natl. Acad. Sci. USA* 90: 1629–1633.

SERRES R., STANG E., 1992. 'Gene transfer using electric discharge particle bombardment and recovery of transformed cranberry plants'. *J. Am. Soc. Hort. Sci.* 117: 174–180.

SHAH D.H., HORSCH R.B., KLEE H.J., KISHORE G.M., WINTER J.A., TURNER N.E., HIRONAKA C.A., SANDERS P.R., GASSER C.S., AYKENT S., SIEGEL N.R., ROGERS S.G., FRALEY R.T., 1986. 'Engineering herbicide tolerance in transgenic plants'. *Science* 233: 478–481.

SHARROCK R.A., QUAIL P.H., 1989. 'Novel phytochrome sequences in *Arabidopsis thaliana*: structure, evolution and differential expression of a plant regulatory photoreceptor family'. *Genes and Development.* 3: 1745–1757.

SHEN Z., CORBIN D.X., PLATE J.T., GREBENOK R.J., GALBRAITH D.W., PURCELL J.P., 1997. 'Studies on the mode of action of cholesterol oxidase on insect midgut membranes'. *Archives of Insect Biochemistry and Physiol.* 34: 429–442.

SHIKANAI T., TAKEDA T., YAMAUCHI H., SANO S., TOMIZAWA K.I., 1998. 'Inhibition of ascorbate peroxidase under oxidative stress in tobacco having bacterial catalase in chloroplasts'. *FEBS Lett.* 428: 47–51.

SIJMONS P.C., ATKINSON H.J., WYSS U., 1994. 'Parasitic strategies of root nematodes and associated host all responses'. *Annu. Rev. Phytopathol.* 32: 235–259.

SIMPSON G.G., GENDALL A.R., DEAN C., 1999. 'When to switch to flowering'. *Ann. Rev. Cell Dev. Biol.*, 15: 519–550.

SINGH Z., SANSAVINI S., 1998. 'Genetic transformation and Fruit Crop Improvement'. In: J. Janik (ed.), *Plant Breeding Reviews*, vol. 16 J. Wiley & Sons Inc, pp. 87–134.

SINGH Z., JONES R.A.C., JONES M.G.K., 1997. 'Effectiveness of coat protein and defective replicase gene mediated resistance against Australian isolates of cucumber mosaic cucumovirus'. *Proc. 11th Biennial Conference Australasian Plant Path. Soc.* p. 271.

SKIRVIN R.M., 1978. 'Natural and induced variation in tissue culture'. *Euphytica* 27, 241–266.

SLIGHTOM J.L., DURANT TARDIF M., JOUANIN L., TEPFER D., 1986. 'Nucleotide sequence analysis of TL-DNA of *Agrobacterium rhizogenes* agropine type plasmid: identification of open reading frames'. *J. Biol. Chem.* 261: 108–121.

SMIGOCKI A.C., 1995. 'Phenotype modification and enhanced tolerance to insect pests by regulated expressing of a cytokinin biosynthesis gene'. *Hort. Sci.* 30: 967–969.

SMIGOCKI A.C., NEAL J.W.J., MCCANA I., DOUGLAS G., 1993. 'Cytokinin mediated insect resistance in *Nicotiana* plants transformed with *ipt* gene'. *Plant*

Mol. Biol. 23: 325–335.

SMITH G.J.S., WATSON C.F., RAY J., BIRD C.R., MORRIS P.C., SCHUCH W., GRIERSON D., 1988a. 'Antisense RNA inhibition of polygalacturonase gene expression in transgenic tomatoes'. *Nature* 334: 724–726.

SMITH G.R., BORG Z., LOCKHART B.E.L., BRAITHWAITE C.S., GIBBS M.J., 2000. 'Sugarcane yellow leaf virus: a novel member of the Luteoviridae that probably arose by inter-specific recombination'. *J. General Virology* 81: 1865–1869.

SMITH H., 1995. 'Physiological and ecological function within the phytochrome family'. *Annu. Rev. Plant Physiol. Plant Mol. Biol.* 46, 289–315.

SMITH I.K., MAUVAIS C.J., KNOWITON S., MAZUR B.J., 1988b. 'Molecular biology of resistance to sulfonylurea herbicides'. *ACS Symp. Biotechnol. Crop Protect.*, Washington, DC. pp. 25–36.

SMITH M.K., HAMIL S.D., LANGDON P.W., PEGG K.G., 1995. 'Genetic improvment of *Musa spp.*' In: D.R. Jones (ed.), *The improvement and Testing of Musa: A Global Partnership* INIBAP, Montpellier (FRA), pp. 233–242.

SMYTH D.R., 1999. 'Gene silencing: plants and viruses fight it out'. *Current Biology* 9: 100–102.

SPENA A., SCHMULLING T., KINCZ C., SCHELL J.S., 1987. 'Independent and synergistic activity of *rol A B* and *C* loci in stimulating abnormal growth in plants'. *EMBO J.* 6: 3891–3899.

SPIEGEL-ROY P., BEN-HAYYIM G., 1985. 'Selection and breeding for salinity tolerance in vitro'. *Plant and Soil* 89: 243–252.

SPIEGEL-ROY P., KOCHBA J., SAAD S., 1983. 'Selection for tolerance to 2,4-Dichlorophenoxyacetic acid in ovular callus or orange (*Citrus sinensis*)'. *Zeitschrift für Pflanzenphysiologie* 109: 41–48.

STARK D.M., TIMMERMAN K.P., BARRY G.F., PREISS J., KISHORE G.M., 1992. 'Regulation of the amount of starch in plant tissues by ADP glucose pyrophosphorylase.' *Science* 258, 287–292.

STEFFENS J.C., HAREL E., HUNT M.D., 1994. 'Polyphenol oxidase'. In: B.E. Ellis, (ed.), *Genetic Engineering of Plant Secondary Metabolism.* Plenum Press, New York, pp. 275–312.

STIEKEMA V.J., VAN DER VOSSEN A.G., ROUPPE VAN DER VOOT J., BAKKER J., KLEIN LANKHORST R.M., 1999. 'Molecular isolation of two cyst nematode resistance genes: the $Hs1^{pro\text{-}1}$ gene of beet and the GPA2 gene of potato'. In: G.T. Scarascia Mugnozza, E. Porceddu and M.A. Pagnotta (eds) *Genetics and Breeding for Crop Quality and Resistance.* Kluwer Academic Publishers, The Netherlands, pp. 185–193.

STORTI E., BOGANI P., BETTINI P., BITTINI P., GUARDIOLA M.L., PELLEGRINI M.G., INZE D., BUIATTI M., 1994. 'Modification of competence for in vitro response to *Fusarium oxysporum* in tomato cells. II. Effect of the integration of *Agrobacterium tumefaciens* genes for auxin an cytokinins synthesis'. *Theor. Appl. Genet.* 88: 89–96.

STOVER R.H., 1987. 'Somaclonal variation in Grand Name and Saba bananas in the nursery and field'. In: G.I. Persley and E.A. De Langhe (eds), *Banana and Plantain Strategies. ACIAR Proceedings, No. 21*, Canberra, pp. 136–139.

STOVER R.H., BUDDENHAGEN I.W., 1986. 'Banana breeding: polyploidy, disease resistance and productivity'. *Fruits* 41, 175–191.

STRIPE F., BARBIERI L., BATTELLI L.G., SORIA M., LAPPI D.A., 1992. 'Ribosome inactivating proteins from plants, present status and future prospects'. *Biotechnol.* 10: 405–412.

STROBEL G.A., NACHMIAS A., 1985. '*Agrobacterium rhizogenes* promotes the initial growth of bare stock almond'. *J. Gen. Microbiol.* 131: 1245–1249.

SWARTS H.J., GALLETTA G.J., ZIMMERMAN R.H., 1983. 'Field performance and phenotypic stability of tissue culture-propagated strawberries'. *J. Am. Soc. Hort. Sci.* 108: 285–290.

SZWACKA M., PALUCHA A., MALEPSZY S., 1996. '*Agrobecterium tumefaciens* – mediated cucumber transformation with thaumatin II cDNA'. *Genet. Pol.* 37A: 58.

TADA Y., NAKASE M., ADACHI T., 1996. 'Reduction of 14–16 kDa allergenic proteins in transgenic rice plants by antisense gene'. *FEBS Lett.* 391: 341–345.

TAKEUCHI Y., YOSHIKAWA M., TAEKBA G., TANAKA K., SHIBATA D., HORINO O., 1990. 'Molecular cloning and ethylene induction of mRNA encoding a phytoalexin elicitor releasing factor, β-1,3 endoglucanase, in soybean'. *Plant Physiol*, 673–682.

TAO R., HANDA T., TAMURA M., SUGIURA A., 1994. *J. Japan Soc. Hort. Sci.* 63: 283–289.

TAO I., DANDEKAR X.M., URATSU S.L., VAIL P.V., TEBBETS I.S., 1997. 'Engineering genetic resistance against insects in Japanese persimmon using the CrylA(c) gene of *Bacillus thuringiensis*'. *J. Am. Soc. Hort. Sci.* 122: 764–771.

TATTERSALL D.B., HEESWIJCK R.V., HOJ P.B., 1997. 'Identification and characterization of a fruit-specific, thaumatin-like protein that accumulates at very high levels in conjection with the onset of sugar accumulation and berry softening in grapes'. *Plant Physiol.* 114: 759–769.

TENNANT P.R., CONSALVES C., LING K.S., FITCH M., MANSHARDT R., SLIGHTOM J.L., GONSALVES D., 1994. 'Differential protection against papaya ringspot virus isolates in coat protein gene transgenic papaya and classically crossprotected papaya'. *Phytopathology* 84: 1359–1366.

TEPFER D., 1984. 'Transformation of several species of higher plants by *Agrobacterium rhizogenes*: sexual transmission of the transformed genotype and phenotype'. *Cell* 37: 959–967.

TEPFER M., CASSE-DELBART F., 1987. '*Agrobacterium rhizogenes* as a vector for transforming higher plants'. *Microbiol. Sci.* 4: 24–28.

TERRAS F.R., PENNINCKX I.A., GODERIS I.J., BROEKAERT W.F., 1998. 'Evidence that the role of plant defensins in radish defense responses is independent of salicylic acid'. *Planta*, 206: 117–124.

TERRAS F.R., TORREKENS S., VAN LEUVEN F., OSBORN R.W., VANDERLEYDEN J., CAMMUE B.P., BROECKAERT W.F., 1993. 'A new family of basic cysteine-rich plant antifungal proteins from Brassicaceae species'. *FEBS Lett.* 316: 233–240.

TORTIGLIONE C., MALVA C., PENNACCHIO F., RAO R., 1999. 'New genes for pest control'. In: G.T. Scarascia Mugnozza, E. Porceddu and M.A. Pagnotta (eds) *Genetics and Breeding for Crop Quality and Resistance.* Kluwer Academic Publishers, The Netherlands, pp. 159–163.

TOYODA H., HORIKOSHI K., YAMANO Y., OUCHI S., 1991. 'Selection of Fusariurn wilt disease resistance from regenerants derived from leaf callus of strawberry'. *Plant Cell Rep.* 10: 167–170.

TRAINOTTI L., SPOLAORE S., FERRARESE L., CASADORO G., 1997. 'Characterisation of ppEG1, a member of a multigene family which encodes endo-beta 1,4 glucanase in peach'. *Plant Mol. Biol.* 34: 791–802.

TRIFONOVA A., SAVOVA D., IVANOVA K., 1994. '*Agrobacterium*-mediated transformation of the apple cultivar Granny Smith.' In: H. Schmidt and M. Kellerhals (eds), *Progress in temperate fruit breeding.* Kluwer, London, UK, pp. 343–347.

TROVATO M., MAURO M.L., CONSTANTINO P., ALTAMURA M.M., 1997. 'The *rolD* gene from *Agrobacterium rhizogenes* is developmentally regulated in transgenic tobacco'. *Protoplasma* 197: 111-120.

TUCKER K.J., 1976. 'Effect of far-red light of the hormonal control of side shoot growth in the tomato'. *Ann. Bot.* 40, 1033–1042.

VAN CAMP W., CAPIAU K., VAN MONTAGU M., INZÈ D., SLOOTEN L., 1996. 'Enhancement of oxidative stress tolerance in transgenic tobacco plants overproducing Fe-superoxide dismutase in chloroplasts'. *Plant Physiol.* 112: 1703–14.

VAN DAMME E.J.M., PEUMANS W.J., BARRE A., ROUGE P., 1998. 'Plant lectins: A composite of several distinct families of structurally and evolutionary related proteins with diverse biological roles'. *Crit. Rev. Pla. Sci.* 17(6): 575–692.

VARDI A., GALUN E., 1988. 'Recent advances in protoplast culture of horticultural crops: Citrus'. *Scientia Hort.* 37: 217–230.

VARDI A., BLEICHMAN S., AVIV D., 1990. 'Genetic transformation of Citrus protoplasts and regeneration of transgenic plants'. *Plant Sci.* 69: 199–206.

VERONESE P., CRINÒ P., TUCCI M., COLUCCI F., YUN D.J., HASEGAWA M.P., BRESSAN R.A., SACCARDO F., 1999. 'Pathogenesis-related proteins for the control of fungal diseases of tomato'. In: G.T. Scarascia Mugnozza, E. Porceddu, M.A. Pagnotta (eds) *Genetics and Breeding for Crop Quality and Resistance.* Kluwer Academic Publishers. Netherlands, pp. 15–24.

VIJN I., VAN DIJKEN A., SPRENGER N., 1997. 'Fructan of the inulin neoseries is synthesised in transgenic chicory plants (*Cichorium intybus* L.) harbouring onion (*Allium cepa* L.) fructan: fructan 6G-fructosyltransferase'. *The Plant Jour.* 11: 387–398.

VILAINE F., CASSE-DELBART F., 1987. 'Independent induction of transformed roots by the TL and TR regions of the Ri plasmid of agropine type *Agrobacterium rhizogenes*'. *Mol. Gen. Genet.* 206: 17–23.

VILAINE F., CHARBONNIER C. CASSE-DELBART F., 1987. 'Further insight concerning the TL region of the Ri plasmid of *Agrobacterium rhizogenes* strain

A4: Transfer of a 1.9 kb fragment is sufficient to induce transformed roots on tobacco leaf fragments'. *Mol. Gen. Genet.*, 210: 111–115.

VINCE-PRUE D., CANHAM A.E., 1983. 'Horticultural significance of photomorphogenesis'. In: Shropshire and Mohr (eds). *Photomorphogenesis. Encyclopedia of Plant Physiology.* Springer Verlag. pp. 518–544.

VISEUR M.J., TAPIA Y FIGUEROA M., 1987. 'In vitro co-culture as a tool for the evaluation of fire blight resistance in pears and apples'. *Acta Hort.* 217, 273–282.

VUYLSTEKE D., SWERMEN R., WILSON G., DE LANGHE E., 1988. 'Phenotypic variation among in vitro propagated plantain (*Musa* sp. cultivar 'AAB')'. *Scientia Hort.* 36, 79–88.

WANGER D., TEPPERMAN J.M., QUAIL H.P., 1991. 'Overexpression of Phytochrome B induces a short hypocotyl phenotype in transgenic Arabidopsis'. *The Plant Cell*, 3; 1275–1288.

WARREN G.J., 1995. Identification and analysis of INA genes and proteins. In: R.E. Lee Jr., C.J. Warren and L.V. Gusta (eds), *Biological Ice Nucleation and Its Applications*, APS Press, St. Paul, MN., pp. 163–181.

WATERWORTH H.E., HADIDI A., 1998. 'Economic losses due to plant viruses'. In: A. Hadidi, R.H. Khetarpal and H. Koganezavra (eds) *Plant Virus Disease Control* 1–13. American Phytopathological Society Press, St. Paul.

WELANDER M., PAWLICKI N., HOLEFORS A., WILSON F., 1998. *Plant. Physiol.* 153: 3–4, 371–380.

WHEELER G. L., JONES M.A., SMIRNOFF N., 1988. 'The biosynthetic pathway of vitamin C in higher plants'. *Nature* 393: 365–369.

WHITELAM G.C. AND HARBERD N.P., 1994. 'Action and function of phytochrome family members revealed through the study of mutant and transgenic plants'. *Plant Cell Environm.* 17: 615–625.

WILKINSON J.Q., LENAHAN M.B., CONNER T.W., KLEE H.D., 1995. 'Identification of mRNAs with enhanced expression in ripening strawberry fruit using polymerase chain reaction differential display'. *Plant Mol. Biol.* 6: 1097–1108.

WILSON T.M.A., 1993. 'Strategies to protect crop plants against viruses'. *Proc. Nat. Acad. Sci. USA* 90: 3134–3141.

WISNIEWSKI M., ARORA R., 1993. 'Adaptation and response of fruit trees to freezing temperatures'. In: A.R. Biggs (ed.) *Cytology, Histology, and Histochemistry of Fruit Tree Diseases.* CRC Press, Boca Raton, FL, pp. 299–320.

WISNIEWSKI M., ARORA R., 2000. 'Structural and biochemical aspects of cold hardiness in woody plants'. *Mol. Biol. of Woody Plants.* 2: 419–437.

WISNIEWSKI M., WEBB R., BALSAMO R., CLOSE T., YU X.M., GRIFFITH M., 1999. 'Purification, immunolocalization, cryoprotective, and antifreeze activity of PCA60: A dehydrin from peach (*Prunus persica* L. Batsch)'. *Physiol. Plant.* 105: 600–608.

WITTY M., HARVEY W.J., 1990. 'Sensory evaluation of transgenic *Solanum tuberosum* producing r-thaumatin II'. *New Zealand J. Crop Hort. Science*, 18: 77–80.

WOLFSON J.L., MURDOCK L.L., 1990. 'Diversity in digestive proteinase activity among insects'. *J. Chem. Ecol.* 16: 1089–1102.

WU G.S., SHORT B.J., LAWRENCE E.B., LEVINE E.B., FITZSIMMONS K.C., SHAH D.M., 1995. 'Disease resistance conferred by expression of a gene encoding H_2O_2-generating glucose oxidase in transegenic potato plants'. *Plant Cell.* 7: 1357–1368.

XU D., DUAN X, WANG B. HONG B., HO T.D., WII R., 1996. 'Expression of a late embryogenesis abundant protein gene, HVAI, from barley confers tolerance to water deficit and salt stress in transgenic rice'. *Plant Physiol.* 110: 249–257.

XUE B., LING K.S., REID C.L., KRASTANOVA S., SEKIYA M., MOMOL E.L., SLUE S., MOZSAR J., GONSALVES D., BURR T.J., 1999. 'Transformation of five grape rootstocks with plant virus genes and a virE2 gene from *Agrobacterium tumefaciens*'. *In Vitro Cell. Develop. Biol. Plant.* 35: 226–231.

YADAV N., MCDEVITT R.E., BENARD S., FALCO S.C., 1986. 'Single amino acid substitutions in the enzyme acetolactate synthase confer resistance to the herbicide sulfonmeturon methyl'. *Proc. Nat. Acad. Sci. USA* 83: 4418–4422.

YAMAKAWA Y., CHEN L.H., 1996. '*Agrobacterium rhizogenes*-mediated transformation of kiwi fruit (*Actinidia deliciosa*) by direct formation of adventitious buds.' *J. Japan Soc. Hort. Sci.* 64: 741–747.

YANG Z.N., INGELBRECHT I.L., LOUZADA E., SVARIA M., MIRKOV T.E., 2000. '*Agrobacterium*-mediated transformation of the commercial important grapefruit cultivar Rio Red (*Citrus paradisi* Macf.)'. *Plant Cell Rep.* 19: 1203–1211.

YANOTSKY M., 1995. 'Floral meristems to floral organs: genes controlling early events in Arabidopsis flower development'. *Annu. Rev. Plant Physiol. Mol. Biol.* 46: 167: 188.

YANOTSKY M.J., ALCONADA-MAGLIANO T.M., MAZZELLA M.A., GATZ C., THOMAS B., CASAL J.J., 1998. 'Phytochrome A affects stem growth, anthocyanin synthesis, sucrose-phosphate-synthase activity and neighbour detection in sunlight-grown potato'. *Planta* 205: 235–241.

YAO J.L., COHEN D., ATKINSON R., RICHARDSON K., MORRIS B., 1995. 'Regeneration of transgenic plants from the commercial apple cultivar Royal Gala. *Plant Cell Rep.* 14: 407–412.

YAZAWA M., SUGINUMA C., ICHIKAWA K., KAMADA H., AKIHAMA T., 1995. Regeneration of transgenic plants from hairy root of kiwi fruit (*Actinidia deliciosa*) induced by *Agrobacterium rhizogenes*'. *Breeding Science* 45, 241–244.

YEH S.D., BAU H.J., CHENG Y.H., YU T.A., YANG J.S., DREW R.A., 1998. 'Greenhouse and field evaluations of coat-protein transgenic papaya resistant to papaya ringspot virus'. *Acta Hort.* 461: 321–328.

YIE Y., TIEN P., 1998. 'Controlling mosaic virus diseases under field conditions using multiple gene strategies in transgenic plants'. In: A. Hadidi, R.H. Khetarpal and H. Koganezawa (eds) *Plant Virus Disease Control.*

American Phytopathological Society Press, St. Paul, pp. 129–141.

YU C.O., MULLINS M.A., WARREN G.W., KOZIEL M.G.W., ESTRUCH J.J., 1997. 'The *Bacillus thuringiensis* vegetative insecticidal protein Vip3A lyses midgut epithelium cells of susceptible insects'. *Appl. Environ. Microbiol.* 63: 532–536.

ZHANG G.J., KLUEVA N.Y., WANG Z.,WU R., HO D., NGUYEN H.T., 2000. 'Genetic engineering for abiotic stress resistance in crop plants'. *In Vitro Cell. Dev. Biol. Plant* 36: 108–11.

ZHANG L., NGUYEN H.T., BLUM A., 1999. 'Genetic analysis of osmotic adjustment in crop plants'. *J. Exp. Bot.* 50: 291–302.

ZHU B., CHEN T.H.H., LI P.H., 1996. 'Analysis of late-blight disease resistance and freezing tolerance in transgenic potato plants expressing sense and antisense genes for an osmotin-like protein'. *Planta* 198, 70–77.

ZHU L.H., 2001. 'Transformation of the apple rootstocks M.9/29 with the *rolB* gene and its influence on rooting and growth'. *Plant Sci.* 160: 433–439.

ZHU Q., MAHLER E.A., MASOUD S., DIXON R., LAMB C.J., 1994. 'Enhanced protection against fungal attack by constitutive co-expression of chitinase and glucanase genes in transgenic tobacco'. *Biotechnol.* 12, 807–812.

ZHU-SALZMAN K., SHADE R.E., KOIWA H., SALZMAN R.A., NARASIMHAN M., BRESSAN R.A., HASEGAWA P.M., MURDOCK L.L., 1998. 'Carbohydrate binding and resistance to proteolysis control insecticidal activity of *Griffonia simplicifolia* lectin II'. *Proc. Natl. Acad. Sci. USA* 25: 15123–15128.

Apple

BROOTHAERTS W., JANSSENS G.A., PROOST P., BROEKAERT W.F., 1995. 'cDNA cloning and molecular analysis of two self-incompatibility alleles from apple'. *Plant Mol. Biol.* 27: 499–511.

CHENG F.S., WEEDEN N.F., BROAN S.K., ALDWINCKLE H.S., GARDINER S.E., BUS V.G., 1998. 'Development of a DNA marker for *Vm*, a gene conferring resistance to apple scab'. *Genome* 41(2): 208–214.

CHEVRAU E., MANGANARIS A.G., GALLET M., 1999. 'Isozyme segregation in five apple progenies and potential use for map construction'. *Theor. Appl. Genet.* 98: 329–336.

CONNER P.J., BROWN S.K., WEEDEN N.F., 1997. 'Randomly amplified polymorphic DNA-based genetic linkage maps of three apple cultivars'. *J. Amer. Soc. Hort. Sci.* 122(3): 350–359.

CONNER P.J., BROWN S.K., WEEDEN N.F., 1998. 'Molecular-marker analysis of quantitative traits for growth and development in juvenile apple trees'. *Theor. Appl. Genet.*, 96(8): 1027–1035.

DONG Y.H., YAO J.L., ATKINSON R.G., MORRIS B.A., GARDNER R.C., 1999. 'Mdh3 encoding a Phalaenopsis O39-like homeodomain protein expressed in ovules of *Malus domestica*'. *J. Exp. Botany* 50(330): 141–142.

DONG Y.H., YAO J.L., ATKINSON R.G., PUTTERILL J.J., MORRIS B.A., GARDNER R.C., 2000. '*MDH1*: an apple homeobox gene belonging to the *BEL1* family'. *Plant Mol. Biol.* 42: 623–633.

DONG Y.H., ZHANG X.C., KVARNHEDEN A., ATKINSON R.G., MORRIS B.A., GARDNER R.C., 1998. 'Expression of a cDNA from apple encoding a homologue of DAD1, an inhibitor of programmed cell death'. *Plant Sci.* 139(2): 165–174.

GIANFRANCESCHI L., SEGLIAS N., TARCHINI R., KOMJANC M., GESSLER C., 1998. 'Simple sequence repeats for the genetic analysis of apple'. *Theor. Appl. Genet.* 96: 1069–1076.

GUILFORD P., PRAKASH S., ZHU J.M., RIKKERINK E., GARDINER S., BASSETT H., FORSTER R., 1997. 'Microsatellites in *Malus X domestica* (apple): abundance, polymorphism and cultivar identification'. *Theor. Appl. Genet.* 94: 249–254.

HARADA T., SUNAKO T., WAKASA Y., SOEJIMA J., SATOH T., NIIZEKI M., 2000. 'An allele of the 1-aminocyclopropane-1–carboxylate synthase gene (*Md-ACS1*) accounts for the low level of ethylene production in climateric fruits of some apple cultivars'. *Theor. Appl. Genet.* 101: 742–746.

HEMMAT M., WEEDEN N.F., ALDWINCKLE H.S., BROWN S.K., 1998. 'Molecular markers for the scab resistance (*Vf*) region in apple'. *J. Am. Soc. Hort. Sci.* 123(6): 992–996.

HEMMAT M., WEEDEN N.F., CONNER P.J., BROWN S.K., 1997. 'A DNA marker for columnar growth habit in apple contains a simple sequence repeat'. *J. Am. Soc. Hort. Sci.*, 122(3): 347–349.

HEMMAT M., WEEDEN N.F., MANGANARIS A.G., LAWSON D.M., 1994. 'Molecular marker linkage map for apple'. *J. Hered.* 85:4–11.

JANSSENS G.A., GODERIS I.J., BROEKAERT W.F., BROOTHAERTS W., 1995. 'A molecular method for S-allele identification in apple based on allele-specific PCR'. *Theor. Appl. Genet.* 91: 691–698.

JOHNSON R.D., JOHNSON L., KOHMOTO K., OTANI H., LANE C.R., KODAMA M., 2000. 'A polymerase chain reaction-based method to specifically detect *Alternaria alternata* apple pathotype (*A. mali*), the causal agent of alternaria blotch of apple'. *Phytopathology* 90(9): 973–976.

KING G.J., ALSTON F.H., BROWN L.M., CHEVREAU F., EVANS K.M., DUNEMANN F., JANSE J., LAURENS F., LYNN J.R., MALIEPAARD C., MANGANARIS A.G., ROCH P., SCHMIDT H., TARTARINI S., VERHAEGH J., VRIELINK R., 1998. 'Multiple field and glasshouse treatments increase the reliability of linkage mapping of the *Vf* source of scab resistance in apple'. *Theor. Appl. Genet.* 96: 699–708.

KITAHARA K., SOEJIMA J., KOMATSU H., FUKUI H., MATSUMOTO S., 2000. 'Complete sequences of the S-genes, Sd- and Sh-Rnase cDNA in apple'. *Hort. Sci.* 35(4): 712–715.

MALIEPAARD C., ALSTON F.H., VAN ARKEL G., BROWN L.M., CHEVRAU E., DUNEMANN F., EVANS K.M., GARDINER S., GUILFORD P., VAN HEUSDEN A.W., JANSE J., LAURENS F., LYNN J.R., MANGANARIS A.G., DEN NIJS A.P.M., PERIAM N., RIKKERINK E., ROCHE P., RYDER C., SANSAVINI S., SCHMIDT H., TARTARICI S., VERHAEGH J.J., VRIELINK-VAN GINKEL M., KING G.J., 1998. 'Aligning male and female linkage maps of apple (*Malus pumila* Mill.) using multi-allelic markers'. *Theor. Appl. Genet.* 97: 60–73.

MATSUMOTO S., KITAHARA K., 2000. 'Discovery of a new self-incompatibility allele in apple'. *Hort Sci.* 35(7): 1329–1332.

PATOCCHI A., GIANFRANCESCHI L., GESSLER C., 1999a. Towards the map-based cloning of *Vf*: fine and physical mapping of the *Vf* Region'. *Theor. Appl. Genet.* 99(6): 1012–1017.

PATOCCHI A., VINATZER B.A., GIANFRANCESCHI L., TARTARICI S., ZHANG H.B., SANSAVINI S., GESSLER C., 1999b. 'Construction of a 550 kb BAC contig spanning the genomic region containing the apple scab resistance gene *Vf*. *Mol. Gen. Genet.* 262(4/5): 884–891.

SUNAKO T., SAKURABA W., SENDA M., AKADA S., ISHIKAWA R., NIIZEKI M., HARADA T., 1999. 'An allele of the ripening-specific 1–aminocyclopropane-1–carboxylic acid synthase gene (ACS1) in apple fruit with a long storage life'. *Plant Physiol.* 119(4): 1297–1303.

SUNG S.K., YU G.H., AN G., 1999. 'Characterization of MdMADS2, a member of the SQUAMOSA subfamily of gene, in apple'. *Plant Physiol.* 120(4): 969–978.

SUNG S.K., YU G.H., NAM J., JEONG D.H., AN G., 2000. 'Developmentally regulated expression of two MADS-box genes, MdMADS3 and MdMADS4, in the morphogenesis of flower buds and fruits in apple'. *Planta* 210(4): 519–528.

TARTARINI S., GIANFRANCESCHI L., SANSAVINI S., GESSLER C., 1999. 'Development of reliable PCR markers for the selection of the *Vf* gene conferring scab resistance in apple'. *Plant Breed.* 118(2): 183–186.

XU M.L., KORBAN S.S., 2000. 'Saturation mapping of the apple scab resistance gene *Vf* using AFLP markers'. *Theor. Appl. Genet.* 101: 844–851.

XU M.L., HUARACHA E., KORBAN S.S., 2001. 'Development of sequence-characterized amplified regions (SCARs) from amplified fragment length polymorphism (AFLP) markers tightly linked to the *Vf* gene in apple'. *Genome* 44(1): 63–70.

YANG H.Y., KORBAN S.S., KRUGER J., SCHMIDT H., 1997. 'The use of a modified bulk segregant analysis to identify a molecular marker linked to a scab resistance gene in apple'. *Euphytica* 94: 175–182.

YAO C.L., CONWAY W.S., REN R.H., SMITH D., ROSS G.S., SAMS C.E., 1999. 'Gene encoding polygalacturonase inhibitor in apple fruit is developmentally regulated and activated by wounding and fungal infection'. *Plant Mol. Biol.* 39(6): 1231–1241.

ZHOU Z.Q., LI Y.N., 2000. 'The RAPD evidence for the phylogenetic relationship of the closely related species of cultivated apple'. *Genet. Resour. Crop Evol.* 47(4): 353–357.

Pear

ISHIMIZU T., SATO Y., YOSHIMURA Y., NARIOKA S., NAKANISHI T., SAKIYAMA F., 1996. 'Identification and partial amino acid sequences of seven S-RNases associated with self-incompatibility of Japanese pear, *Pyrus pyrifolia*'. *Nakai J. Biochem.* 120(2): 326–334.

ISHIMIZU T., INOUE K., SHIMONAKA M., SAITO T., TERAI O., NORIOKA S., 1999. 'PCR-based method for identifying the S-genotypes of Japanese pear cultivars'. *Theor. Appl. Genet.* 98(6/7): 961–967.

ITAI A., KAWATA T., TANABE K., TAMURA F., UCHIYAMA M., TOMOMITSU M., SHIRAIWA N., 1999a. 'Identification of 1–aminocyclopropane-1–carboxylic acid synthase genes controlling the ethylene level of ripening fruit in Japanese pear (*Pyrus pyrifolia* Nakai)'. *Mol. Gen. Genet.* 261(1): 42–49.

ITAI A., TANABE K., TAMURA F., TANAKA T., 2000. 'Isolation of cDNA clones corresponding to genes expressed during fruit ripening in Japanese pear (*Pyrus pyrifolia* Nakai): involvement of the ethylene signal transduction pathway in their expression'. *J. Exp. Bot.* 51(347): 1163–1166.

ITAI A., YOSHIDA K., TANABE K., TAMURA F., 1999b. 'A beta-D-xylosidase-like gene is expressed during fruit ripening in Japanese pear (*Pyrus pyrifolia* Nakai)'. *J. Exp. Bot.* 50(335): 877–878.

MONTE-CORVO L., CABRATA L., OLIVEIRA C., LEITAO J., 2000. 'Assessment of genetic relationships among *Pyrus* species and cultivars using AFLP and RAPD markers'. *Gen. Resour. Crop Evol.* 47: 257–265.

NORIOKA N., OHNISHI Y., NORIOKA S., ISHIMIZU T., NAKANISHI T., SAKIYAMA F., 1995. 'Nucleotide sequences of cDNAs encoding S_2- and S_4-Rnases (D49527 and D49528 for EMBL) from Japanese pear (*Pyrus pyrifolia* Nakai) (PGR95-020)'. *Plant Physiol.* 108: 1343.

NORIOKA N., NORIOKA S., OHNISHI Y., ISHIMIZU T., ONEYAMA C., NAKANISHI T., SAKIYAMA F., 1996. 'Molecular cloning and nucleotide sequences of cDNAs encoding S-allele-specific stylar Rnases in a self-incompatible cultivar and its self-compatible mutant of Japanese pear, *Pyrus pyrifolia* Nakai'. *J. Biochem.* 120: 335–345.

OLIVEIRA C., MOTA M., MONTE-CORVO L., GOULAO L., SILVA D., 1999. 'Molecular typing of *Pyrus* based on RAPD markers'. *Scientia Hort.* 79: 163–174.

SASSA H., HIRANO H., NISHIO T., KOBA T., 1997. 'Style-specific self-incompatible mutation caused by deletion of the S-RNase gene in Japanese pear (*Pyrus serotina*)'. *Plant J.* 12: 223–227.

TATEISHI A., INOUE H., SHIBA H., YAMAKI S., 2001. 'Molecular cloning of beta-galactosidase from Japanese pear (*Pyrus pyrifolia*) and its gene expression with fruit ripening'. *Plant Cell Physiol.* 42(5): 492–498.

YAMAMOTO T., KIMURA T., SAWAMURA Y., KOTOBUKI K., BAN Y., HAYASHI T., MATSUTA N., 2001. 'SSRs isolated from apple can identify polymorphism and genetic diversity in pear'. *Theor. Appl. Genet.* 102: 865–870.

Peach

BASSETT C., CALLAHAN A., DUNN L., 1998. 'Characterization of a type II gene chlorophyll a/b-binding protein gene (*Lhcb*Pp1*) in peach. I. Isolation, identification, and abundance in developing leaves of mature 'Loring' trees in the absence of flowering'. *J. Amer. Soc. Hort. Sci.* 123(4): 486–492.

CHUNG K.H., BUETOW D.E., KORBAN S.S., 1998. 'Isolation and characterization of a type I gene encoding a light-harvesting chlorophyll a/b-binding protein of photosystem II in peach'. *J. Amer. Soc. Hort. Sci.* 123(4): 493–499.

CIPRIANI G., LOT G., HUANG W.G., MARRAZZO M.T., PETERLUNGER E., TESTOLIN R., 1999. 'AC/GT and AG/CT microsatellite repeats in peach [*Prunus persica* (L.) Batsch]: isolation, characterisation and cross-species amplification in *Prunus*'. *Theor. Appl. Genet.* 99(1/2): 65–72.

DIRLEWANGER E., PRONIER V., PARVERY C., ROTHAN C., GUYE A., MONET R., 1998. 'Genetic linkage map of peach [*Prunus persica* (L.) Batsch] using morphological and molecular markers'. *Theor. Appl. Genet.* 97(5/6): 888–895.

DIRLEWANGER E:, MOING A., ROTHAN C., SVANELLA L., PRONIER V., GUYE A., PLOMION C., MONET R., 1999. 'Mapping QTLs controlling fruit quality in peach [*Prunus persica* (L.) Batsch]'. *Theor. Appl. Genet.* 98(1): 18–31.

FOOLAD M.R., ARULSEKAR S., BECERRA V., BLISS F.A., 1995. 'A genetic linkage map of *Prunus* based on an interspecific cross between peach and almond'. *Theor. Appl. Genet.* 91: 262–269.

GREEN M.J., THOMPSON D.A., MACKENZIE D.J., 1999. 'Easy and efficient DNA extraction from woody plants for the detection of phytoplasmas by polymerase chain reaction'. *Plant Dis.* 83(5): 482–485.

HEUSS K., LIU Q., HAMMERSCHLAG F.A., HAMMOND R.W., 1999. 'A cRNA probe detects prunus necrotic ringspot virus in three peach cultivars after micrografting and in peach shoots following long-term culture at 4 degrees C'. *Hort. Sci.* 34(2): 346–347.

JAUREGUI B., DE VICENTE M.C., MESSEGUER R., FELIPE A., BONNET A., SALESSES G., ARUS P., 2001. 'A reciprocal translocation between 'Garfi' almond and 'Nemared' peach'. *Theor. Appl. Genet.* 1023: 1169–1176.

JOOBEUR T., VIRUEL M.A., DE VICENTE M.C., JAUREGUI B., BALLESTER J., DETTORI M.T., VERDE I., TRUCO M.J., MESSEGUER R., BATELLE I., QUARTA R., DIRLWANGER E., ARUS P., 1998. 'Construction of a saturated linkage map for *Prunus* using an almond x peach F_2 progeny'. *Theor. Appl. Genet.* 97: 1034–1041.

LU Z.X., REIGHARD G.L:, BAIRD W.V., RAJAPAKSE S., ABBOTT A.G., 1996. 'Identification of peach rootstock cultivars by RAPD markers'. *Hort. Sci.* 31: 127–129.

LU Z.X., SOSINSKI B., REIGHARD G.L., BAIRD W.V., ABBOTT A.G., 1998. 'Construction of a genetic linkage map and identification of AFLP markers for resistance to root-knot nematodes in peach rootstocks'. *Genome* 41: 199–207.

LU Z.X., SOSSEY-ALOUI K., REIGHARD G.L., BAIRD W.V., ABBOTT G.A., 1999. 'Development and characterization of a codominant marker linked to root-knot nematode resistance, and its application to peach rootstock breeding'. *Theor. Appl. Genet.* 99: 115–122.

MANUBENS A., LOBOS S., JADUE Y., TORO M., MESSINA R., LLADSER M., SEELENFREUND D., 1999. 'DNA isolation and AFLP fingerprinting of nectarine and peach varieties (*Prunus persica*)'. *Plant Mol. Biol. Rep.* 17(3): 255–267.

MATHOOKO F.M., TSUNASHIMA Y., OWINO W.Z.O., KUBO Y., INABA A., 2001.

'Regulation of genes encoding ethylene biosynthetic enzymes in peach (*Prunus persica* L.) fruit by carbon dioxide and 1-methylcyclopropane'. *Postharvest Biol. Technol.* 21(3): 265–281.

OHMIYA A., TANAKA Y., KADOWAKI K., HAYASHI T., 1998. 'Cloning of genes encoding auxin-binding proteins (ABP19/20) from peach: significant peptide sequence similarity with germin-like proteins'. *Plant Cell Physiol.* 39(5): 492–499.

SAKANISHI K., KANAYAMA Y., MORI H., YAMADA K., YAMAKI S., 1998. 'Expression of the gene for NADP-dependent sorbitol-6-phosphate dehydrogenase in peach leaves of various developmental stages'. *Plant Cell Physiol.* 39(12): 1372–1374.

SOSINSKI B., GANNAVARAPU M., HAGER L.D., BECK L.E., KING G.J., RYDER C.D., RAJAPAKSE S., BAIRD W.V., BALLARD R.E., ABBOTT A.G., 2000. 'Characterization of microsatellite markers in peach [*Prunus persica* (L.) Batsch]'. *Theor. Appl. Genet.* 101: 421–428.

TESTOLIN R., MARRAZZO T., CIPRIANI G., QUARTA R., VERDE I., DETTORI M.T., 2000. 'Microsatellite DNA in peach (*Prunus persica* L. Batsch) and its use in fingerprinting and testing the genetic origin of cultivars'. *Genome* 43(3): 512–520.

Apricot

BADENES M.L., HURTADO M.A., SANZ F., ARCHELOS D.M., BURGOS L., EGEA J., LLACER G., 2000. 'Searching for molecular markers linked to male sterility and self-compatibility in apricot'. *Plant Breed.* 119(2): 157–160.

CHEVALIER T., DE RIGAL D., MBEGUIE-A-MBEGUIE D., GAUILLARD F., RICHARD-FORGET F., FILS-LYCAON B.R., 1999. 'Molecular cloning and characterization of apricot fruit polyphenol oxidase'. *Plant Physiol.* 119(4): 1261–1269.

DE VICENTE M.C., TRUCO M.J., EGEA J., BURGOS L., ARUS P., 1998. 'RFLP variability in apricot (*Prunus armeniaca* L.)'. *Plant Breed.* 117(2): 153–158.

TAO R., HABU T., YAMANE H., SUGIURA A., IWAMOTO K., 2000. 'Molecular markers for self-compatibility in Japanese apricot (*Prunus mume*)'. *Hort. Sci.* 35(6): 1121–1123.

Cherry

BOSKOVIC R., TOBUTT K.R., 1998. 'Inheritance and linkage relationships of isozymes in two interspecific cherry progenies'. *Euphytica* 103: 273–286.

BOSKOVIC R., TOBUTT K.R., NICOLL F.J., 1997. 'Inheritance of isozymes and their linkage relationships in two interspecific cherry progenies'. *Euphytica* 93: 129–143.

BOSKOVIC R., RUSSELL K., TOBUTT K.R., RIDOUT M.S., 2000. 'An isozyme marker linked to the incompatibility locus in cherry'. *Theor. Appl. Genet.* 100: 512–518.

CANTINI C., IEZZONI A.F., LAMBOY W.F., BORIZKI M., STRUSS D., 2001. 'DNA fingerprinting of tetraploid cherry germplasm using simple sequence repeats'. *J. Am. Soc. Hort. Sci.* 126(2): 205–209.

DOWNEY S.L., IEZZONI A.F., 2000. 'Polymorphic DNA markers in black cherry (*Prunus serotina*) are identified using sequences from sweet cherry, peach, and sour cherry'. *J. Am. Soc. Hort. Sci.* 125(1): 76–80.

FILS-LYCAON B.R., WIERSMA P.A., EASTWELL K.C., SAUTIERE P., 1996. 'A cherry protein and its gene, abundantly expressed in ripening fruit, have been identified as thaumatin-like'. *Plant Physiol.* 111: 269–273.

LECOULS A.C., RUBIO-CABETAS M.J., MINOT J.C., VOISIN R., BONNET A., SALESSES G., DIRLEWANGER E., ESMENJAUD D., 1999. 'RAPD and SCAR markers linked to the Ma1 root-knot nematode resistance gene in Myrabolan plum (*Prunus cerasifera* Her.)'. *Theor. Appl. Genet.* 99(1/2): 328–335.

TAO R., YAMANE H., SUGIURA A., MURAYAMA H., SASSA H., MORI H., 1999. 'Molecular typing of S-alleles through identification, characterization and cDNA cloning for S-Rnases in sweet cherry'. *J. Am. Soc. Hort. Sci.* 124: 224–233.

WANG D., KARLE R., IEZZONI A.F., 2000. 'QTL analysis of flower and fruit traits in sour cherry'. *Theor. Appl. Genet.* 100: 535–544.

WANG D., KARLE R., BRETTIN T.S., IEZZONI A.F., 1998. 'Genetic linkage map in sour cherry using RFLP markers'. *Theor. Appl. Genet.* 97(8): 1217–1224.

WIERSMA P.A., WU Z., ZHOU L., HAMPSON C., KAPPEL F., 2001. 'Identification of new self-incompatibility alleles in sweet cherry (*Prunus avium* L.) and clarification of compatibility groups by PCR and sequencing analysis'. *Theor. Appl. Genet.* 102: 700–708.

Citrus

ASINS M.J., MONFORTE A.J., MESTRE P.F., CARBONELL E.A., 1999. '*Citrus* and *Prunus copia*-like retrotransposons'. *Theor. Appl. Genet.* 99(3/4): 503–510.

CAI Q., GUY C.L., MOORE G.A., 1994. 'Extension of the genetic linkage map in Citrus using random amplified polymorphic DNA (RAPD) markers and RFLP mapping of cold-acclimation-responsive loci'. *Theor. Appl. Genet.* 89: 606–614.

CRISTOFANI M., MACHADO M.A., GRATTAPAGLIA D., 1999. 'Genetic linkage maps of *Citrus sunki* Hort. Ex Tan. and *Poncirus trifoliata* (L.) Raf. and mapping of citrus tristeza virus resistance gene'. *Euphytica* 109(1): 25–32.

DENG Z., HUANG S., XIAO S., GMITTER F.G., 1997. 'Development and characterization of SCAR markers linked to the citrus tristeza virus resistance gene from *Poncirus trifoliata*'. *Genome* 40: 697–704.

DENG Z., HUANG S., LING P., CHEN C., YU C., WEBER C.A., MOORE G.A., GMITTER JR. F.G., 2000. 'Cloning and characterization of NBS-LRR class resistance gene candidate sequences in *Citrus*'. *Theor. Appl. Genet.* 101: 814–822.

DENG Z., HUANG S., LING P., YU C., TAO Q., CHEN C., WENDELL M.K., ZHANG H.B., GMITTER JR. F.G., 2001a. 'Fine genetic mapping and BAC contig development for the citrus tristeza virus resistance gene locus in *Poncirus trifoliata* (Raf.)'. *Mol. Genet. Genomics* 265: 739–747.

DENG Z., TAO Q., CHANG Y.L., HUANG S., LING P., YU C., CHEN C., GMITTER JR. F.G., ZHANG H.B., 2001b. 'Construction of a bacterial artificial chromosome (BAC) library for citrus and identification of BAC contigs containing resistance gene candidates'. *Theor. Appl. Genet.* 102: 1177–1184.

FANG D.Q., ROOSE M.L., 1997. 'Identification of closely related citrus cultivars with inter-simple sequence repeat markers'. *Theor. Appl. Genet.* 95: 408–417.

FANG D.Q., FEDERICI C.T., ROOSE M.L., 1998. 'A high-resolution linkage map of the citrus tristeza virus resistance gene region in *Poncirus trifoliata* (L.) Raf'. *Genetics* 150: 883–890.

FANG D.Q., ROOSE M.L., KRUEGER R.R., FEDERICI C.T., 1997. 'Fingerprinting trifoliate orange germplasm accessions with isozymes, RFLP, and inter-simple sequence repeat markers'. *Theor. Appl. Genet.* 95: 211–219.

GARCIA M.R., ASINS M.J., CARBONELL E.A., 2000. 'QTL analysis of yield and seed number in *Citrus*'. *Theor. Appl. Genet.* 101: 487–493.

GARCIA R., ASINS M.J., FORNER J., CARBONELL E.A., 1999. 'Genetic analysis of apomixes in *Citrus* and *Poncirus* by molecular markers'. *Theor. Appl. Genet.* 99(3/4): 511–518.

GMITTER. F.G. Jr., XIAO S.,, HUANG S., HU X., GARNSEY S.M., DENG Z., DENG Z., 1996. 'A localized linkage map of the citrus tristeza virus resistance gene region'. *Theor. Appl. Genet.* 92: 688–695.

KIJAS J.M.H., THOMAS M.R., FOWLER J.C.S., ROOSE M.L., 1997. 'Integration of trinucleotide microsatellites into a linkage map of *Citrus*'. *Theor. Appl. Genet.* 94: 701–706.

LING P., DUNCAN L.W., DENG Z., DUNN D., HU X., HUANG S., GMITTER. F.G. Jr., 2000. 'Inheritance of citrus nematode resistance and its linkage with molecular markers'. *Theor. Appl. Genet.* 100: 1010–1017.

LURO F., LAIGRET F., BOVE' J.M., OLLITRAULT P., 1995. 'DNA amplified fingerprinting, a useful tool for determination of genetic origin and diversity analysis in *Citrus*'. *Hort. Sci.* 30(5): 1063–1067.

MESTRE P.F., ASINS M.J., PINA J.A., CARBONELL E.A., NAVARRO L., 1997. 'Molecular markers flanking a citrus tristeza virus resistance gene from *Poncirus trifoliata* (L) Raf'. *Theor. Appl. Genet.* 94: 458–464.

NICOLOSI E., DENG Z.N., GENTILE A., LA MALFA S., CONTINELLA G., TRIBULATO E., 2000. '*Citrus* phylogeny and genetic origin of important species as investigated by molecular markers'. *Theor. Appl. Genet.* 100: 1155–1166.

SANKAR A.A., MOORE G.A., 2001. 'Evaluation of inter-simple sequence repeat analysis for mapping in *Citrus* and extension of the genetic linkage map'. *Theor. Appl. Genet.* 102(2/3): 206–214.

Grape

ABLETT E., SEATON G., SCOTT K., SHELTON D., GRAHAM M.W., BAVERSTOCK P., SLADE LEE L., HENRY R., 2000. 'Analysis of grape ESTs: global gene expression patterns in leaf and berry'. *Plant Sci.* 159(1): 87–95.

BOSS P.K., DAVIES C., ROBINSON S.P., 1996. 'Analysis of the expression of anthocyanin pathway genes in developing *Vitis vinifera* L. cv Shiraz grape berries and the implications for pathway regulation'. *Plant Physiol.* 111: 1059–1066.

BOSS P.K., VIVIER M., MATSUMOTO S., DRY I.B., THOMAS M.R., 2001. 'A cDNA from grapevine (*Vitis vinifera* L.), which shows homology to Agamous and Shatterproof, is not only expressed in flowers but also throughout berry development'. *Plant Mol. Biol.* 45: 541–553.

BOTTA R., SCOTT N.S., EYNARD I., THOMAS M.R., 1995. 'Evaluation of microsatellite sequence-tagged site markers for characterising *Vitis vinifera* cultivars'. *Vitis* 34: 99–112.

BOURQUIN J.C., SONKO A., OTTEN L., WALTER B., 1993. 'Restriction fragment length polymorphism and molecular taxonomy in *Vitis vinifera* L'. *Theor. Appl. Genet.* 87: 431–438.

BOWERS J.E., MEREDITH C.P., 1997. 'The parentage of a classic wine grape, Cabernet Sauvignon'. *Nature Genet.* 16: 84–87.

BOWERS J.E., DANGL G.S., MEREDITH C.P., 1999a. 'Development and characterisation of additional microsatellite DNA markers for grape'. *Am. J. Enol. Vitic.* 50: 243–247.

BOWERS J.E., DANGL G.S., VIGNAIN R., MEREDITH C.P., 1996. 'Isolation and characterisation of new polymorphic simple sequence repeat loci in grape (*Vitis vinifera* L.)'. *Genome* 39: 628–633.

BOWERS J., BOURSIQUOT J.M., THIS P., CHU K., JOHANSSON H., MEREDITH C., 1999b. 'Historical genetics: the parentage of Chardonnay, Gamay, and other wine grapes of Northeastern France'. *Science* 285: 1562–1565.

CERVERA M.T., CABEZAS J.A., SANCHA J.C., MARTINEZ DE TODA F., MARTINEZ-ZAPATER J.M., 1998. 'Application of AFLPs to the characterization of grapevine *Vitis vinifera* L. genetic resources. A case of study with accessions from Rioja (Spain)'. *Theor. Appl. Genet.* 97: 51–59.

CIPRIANI G., FRAZZA G., PETERLUNGER E., TESTOLIN R., 1994. 'Grapevine fingerprinting using microsatellite repeats'. *Vitis* 33: 211–215.

DALBO M.A., YE G.N., WEEDEN N.F., STEINKELLNER H., SEFC K.M., REISCH B.I., 2000. 'A gene controlling sex in grapevines placed on a molecular marker-based genetic map'. *Genome* 43(2): 333–340.

DAVIES C., ROBINSON S.P., 1996. 'Sugar accumulation in grape berries: cloning of two putative vacuolar invertase cDNAs and their expression in grapevine tissue'. *Plant Physiol.* 111: 275–283.

DAVIES C., ROBINSON S.P., 2000. 'Differential screening indicates a dramatic change in mRNA profiles during grape berry ripening. Cloning and characterization of cDNAs encoding putative cell wall and stress response proteins'. *Plant Physiol.* 122: 803–812.

DI GASPERO G., PETERLUNGER E., TESTOLIN R., EDWARDS K.J., CIPRIANI G., 2000. 'Conservation of microsatellite loci within the genus *Vitis*'. *Theor. Appl. Genet.* 101: 301–308.

FOSSATI T., LABRA M., CASTIGLIONE S., FAILLA O., SCIENZA A:, SALA F., 2001. 'The use of AFLP and SSR molecular markers to decipher homonyms and synonyms in grapevine cultivars: the case of the varietal group known as "Schiave"'. *Theor. Appl. Genet.* 102(2/3): 200–205.

JACOBS A.K., DRY I.B., ROBINSON S.P., 1999. 'Induction of different pathogenesis-related cDNAs in grapevine infected with powdery mildew and treatment with ethephon'. *Plant Pathol.* 48: 325–336.

LAMBOY W.F., ALPHA C.G., 1998. 'Using simple sequence repeats (SSRs) for DNA fingerprinting germplasm accessions of grape (*Vitis*) species'. *J. Am. Soc. Hort. Sci.* 123: 182–188.

LODHI M.A., DALY M.J., YE G.N., WEEDEN N.F., REISCH B.I., 1995. 'A molecular marker based linkage map of *Vitis*'. *Genome* 38(4): 786–794.

SARNI-MANCHADO P., VERRIES C., TESNIERE C., 1997. 'Molecular characterisation and structural analysis of one alcohol dehydrogenase gene (GV-*Adh*1) expressed during ripening of grapevine (*Vitis vinifera* L.) berry'. *Plant Sci.* 125: 177–187.

SCOTT K.D., EGGLER P., SEATON G., ROSSETTO M., ABLETT E.M., LEE L.S., HENRY R.J., 2000. 'Analysis of SSRs derived from grape ESTs'. *Theor. Appl. Genet.* 100: 723–726.

SEFC K.M., REGNER F., TURETSCHET E., GLOSSL J., STEINKELLNER H., 1999. 'Identification of microsatellite sequences in *Vitis riparia* and their applicability for genotyping of different *Vitis* species'. *Genome* 42: 367–373.

SEFC K.M., STEINKELLNER H., WAGNER H.W., GLOSSL J., REGNER F., 1997. 'Application of microsatellite markers to parentage studies in grapevine'. *Vitis* 36: 179–183.

SEFC K.M., LOPES M.S., LEFORT F., BOTTA R., ROUBELAKIS-ANGELAKIS K.A., INANEZ J., PEJIC I., WAGNER, H.W., GLOSSL J., STEINKELLNER H., 2000. 'Microsatellite variability in grapevine cultivars from different European regions and evaluation of assignment testing to assess the geographic origin of cultivars'. *Theor. Appl. Genet.* 100: 498–505.

STRAVRAKAKIS M.N., BINIARI K., 1998. 'Genetic study of grape cultivars belonging to the Muscat family by random amplified DNA markers'. *Vitis* 37: 119–122.

TASSERSALL D.B., VAN HEESWIJCK R., HOJ P.B., 1997. 'Identification and characterization of a fruit-specific, thaumatin-like protein that accumulates at very high levels in conjunction with the onset of sugar accumulation and berry softening in grapes'. *Plant Physiol.* 114: 759–769.

TESNIERE C., VERRIES C., 2000. 'Molecular cloning and expression of cDNAs encoding alcohol dehydrogenases from *Vitis vinifera* L., during berry development'. *Plant Sci.* 157(1): 77–88.

TESSIER C., DAVID J., THIS P., BOURSIQUOT J.M., CHARRIER A., 1999. 'Optimization of the choice of molecular markers for varietal identification in *Vitis vinifera* L.' *Theor. Appl. Genet.* 98(1): 171–177.

THOMAS M.R., SCOTT N.S., 1993. 'Microsatellite repeats in grapevine reveals DNA polymorhisms when analysed as sequence-tagged sites (STSs)'. *Theor. Appl. Genet.* 86: 985–990.

THOMAS M.R., CAIN P., SCOTT N.S., 1994. 'DNA typing of grapevine: a universal methodology and database for describing cultivars and evaluating genetic relatedness'. *Plant Mol. Biol.* 25: 939–949.

VIDAL J.R., DELAVAULT P., COARER M., DEFONTAINE A., 2000. 'Design of grapevine (*Vitis vinifera* L.) cultivar-specific SCAR primers for PCR fingerprinting'. *Theor. Appl. Genet.* 101: 1194–1201.

XU H., BAKALINSKI A.T., 1996. 'Identification of grape (*Vitis*) rootstocks using sequence characterised amplified region DNA markers'. *Hort. Sci.* 31: 267–268.

4

Genes involved in plant defence mechanisms

M.A. Gómez Lim, CINVESTAV-Irapuato

4.1 Introduction

Historically, pathogen infection and insect infestation of staple crops have led to food shortages and considerable economic losses. Resistant varieties have, therefore, been developed by plant breeders for a number of years to reduce such losses, but pathogens are eventually able to overcome this resistance. Many pesticides have been developed to combat crop losses, with the consequence that plant disease control has become heavily dependent on these compounds. Yet the use of pesticides has also resulted in significant costs to public health and the environment. Plant biotechnology (gene isolation and plant transformation techniques), together with conventional breeding programmes, could make significant contributions to sustainable agriculture. In this regard, there has been intensive research in agricultural biotechnology aimed at plant protection. This chapter will describe the main defence-related mechanisms that plants display to cope with pathogen infection. Subsequently, the current status of the genes identified for resistance against virus, fungi, insects and nematodes, with an emphasis on their role in resistance to pathogens in transgenic plants, will be discussed. Reference will be made to the use of species of *Trichoderma* as a biological control agent.

4.2 Mechanisms of plant response to pathogens

Plant disease resistance is dependent on the genetic background of both host and pathogen and relies on a series of complex mechanisms of molecular recognition and signal transduction (Crute 1985). In general, plant resistance occurs in the following circumstances:

1. The pathogen fails to infect the plant because it belongs to a taxonomic group outside the host range of the pathogen (nonhost resistance). This is the most common form of resistance exhibited by plants.
2. The plant contains preformed physical and chemical barriers, which prevent pathogen penetration.
3. The plant recognises the presence of the pathogen and rapidly triggers an array of defence mechanisms, which involve differential gene expression (host resistance).

It is now clear that, in the latter type of resistance, disease susceptibility frequently results from poor pathogen perception, rather than a lack of host resistance machinery. Therefore, early recognition of the pathogen at the level of single cells is essential to mount an efficient defence response. Successful pathogen recognition triggers the activation of several and diverse defence responses. Sometimes, resistance is manifested at the macroscopic level by the appearance of necrotic lesions at the site of infection. This is the result of rapid localised cell death termed hypersensitive response (HR) which is thought to limit pathogen growth and spread. Early and local responses associated with the HR include the transient opening of ion channels, production of reactive oxygen species, cell wall fortification, production of antimicrobial pytoalexins, host cell death and synthesis of pathogenesis-related proteins (PRP), which are thought collectively to confer the observed resistance to bacterial, fungal and viral pathogens (Hammond-Kosack and Jones 1996).

In addition to localised responses, plants often induce defence mechanisms in uninfected areas. Defence responses at such secondary sites are collectively referred to as systemic acquired resistance (SAR). SAR can be distinguished from other inducible resistances based upon the spectrum of pathogen protection and the associated changes in gene expression. SAR is induced following infection by necrotising pathogens (e.g. *Colletotrichum lagenarium*, tobacco mosaic virus, etc.) or experimentally by treatments with salicylic acid (SA) (Stichter *et al.* 1997). SAR leads to induction of pathogenesis related (PR)s genes, such as glucanases and chitinases (Stichter *et al.* 1997) and confers a long lasting, broad-spectrum disease resistance that is dependent upon SA accumulation (Stichter *et al.* 1997). SA has been shown to have multiple roles and appears to be a common signalling molecule in both the HR and SAR responses (Malek and Lawton 1998).

A series of *Arabidopsis* mutants exhibiting a constitutive SAR have been identified (Bowling *et al.* 1997). They display high levels of PR gene expression, and broad-spectrum pathogen resistance (Bowling *et al.* 1997). Nevertheless, these mutants also displayed phenotypic alterations such as reduced size or altered morphology, which suggests that genetic manipulation for constitutive SAR in crop plants may result in yield losses. However, transfer of two bacterial genes coding for enzymes that convert chorismate into SA in tobacco plants resulted in overproduction of salicylic acid and enhanced resistance to viral and fungal infection (Verberne *et al.* 2000). The plants did not present any phenotypic alteration but genes encoding acidic pathogenesis-related (PR) proteins, were constitutively expressed.

In addition to the SA-dependent pathway, an SA-independent pathway has been identified and termed induced systemic resistance (ISR) (Pieterse *et al.* 1998). ISR is not associated with SAR gene expression and confers quantitative resistance (40–60% protection) to fungal and bacterial pathogens. Moreover, ISR is dependent on jasmonic acid (JA) and ethylene signalling (Knoester *et al.* 1999). The necrotrophic bacteria *Erwinia carotovora*, has been shown to induce expression of certain PR genes via an SA-independent, and potentially even SA-antagonistic, pathway during an early phase of infection (Vidal *et al.* 1997). Similarly, infection of *Arabidopsis* with necrotrophs such as *Alternaria brassicola* leads to induction of thionin and defensin-like genes such as *PDF1.2* (whose expression is SA-independent) but does not result in PR-1 induction (Penninckx *et al.* 1996). It is likely that the plant response to pathogen invasion involves a combination of the different mechanisms described (Somssich and Hahlbrock 1998). That may explain why different pathogens may induce the same defence mechanism (i.e. synthesis of PRP).

4.3 Genes in the defence against virus

Viral diseases cause serious losses world wide in horticultural and agricultural crops in many parts of the world. They have been difficult to control and resistance has traditionally relied on either the use of pesticides to kill viral insect vectors or the introduction of natural resistance genes through conventional breeding programmes. More recently, viral resistance has been engineered by transforming susceptible plants with genes or sequences derived from viral sequences, the so-called pathogen-derived resistance (PDR) (Lomonossoff 1995). The mechanisms by which PDR is activated have not been well defined, but resistance may result from the expression of a viral protein (native or mutated) or an RNA-mediated mechanism that appears to be analogous to gene silencing (Baulcombe 1999). Interestingly, gene silencing appears to be the natural mechanism by which kohlrabi and *Nicotiana clevelandii* resist infection by cauliflower mosaic virus and tomato black ring nepovirus respectively (Ratcliff *et al.* 1997; Covey *et al.* 1997).

The first demonstration of PDR came from transgenic tobacco expressing the coat protein (CP) of tobacco mosaic virus (TMV) (Lomonossoff 1995). Since then, CP-mediated resistance (CPMR) has been engineered against a wide variety of viruses in many plant species. It is still unclear why CPMR sometimes confers resistance only to the viral strain from which the CP was derived, whereas in other cases it provides protection against related viruses. Nevertheless, as CPMR is effective in many plants against diverse viruses, it has been employed in different crops of agronomic importance. Transgenic lines of squash and papaya exhibiting CPMR have already been approved for commercial release and other crops are under development (Kaniewski *et al.*

1999). As more is known about the mechanisms associated with CPMR, it will be possible to manipulate the level and/or breadth of resistance, as in the case of TMV where mutant CPs that confer much greater resistance than the wild-type protein have been generated (Beachy, 1997).

Viral replicase genes in either wild-type or defective forms have also been employed to engineer PDR (Baulcombe 1996). Replicase-mediated resistance (RMR) can confer nearly full immunity to infection, but similarly to CPMR, it tends to be effective only against the virus strain from which the gene was derived and it is even more limited than CPMR (Baulcombe 1996). The mechanisms by which RMR is activated are not well understood but the high degree of resistance associated with RMR makes it attractive for engineering virus-resistant crops. Interestingly, two cases of broader-base RMR have been reported (Beachy 1997). Clearly, a better understanding of the mechanisms by which RMR is achieved is needed before immunity can be conferred against more than a single viral strain.

PDR has also been engineered by expressing mutant forms of viral movement proteins (MP). MP-mediated resistance (MPMR) represents a very interesting approach because it confers delayed symptoms and/or decreased systemic viral accumulation against a much broader spectrum of viruses than either CPMR or RMR (Baulcombe 1996). Furthermore, MPMR can also protect against viruses thought to move through plasmodesmata as well as tubules (Beachy 1997). It has been hypothesised that MPMR results from a dominant negative mutation in the MP, which obstructs the wild-type MP from either binding the viral genome or facilitating the cell-to-cell or systemic movement (Beachy 1997).

In addition to the above examples, PDR has been engineered using other viral genes as well as entire viral genomes (Baulcombe 1996; Beachy 1997; Song *et al.* 1999). The mechanisms through which viral genomes confer resistance are not well defined but it has been shown that a potato virus X (PVX) replicon provides high-level, strain specific resistance via an RNA-dependent silencing mechanism (Angell and Baulcombe 1997). Several strategies for engineering virus-resistant plants independent of PDR have also been developed. One employs ribosome-inactivating proteins (RIP), which cleave the N-glycosidic bond of adenine in a specific ribosomal RNA sequence thereby rendering them incapable of protein synthesis. One of the best studied RIPs is pokeweed antiviral protein (PAP), which exhibits potent nonspecific antiviral activity and has been used to engineer resistance in tobacco to a broad spectrum of plant viruses (Tumer *et al.* 1999). Unfortunately, construction of these transgenic plants has been difficult probably due to the toxic effects of PAP. For that reason, the use of a carboxy-terminal deletion mutant of PAP (Zoubenko *et al.* 1997) and a recently isolated PAP isoform (PAP II) have resulted in reduced toxicity and a broad spectrum resistance to viral infection (Tumer *et al.* 1997; Wang *et al.* 1998). Interestingly, PAP overexpression also confers resistance to fungal infection (Wang *et al.* 1998).

Proteins that recognise and degrade double-stranded RNA, the replication intermediate for most plant viruses, have also been employed to engineer resistance. Tobacco plants expressing *pac1*, a yeast double-strand RNA-specific ribonuclease, exhibit modest levels of resistance after infection by several unrelated viruses (Watanabe *et al.* 1995), whereas plants expressing 2′,5′-oligoadenylate system are either completely or partially resistant to various unrelated viruses (Ogawa *et al.* 1996).

Recently there have been a number of reports on the use of plant genes to confer resistance against viral diseases. They have used natural *R* genes (Van Der Vossen *et al.* 1997; Bendahmane *et al.* 1999; Cooley *et al.* 2000), genes coding for heat shock-like proteins (Whitham *et al.* 2000) or genes coding for unknown proteins (Kachroo *et al.* 2000; Mourrain *et al.* 2000).

A novel approach to engineer resistance against viral diseases has been the use of proteinase inhibitors. Proteinase activity is an important component of the processing mechanism of several groups of plant viruses (Spall *et al.* 1997). Therefore, introduction of the corresponding proteinase inhibitor may be one way to inhibit viral replication and confer resistance. Gutiérrez-Campos *et al.* (1999) have shown that transgenic tobacco plants containing a cysteine proteinase inhibitor (oryzacystatin I) exhibited immunity against more than one potyvirus (the most important virus group, from the standpoint of the economic losses that they cause) and presented several beneficial pleiotropic effects (Gutiérrez-Campos *et al.* 2000). Since the proteinases employed by the different groups of plant viruses may vary (serine, cysteine, etc.), an adequate choice of the inhibitor may provide resistance to a particular plant virus.

In addition to providing broad-based resistance, strategies involving non-viral, plant genes may be advantageous because they eliminate concerns that recombination between a virus and a related transgene or transcapsidation of a viral RNA might create a novel 'superpathogen' capable of massive crop infestation.

4.4 Genes in the defence against fungi

Fungal diseases have been one of the main causes of crop losses for years. Control of fungal diseases has traditionally involved three strategies: husbandry techniques, such as crop rotation and confinement of contaminated soil and plant material, breeding of resistant crop cultivars and application of fungicides. Although conventional plant breeding has made a significant impact by improving crop resistance to many important diseases, it still requires extended periods of time to develop a new variety and there may not be any natural sources of resistance to major diseases available to the breeder (Oerke 1994). In modern agriculture farmers employ a variety of fungicides, but their use is being restricted because of their high costs and growing concerns about the degradation of the environment. Furthermore, the excessive use of fungicides frequently results in the development of resistant fungal strains.

The advent of advanced molecular techniques for plant breeding has allowed the development of crops resistant to a number of fungal pathogens. There have been mainly two approaches to generate broad-spectrum, fungal-resistant crops: one relies on the overexpression of genes encoding antifungal proteins (AFP) and the other aims to enhance pathogen perception, by manipulation of certain pathogen and plant *R* genes.

4.4.1 Antifungal proteins

Many of the genes induced by the plant disease-resistance responses described in Section 4.4 encode proteins with antifungal activity, which probably have an important role against fungal infection. The AFP strategy involves the constitutive expression in transgenic plants of genes encoding proteins with a fungitoxic or fungistatic capacity. This is an extension of the paradigm that has worked so well for insect control genes based on insecticidal proteins from *Bacillus thuringiensis* (see below). Table 4.1 presents an overview of the reports on transgenic plants obtained using this approach.

Plant defence mechanisms are usually accompanied by the expression of a large set of genes termed PR. At least ten families of PRPs have been identified (Dempsey *et al.* 1998) (Table 4.2) and the two most prominent members have been the hydrolytic enzymes chitinase and β-1,3 glucanase, which are capable of degrading the major cell wall constituents (i.e. chitin and β-1,3 glucan) of most filamentous fungi. Expression of either enzyme individually in a number of plants has conferred resistance against a particular pathogen, but constitutive coexpression of both enzymes confers even higher levels of resistance (Table 4.1), suggesting a synergistic interaction between the two enzymes. Other genes coding for different PRPs have yielded comparable results. Constitutive expression in tobacco of PR1a, a protein with unknown biological function, increased resistance to two oomycete pathogens (Alexander *et al.* 1993). However, fungal resistance is significantly enhanced when more than one gene is employed. Future studies are necessary to identify different combinations of PR proteins that might confer effective broad-spectrum protection.

In addition to PR proteins, a broad family of small, cysteine-rich AFP has been characterised (Broekaert *et al.* 1997). This family include plant defensins, thionins, lipid transfer proteins (LTP) and hevein- and knottin-type peptides and have been shown to posses antifungal activity *in vitro* against a broad spectrum of fungal pathogens (Broekaert *et al.* 1995; 1997). They seem to inhibit fungal growth by permeabilisation of fungal membranes (Thevissen *et al.* 1999). Plant defensins have been isolated from seeds of a variety of plants and shown to be induced upon pathogen infection in an SA-independent manner (Terras *et al.* 1998). They have been used to confer resistance against *Alternaria longipes* in tobacco (Broekaert *et al.* 1995). Thionins and LTP are also induced in an SA-independent manner by infection of a variety of plant pathogens in many plant tissues (Broekaert *et al.* 1997). Overexpression of an endogenous thionin gene in *Arabidopsis* conferred protection against *Fusarium oxysporum* (Epple *et al.* 1997) whereas expression of a barley non-specific LTP in *Arabidopsis* and tobacco conferred enhanced resistance to the bacterial pathogen *Pseudomonas*

Table 4.1 Fungal resistance in transgenic plants

Plant	Transgene(s)	Pathogen	Reference
Alfalfa	Peanut Resveratrol synthase	*Phoma medicaginis*	Hipskind and Paiva 2000
Brassica napus	Bean chitinase	*Rhizoctonia solani*	Broglie *et al.* 1991
	Tomato/tobacco Chitinase	*Cylindrosporium conc.*	Grison *et al.* 1996
		Sclerotinia sclerotiorum	Grison *et al.* 1996
		Phoma lingam	Grison *et al.* 1996
Carrot	Tobacco Chitinase + β-1,3 Glucanase	*Alternaria dauci*	Melchers and Stuiver 2000
	Tobacco AP24	*Alternaria radicina*	Melchers and Stuiver 2000
		Cercospora carotae	Melchers and Stuiver 2000
		Erysiphe heracleï	Melchers and Stuiver, 2000
Potato	AP24	*Phytophthora infestans*	Liu *et al.* 1994
	Glucose oxidase	*Phytophthora infestans*	Wu *et al.* 1995
		Verticillium dahliae	Wu *et al.* 1995
	Osmotin	*Phytophthora infestans*	Li *et al.* 1999
	Tobacco class II catalase	*Phytophthora infestans*	Yu *et al.* 1999
	Aly AFP	*Verticillium* sp.	Liang *et al.* 1998
	Soybean β-1,3-Glucanase	*Phytophthora infestans*	Borkowska *et al.* 1998
	PR-1a, SAR 8.2	*Peronospora tabacina*	Alexander *et al.* 1993
Tobacco		*Phytophthora parasitica, Pythium* sp.	Alexander *et al.* 1993
	Class III Chitinase	*Phytophthora parasitica*	Alexander *et al.* 1993
	Class I Chitinase	*Rhizoctonia solani*	Alexander *et al.* 1993
	Bean Chitinase	*Rhizoctonia solani*	Broglie *et al.* 1991
	Barley RIP	*Rhizoctonia solani*	Logemann *et al.* 1992
	Serratia marcescens Chitinase	*Rhizoctonia solani*	Logemann *et al.* 1992
	Barley Chitinase + β-1, 3 Glucanase	*Rhizoctonia solani*	Jach *et al.* 1995
	Barley Chitinase + RIP	*Rhizoctonia solani*	Jach *et al.* 1995
	Rice Chitinase + Alfalfa Glucanase	*Cercospora nicotianae*	Zhu *et al.* 1994
	Sarcotoxin IA	*Rhizoctonia solani*	Mitsuhara *et al.* 2000
		Pythium aphanidermatum	Mitsuhara *et al.* 2000
	Pseudomonas syringae hrmA	*Phytophthora parasitica*	Shen *et al.* 2000
	Oxalate decarboxylase	*Sclerotinia sclerotiorum*	Kesarwani *et al.* 2000
	Radish AFP	*Alternaria longipes*	Terras *et al.* 1995
Tomato	Tobacco Chitinase + β-1, 3 Glucanase	*Fusarium oxysporum*	Jongedijk *et al.* 1995
Rice	Rice Chitinase	*Rhizoctonia solani*	Lin *et al.* 1995
Wheat	Aly AFP	*Fusarium* sp.	Liang *et al.* 1998

Table 4.2 Pathogenesis-related proteins in plants

PR protein family	Enzymatic activity	Target in pathogen
PR-1	Unknown	Membrane?
PR-2	1, 3-β-glucanase	Cell wall glucan
PR-3	Endochitinase	Cell wall chitin
PR-4	Endochitinase	Cell wall chitin
PR-5	Osmotin-like	Membrane
PR-6	Proteinase inhibitor	Proteinase
PR-7	Proteinase	Unknown
PR-8	Endochitinase	Cell wall chitin
PR-9	Peroxidase	Plant cell wall
PR-10	RNAase?	Unknown
PR-11	Endochitinase	Cell wall chitin
Unclassified	α-Amylase	Cell wall α-glucan
	Polygalacturonase	Unknown
	inhibitor protein (PGIP)	Polygalacturonase

syringae (Molina and Garcia-Olmedo 1997). In the only example where hevein- and knottin-type peptides were overexpressed in transgenic plants, the resultant tobacco plants were not any more resistant to infection by *Alternaria longipes* than control plants (De Bolle *et al.* 1993). This might be explained by the high susceptibility of the antifungal activity of hevein- and knottin-type peptides to the presence of inorganic cations (De Bolle *et al.* 1993).

Finally, there are several AFP that do not fall into any of the classes described above. For example, the H_2O_2-producing enzyme oxalate oxidase has been shown to accumulate in barley infected by *Erysiphe gramminis* (Zhou *et al.* 1995). Interestingly, Wu *et al.* (1995) demonstrated that constitutive expression of another H_2O_2-producing enzyme, glucose oxidase provided disease resistance to a range of plant pathogens, including *Phytophtora infestans*, *Erwinia carotovora* and *Verticillium* wilt disease.

Most of the reports described above are based exclusively on observations of the increased fungal resistance of transgenic plants tested in climate-controlled growth chambers or greenhouse facilities. The challenge now is to translate such results into a significant outcome in the field.

4.4.2 Plant *R* genes

The genetically controlled induction of HR is triggered in plant-pathogen interactions only if the plant contains a disease-resistance protein (R) that recognises the correspondent avirulence (Avr) protein from the pathogen. In the absence of a functional *R* gene or avirulence gene product, no recognition occurs and disease ensues. This indicates that the factors controlling HR are quite specific and that they do not provide resistance to more than a limited number of races or pathotypes. To engineer broad-spectrum disease resistance relying on

HR, two approaches have been employed, although they are still in the early stages: one is based on the transfer of an avirulence gene (i.e. the *Cladosporium fulvum avr9* gene) into a plant containing the corresponding resistance gene (i.e. the tomato *Cf9* gene) and its subsequent expression under the control of a promoter inducible by fungal pathogens. Pathogen-induced expression of the *Avr* gene will then provoke a resistance reaction manifested by a HR. A localised HR will be induced preventing further spread of any invading pathogen, followed by a general defence response. The other approach is based on the overexpression of *R* genes.

The first approach has been tested experimentally three times, twice in tomato using the *AvrPto* gene (Tobias *et al.* 1999; Melchers and Stuiver 2000) and the other in tobacco using an elicitin gene coupled to a pathogen-inducible promoter (Keller *et al.* 1999). In both cases, increased resistance to a broad spectrum of diseases, including both fungal and viral, was obtained. Nevertheless, the main limitation of this approach is still the limited number of *Avr* genes available.

The second approach has benefited from the cloning and analysis of over 20 *R* genes isolated from seven plant species, including both monocots and dicots (Martin 1999). Resistance genes involved in race-specific interactions often provide full disease resistance and are well-known from conventional breeding programmes (Rommens and Kishore 2000). In spite of the great diversity in lifestyles and pathogenic mechanisms of disease-causing organisms, it was somewhat surprising that *R* genes were found to encode proteins with sequence similarities and conserved motifs (Martin 1999). They have been classified into five classes according to the structural characteristics of their predicted protein products: intracellular protein kinases; receptor-like protein kinases with an extracellular leucine-rich repeat (LRR) domain; intracellular LRR proteins with a nucleotide binding site (NBS) and a leucine zipper motif; intracellular NBS-LRR proteins with a region with similarity to the Toll and interleukin-1 receptor (TIR) proteins; and LRR proteins that encode membrane-bound extracellular proteins.

Overexpression of *Pto* (an *R* gene) in tomato elicited an array of defence responses including microscopic cell death, SA accumulation and PR gene expression, and the plants showed increased resistance to several pathogenic bacteria and fungi (Tang *et al.* 1999). Similarly, overexpression of the *Bs2* pepper *R* gene in tomato allowed enhance resistance against *Xantomonas campestris* (Oldroyd and Staskawicz 1998) and bacterial spot disease (Tai *et al.* 1999). Isolation and analysis of plant *R* genes is an extremely active area of research and considering the large number of genes already available and the current work to isolate more genes in many laboratories around the world, there will be substantial progress in this field in the short term.

Overexpression of signalling components that lie downstream of *R* genes is another possible strategy to increase disease resistance. In the first successful example of this approach, the *NPR1* gene was overexpressed in *Arabidopsis* and the resultant transgenic plants exhibited significant increases in resistance to

Pseudomonas and *Peronospora* pathogens (Cao *et al.* 1998). Manipulation of downstream components such as *NPR1* potentially allows activation of only specific defence pathways. This might be one way to avoid agronomic problems associated with constitutive activation of *R* gene-mediated pathways such as HR (see Section 4.1).

4.4.3 Other strategies

In addition to the strategies described above, other approaches to control fungal disease are based on the manipulation of the levels of phytoalexins which are small, broad-spectrum antimicrobial compounds whose synthesis and accumulation is frequently associated with HR. For example, increased resistance to *Botrytis cinerea* has been observed in tobacco expressing a grapevine stilbene synthase gene, which increased the level of the phytoalexin resveratrol (Hain *et al.* 1993). However, there are several caveats to engineering phytoalexin-mediated resistance. Many of these compounds are synthesised via complex pathways, which in order to alter existing phytoalexin structure or content, would require the manipulation of several genes, making the whole process technically more demanding. In addition, phytoalexins are often toxic to the pathogen as well as to the plant. Thus, some type of inducible expression system may be required. A novel and promising approach to engineer broad-spectrum resistance relies on the use of antimicrobial peptides. These peptides are ubiquitous (Gabay 1994) and show a strong antimicrobial activity. Expression in potato plants of a synthetic gene encoding an N-terminus-modified, cecropinmelittin cationic peptide chimera resulted in significant resistance against several bacterial and fungal phytopathogens (Osusky *et al.* 2000). It is likely that this type of approach will be employed considerably in the near future to engineer a range of disease-resistant plants.

4.4.4 *Trichoderma harzianum* as a biological control agent

Biological control of soil-borne plant pathogens is an attractive approach to control fungal diseases (Chet 1990). When effective, it is a significant method of pest control not only because it eliminates the use of fungicides, but also because if the introduced biocontrol agent becomes properly established, it does not require repeated applications. *Trichoderma* spp. is among the most studied biocontrol agents (Chet 1990). Several species of *Trichoderma* spp have been isolated and found to be effective biocontrol agents of various soil-borne plant pathogenic fungi under greenhouse and field conditions (Chet and Inbar 1994). *Trichoderma harzianum* has proved to be the most effective species, and it has been shown to attack a range of economically important soil-borne plant-pathogenic fungi (Chet 1990). *Trichoderma* can be added to the soil as a powder, wheat bran or a peat-wheat bran preparation. It can be sprayed or injected; painted on tree wounds; inserted in pellets in holes drilled in trees; and conidia can be applied directly to the ground or as seed coating (Chet 1990). In addition, *Trichoderma* may contribute to the overall plant defence response as it has been recently shown that *Trichoderma* application induces systemic resistance mechanisms in cucumber plants (Yedidia *et al.* 1999).

Trichoderma may have three modes of action as part of its antagonic activity, antibiosis, competition and mycoparasitism. However, mycoparasitism has been suggested as the main mechanism involved in the antagonism as a biocontrol agent (Haran *et al.* 1996). In order to attack a fungal cell, *Trichoderma* must degrade the cell wall. Given the composition of most fungal cell walls, it has been suggested that chitinases, proteases and β-1,3 glucanases are the main enzymes involved in the mycoparasitic process (Elad *et al.* 1982; Haran *et al.* 1996). However, it is likely that the co-ordinated action of all hydrolases produced by *Trichoderma* is required for a complete dissolution of the cell wall. Indeed, a number of *Trichoderma* isolates are able to secrete different kinds of hydrolytic enzymes into the medium when grown in the presence of cell walls of phytopathogenic fungi (Geremia *et al.* 1993). In addition, the production of several of these lytic enzymes by *Trichoderma* is induced during the parasitic interaction (Haran *et al.* 1996; Flores *et al.* 1997).

As of late, people have been trying to increase the effectiveness, stability, and biocontrol capacity of *Trichoderma* spp by altering the levels of different hydrolytic enzymes. Several improved *Trichoderma* strains have, therefore, been obtained which display an increased antifungal activity by overexpression of a proteinase (Flores *et al.* 1997) or a 33 kDa chitinase (Limon *et al.* 1999). For that reason, considering the activity and specificity of many fungal enzymes, mycoparasitic fungi may serve as excellent sources of genes for disease resistance (Lorito *et al.* 1998). In spite of the success achieved, treatments with *Trichoderma harzianum* have usually not been as effective as the use of some fungicides. Treatment with *T. harzianum* has, therefore, been combined with other cultural practices to implement integrated pest management (Hall 1991).

4.5 Genes in the defence against insects and nematodes

4.5.1 Insects

Insect infestation has caused heavy losses in many agricultural crops for years. Depending on the crop, it is estimated that losses range from 5–30% (FAOSTAT 2000). Control of insects has traditionally employed application of pesticides and to some extent biocontrol agents such as *Bacillus thuringiensis* (*Bt*). The cloning of the δ-endotoxin of *Bt* has allowed the generation of transgenic plants containing the gene. For the past ten years, genes coding for δ-endotoxins from different *Bacillus thuringiensis* (*Bt*) subspecies individually or in combination, have been used to protect crops against insects (Sanchis and Lereclus 1999). Currently, corn, potato and cotton plants expressing different synthetic *Bt* are commercially available; they show meaningful protection again different insects such as European corn borer, Colorado potato beetle and bollworm infestations respectively (Dempsey *et al.* 1998) Recently transgenic rice containing a fusion gene derived from cryIA(b) and cryIA(c) was field-tested in natural and repeated heavy manual infestation of two lepidopteran insects, leaffolder and yellow stem borer (Tu *et al.* 2000). The transgenic hybrid plants showed high resistance against both insect pests without reduced yield.

In spite of these successes, it is worth mentioning that *Bt* δ-endotoxins have not been effective against all insects and, most importantly, that insects have developed resistance against different δ-endotoxins (Tabashnik *et al.* 2000). For that reason alternative insecticidal proteins are being actively pursued. Several such proteins have been identified, including Vip3A and cholesterol oxidases (Dempsey *et al.* 1998). Various endogenous proteins, which are synthesised in response to insect attack, could potentially be used to engineer pest-resistant plants. One such protein is systemin, the first plant polypeptide hormone discovered. Systemin is phloem-mobile and is an essential component of the wound-inducible systemic signal transduction system leading to the transcriptional activation of the defensive genes (Ryan 2000). Systemin is processed from a larger prohormone protein, called prosystemin, by proteolytic cleavages and it has being suggested that overexpression of prosystemin in transgenic plants may confer protection against insect invasion (Schaller 1999).

A quite different alternative to control insect infestation has been the use of proteinase inhibitors (PIs). These compounds can inhibit various digestive enzymes (proteinases) found in the gut of many insects. The synthesis of some PI is stimulated by wounding, including insect attack, whereas others are induced by pathogen infection (Ryan 1990). The wound-inducible serine PIs from tomato have been studied the most extensively. They were divided into two groups based on sequence and molecular weight (PI–1 = 8kDa; PI–I = 12kDa) (Ryan 1990). Several components in the induction pathway leading to PI synthesis have been identified, including systemin, various intermediates of the octadecanoid pathway and jasmonic acid (Ryan and Pearce 1998). Constitutive expression of different types of PI, including serine (Duan *et al.* 1996; Hilder and Boulter 1999) and cysteine (Irie *et al.* 1996) in transgenic plants reduces predation by inhibiting important digestive enzymes in the insect gut. Nevertheless, similarly to the *Bt* δ-endotoxin, insects have developed resistance against PI (Girard *et al.* 1998). For a more effective and durable resistance, it may be necessary to combine different strategies.

4.5.2 Nematodes

Plant parasitic nematodes are obligate parasites that cause billions of dollars in losses annually to the world's farmers (Williamson 1998). They have been divided into ectoparasites and endoparasites. The sedentary endoparasites of the family Heteroderidae cause the most economic damage and include two groups: the cyst nematodes, which include the groups *Heterodera* and *Globodera*, and the root-knot nematodes (genus *Meloidogyne*). Root-knot nematodes infect a broad range of plant species whereas cyst nematodes have a narrower host range.

With increasing restrictions on the use of chemical pesticides, the use of host resistance for nematode control has grown in importance. A number of genes that mediate nematode resistance have now been or soon will be cloned from a variety of plant species. Nematode resistance genes are present in several crop species and are an important component of many breeding programmes including those for tomato, potato, soybeans and cereals (Trudgill 1991). Several resistance genes have been mapped to chromosomal locations or linkage groups and some of them have been cloned. The first nematode resistance gene to be cloned was $Hs1^{pro\text{-}1}$, a gene from a wild relative of sugar beet conferring resistance to *Heterodera schachtii* (Cai *et al.* 1997). The cDNA, under the control of the CaMV 35S promoter, was able to confer nematode resistance to sugar beets transformed with *Agrobacterium rizhogenes* in an *in vitro* assay (Cai *et al.* 1997). The *Mi* gene from tomato conferred resistance against a root-knot nematode and an aphid in transgenic potato (Rossi *et al.* 1998). The gene *Mi* is a true *R* gene, characterised by the presence of NBS and LRR domains (see Section 4.4.2). Recently, *Gpa2*, a gene that confers resistance against some isolates of the potato cyst nematode *Globodera pallida* was identified (Van Der Vossen *et al.* 2000). This gene shares extensive homology with the Rx1 gene that confers resistance to potato virus X suggesting a similarity in function (Van Der Vossen *et al.* 2000).

Although an important constituent of current nematode management strategies is the incorporation of natural resistance, one must be aware of the fact that there may not be appropriate resistance loci available for many crops. In addition, it is not a given fact that a particular gene will function effectively in heterologous hosts. Attempts to transfer *Mi*-mediated resistance into tobacco have not so far been successful (Williamson 1998). Furthermore, nematodes can eventually develop virulence, which may limit the effectiveness of this approach. Clearly, novel strategies to control nematode infestation are required combining existing with new approaches (Gheysen *et al.* 1996).

4.6 Long-term impact of genetically modified plants in their response to pathogens

There are three main concerns about the long-term impact of engineering genes for disease resistance in transgenic plants: the resistance of the plants to pathogen attack once they are grown in large scale, the development of resistance by the pathogen and the possible phenotypic alterations of the plant. The application of biotechnology in agriculture has had great success in the generation of commercially useful insect- and virus-resistant crops (Dunwell 2000). However, the first commercially available *Bt*-expressing cotton crop (grown in 1996) showed mixed success (Dempsey *et al.* 1998). Clearly, detailed and sufficiently extensive studies about the large-scale agronomic performance of each new variety grown in different conditions are necessary. However, until the new variety is grown on a large scale in the appropriate place, one may not determine precisely the actual resistance to pathogen attack.

Resistant plants can impose a selective pressure that results in development of resistance in the pathogen, which is not an uncommon event (Tabashnik *et al.* 2000). There exists an antagonistic coevolution between plants and pathogens that is constantly selecting for genotypes that can overcome the other's defences (Stahl and Bishop 2000). Clearly, the introduction of a single transgene, therefore, may not be sufficient to achieve durable and broad-spectrum disease resistance. A combination of transgenic strategies will be needed to ensure durability of resistance. By combining several methods for pest control, one may reduce the probability that any of these methods will soon become obsolete as a result of adaptation by the pest or disease causing agent. The increasing availability of resistance genes is allowing the generation of transgenic plants with resistance to various pests. However, since constitutive expression of certain *R* and *Avr* genes may have deleterious effects on the plant (Honée *et al.* 1995), a judicious choice of the genes to be transferred, combined with detailed molecular studies, will be necessary to achieve durable resistance together with an optimal field performance (Rommens and Kishore 2000). Furthermore, multiple *R* genes may compete for the signalling components during a mixed infection, thereby interfering with the activation of defences (Bent 1996).

There still remains a series of questions that need to be addressed concerning the response to pathogens of the engineered plants. For instance, what is the likelihood that plants engineered for disease resistance will provide an environment for the development of a novel pathogen that exhibits an increased host range or is resistant to currently available control methods? How durable will the engineered resistance prove to be once crops are grown on a large scale? Will the different *R* genes, cloned and those to be cloned, prove to be functional in heterologous plant species?

4.7 Future trends

The field of plant resistance control is undergoing a very active and exciting period, during which major breakthroughs are being made. Increased availability of cloned *R* genes will permit their testing in different plant backgrounds. In contrast with the success in the production of insect- and virus-resistant crops, the production of fungi-resistant crops with commercially useful levels of resistance has not been achieved. However, it is likely that commercial introduction of fungi-resistant crops can be expected within 4–8 years.

In the same way that plant breeders are continually developing new varieties that contain the most effective combination of existing characteristics, there is a similar trend with transgenic crops. Many laboratories are experimenting with 'pyramiding of genes', which consists of the introduction of multiple genes conferring different characters. A good example of this is a potato line containing seven transgenes that will confer resistance to insects, fungi, virus, and will alter other phenotypic characteristics (Dunwell 2000).

Overexpression of signalling components that lie downstream of *R* genes is an interesting approach that is currently being tested. This may allow activation of only certain defence pathways and may avoid agronomic problems associated with constitutive activation of some *R*-mediated pathways (see Section 4.4.2). The use of antimicrobial peptides to engineer broad-spectrum resistance is a promising and powerful approach that will be used considerably in the near future.

The increased use of inducible promoters is another current trend to manipulate specific pathways or to express *R* genes in a much more controlled manner (Shen *et al.* 2000). Finally, in addition to the progress being made on the plant side of the equation, an understanding of the genetic make-up of pathogens and the critical genes involved in the pathogenesis process are expected to open new avenues in crop protection.

4.8 Sources of further information and advice

Many biotechnology companies and universities are evaluating the performance of transgenic plants containing different disease resistance genes in the US. The interested reader may consult http://www.nbiap.vt.edu/cfdocs/fieldtests1.cfm. For general information on growth of transgenic crops all over the world and current trends it is recommended that the reader consult http://www.isaaa.cornell.edu. Useful sites with information on patents include the US Patent office (http://pctgazette.wipo.int/) and an extensive site supported by IBM ((http://www.patents.ibm.com/). For information on losses caused by pests and in general on world agriculture, the reader is referred to the excellent web site of FAO (http://apps.fao.org/cgi-bin/nph-db.pl?subset=agriculture). The books by D. Hornby and E.C. Oerke included in the list of references are particularly recommended for further reading.

4.9 References

ALEXANDER D, GOODMAN RM, GUT-RELLA M, GLASCOCK C, WEYMANN K, FRIEDRICH L, MADDOX D, AHL-GOY P, LUNTZ T, WARD E and RYALS J (1993), 'Increased Tolerance to Two Oomycete Pathogens in Transgenic Tobacco Expressing Pathogenesis-Related Protein 1a', *Proc Natl Acad Sci USA*, 90: 7327–7331.

ANGELL SM and BAULCOMBE DC (1997), 'Consistent gene silencing in transgenic plants expressing a replicating potato virus X RNA', *EMBO J*, 16, 3675–3684.

BAULCOMBE DC (1996), 'Mechanisms of pathogen derived resistance to viruses in transgenic plants', *Plant Cell*, 8, 1833–1844.

BAULCOMBE DC (1999), 'Viruses and gene silencing in plants', *Arch Virol Suppl*, 15, 189–201.

BEACHY RN (1997), 'Mechanisms and applications of pathogen-derived resistance in transgenic plants', *Curr Opin Biotechnol*, 8, 215–220.

BENDAHMANE A, KANYUKA K and BAULCOMBE DC (1999), 'The Rx gene from potato controls separate virus resistance and cell death responses', *Plant Cell*, 11, 781–792.

BENT AF (1996), 'Plant disease resistance genes: Function meets structure', *Plant Cell*, 8, 1757–1771.

BORKOWSKA M, KRZYMOWSKA M, TALARCZYK A, AWAN MF, YAKOVLEVA L, KLECZKOWSKI K and WIELGAT B (1998), 'Transgenic potato plants expressing soybean beta-1,3-endoglucanase gene exhibit an increased resistance to *Phytophthora infestans*', *Z Naturforsch*, 53, 1012–1016.

BOWLING SA, CLARKE JD, LIU Y, KLESSIG DF and DONG X (1997), 'The *cpr5* mutant of *Arabidopsis* expressed both *NPR1*-dependent and *NPR1*-independent resistance', *Plant Cell*, 9, 1573–1584.

BROEKAERT WF, TERRAS FR, CAMMUE BP and OSBORN RW (1995), 'Plant defensins: novel antimicrobial peptides as components of the host defense system', *Plant Physiol*, 108, 1353–1358.

BROEKAERT WF, CAMMUE BP, DE BOLLE MFC, THEVISSEN K, DESAMBLANX GW and OSBORN RW (1997), 'Antimicrobial peptides from plants', *Crit Rev Plant Sci*, 16, 297–323.

BROGLIE K, CHET I, HOLIDAY M, CRESSMAN R, BIDDLE P, KNOWLTON S, MAUVIS CJ and BROGLIE R (1991), 'Transgenic plants with enhanced resistance to the fungal pathogen *Rhizoctonia solani*', *Science*, 254, 1194–1197.

CAI D, KLEINE M, KIFLE S, HARLOFF HJ, SANDAL NN, MARCKER KA, KLEIN-LANKHORST RM, SALENTIJN EMJ, LANGE W, STIEKEMA WJ, WYSS U, GRUNDLER FMW and JUNG C (1997), 'Positional cloning of a gene for nematode resistance in sugar beet', *Science*, 275, 832–834.

CAO H, LI X and DONG X (1998), 'Generation of broad-spectrum disease resistance by overexpression of an essential regulatory gene in systemic acquired resistance', *Proc Natl Acad Sci USA*, 95, 6531–6536.

CHET I (1990), 'Biological control of soil-borne plant pathogens with fungal antagonists in combination with soil treatments', in Hornby D, *Biological control of soil-borne plant pathogens*, Wallingford, CAB International, 15–25.

CHET I and INBAR J (1994), 'Biological control of fungal pathogens', *Appl Biochem Biotechnol*, 48, 37–43.

COOLEY MB, PATHIRANA S, WU HJ, KACHROO P and KLESSIG DF (2000), 'Members of the Arabidopsis HRT/RPP8 family of resistance genes confer resistance to both viral and oomycete pathogens', *Plant Cell*, 12, 663–676.

COVEY SN, AL-KAFF NS, LANGARA I and TURNER DS (1997), 'Plants combat infection by gene silencing', *Nature*, 385, 781–782.

CRUTE IR (1985), 'The genetic bases of relationships between microbial parasites and their hosts', in Fraser RSS, *Mechanisms of resistance to plant diseases*, Dordrecht, Martinus Nijhoff/Dr. W Junk, 80–142.

DE BOLLE MF, DAVID KM, REES SB, VANDERLEYDEN J, CAMMUE BP and BROEKAERT WF (1993), 'Cloning and characterization of a cDNA encoding an antimicrobial chitin-binding protein from amaranth, *Amaranthus caudatus*', *Plant Mol Biol*, 22, 1187–1190.

DEMPSEY DA, SILVA H and KLESSIG DF (1998), 'Engineering disease and pest resistance in plants', *Trends Microbiol*, 6, 54–61.

DUAN X, LI X, XUE Q, ABO-EL-SAAD M, XU D and WU R (1996), 'Transgenic rice plants harboring an introduced potato proteinase inhibitor II gene are insect resistant', *Nat Biotechnol*, 14, 494–498.

DUNWELL JM (2000), 'Transgenic approaches to crop improvement', *J Exp Bot*, 51, 487–496.

ELAD Y, CHET I and HENIS Y (1982), 'Degradation of plant pathogenic fungi by *Trichoderma harzianum*', *Can J Microbiol*, 28, 719–725.

EPPLE P, APEL K and BOHLMANN H (1997), 'Overexpression of an endogenous thionin enhances resistance of Arabidopsis against *Fusarium oxysporum*', *Plant Cell*, 1997 Apr 9, 509–520.

FAOSTAT (2000). http://apps.fao.org/cgi-bin/nph-db.pl?subset=agriculture

FLORES A, CHET I and HERRERA-ESTRELLA A (1997), 'Improved biocontrol activity of *Trichoderma harzianum* by over-expression of the proteinase-encoding gene *prb1*', *Curr Genet*, 31, 30–37.

GABAY JE (1994), 'Ubiquitous natural antibiotics', *Science*, 264, 373–374.

GEREMIA RA, GOLDMAN GH, JACOBS D, ARDILES W, VILA SB, VAN MONTAGU M and HERRERA-ESTRELLA A (1993), 'Molecular characterization of the proteinase-encoding gene, *prb1*, related to mycoparasitism by *Trichoderma harzianum*', *Mol Microbiol*, 8, 603–613.

GHEYSEN G, VAN DER EYCKEN W, BARTHELS N, KARIMI M and VAN MONTAGU M (1996), 'The exploitation of nematode-responsive plant genes in novel nematode control methods', *Pest Sci*, 47, 95–101.

GIRARD C, LE METAYER M, BONADE-BOTTINO M, PHAM-DELEGUE MH and JOUANIN L (1998), 'High level of resistance to proteinase inhibitors may be conferred by proteolytic cleavage in beetle larvae', *Insect Biochem Mol Biol*, 28, 229–237.

GRISON R, GREZES-BESSET B, SCHNEIDER M, LUCANTE N, OLSEN L, LEGUAY JJ and TOPPAN A (1996), 'Field tolerance to fungal pathogens of Brassica napus constitutively expressing a chimeric chitinasegene', *Nat Biotechnol,* 14, 643–646.

GUTIÉRREZ-CAMPOS R, TORRES-ACOSTA JA, PÉREZ-MARTÍNEZ JJ and GÓMEZ-LIM MA (2000), 'Pleiotropic effects in transgenic tobacco plants expressing the oryzacystatin 1 Gene', *Hortsci*, (in press).

GUTIÉRREZ-CAMPOS R, TORRES-ACOSTA JA, SAUCEDO-ARIAS LJ and GÓMEZ-LIM MA (1999), 'A novel method to engineer resistance to tobacco etch virus in transgenic tobacco plants: the use of cysteine proteinase inhibitors for plant virus control', *Nat Biotechnol* 17: 1223–1226.

HAIN R, REIF HJ, KRAUSE E, LANGEBARTELS R, KINDL H, VORNAM B, WIESE W, SCHMELZER E, SCHREIER, STOCKER RH, et al. (1993), 'Disease resistance results from foreign phytoalexin expression in a novel plant', *Nature*, 361, 153–156.

HALL FR (1991), 'Pesticide application technology and integrated pest management', in Pimentel D, *Handbook of pest management in agriculture*, Boca Raton, CRC Press, 135–163.

HAMMOND-KOSACK KE and JONES JD (1996), 'Resistance gene-dependent plant defense responses', *Plant Cell* 8, 1773–1791.

HARAN S, SCHICKLER H and CHET I (1996), 'Molecular mechanisms of lytic enzymes involved in the biocontrol activity of *Trichoderma harzianum*', *Microbiol* 142, 2321–2331.

HILDER VA and BOULTER D (1999) 'Genetic engineering of crop plants for insect resistance – a critical review', *Crop Prot*, 18, 177–191.

HIPSKIND JD and PAIVA NL (2000), 'Constitutive accumulation of a resveratrol-glucoside in transgenic alfalfa increases resistance to *Phoma medicaginis*', *Mol Plant Microbe Interact*, 13, 551–562.

HONÉE G, MELCHERS LS, VLEESHOUWERS VGAA, VAB ROEKEL JSC and DE WIT PJGM (1995), 'Production of the Avr9 elicitor from the fungal pathogen *Cladosporium fulvum* in transgenic tobacco and tomato plants', *Plant Biol Mol*, 29, 909–920.

IRIE K, HOSOYAMA H, TAKEUCHI T, IWABUCHI K, WATANABE H, ABE M, ABE K and ARAI S (1996), 'Transgenic rice established to express corn cystatin exhibits strong inhibitory activity against insect gut proteinases', *Plant Mol Biol*, 30, 149–157.

JACH G, GORNHARDT B, MUNDY J, LOGEMANN J, PINSDORF E, LEAH R, SCHELL J and MAAS C (1995), 'Enhanced quantitative resistance against fungal disease by combinatorial expression of different barley antifungal proteins in transgenic tobacco', *Plant J*, 8, 97–109.

JONGEDIJK E, TIGELAAR H, VAN ROEKEL JSC, BRES-VLOEMANS SA, DEKKER I, VAN DEN ELZEN PJM, CORNELISSEN BJC and MELCHERS LS (1995), 'Synergistic activity of chitinases and β–1,3 glucanases enhances fungal resistance in transgenic tomato plants', *Euphytica*, 85, 173–180.

KACHROO P, YOSHIOKA K, SHAH J, DOONER HK and KLESSIG DF (2000), 'Resistance to turnip crinkle virus in Arabidopsis is regulated by two host genes and is salicylic acid dependent but NPR1, ethylene, and jasmonate independent', *Plant Cell*, 12, 677–690.

KANIEWSKI WK and THOMAS PE (1999), 'Field testing for virus resistance and agronomic performance in transgenic plants', *Mol Biotechnol*, 12, 101–115.

KELLER H, PAMBOUKDJIAN N, POCHET M, POUPET A, DELON R, VERRIER JL, ROBY D and RICCI P (1999), 'Pathogen-induced elicitin production in transgenic tobacco generates a hypersensitive response and nonspecific disease resistance', *Plant Cell*, 11, 223–235.

KESARWANI M, AZAM M, NATARAJAN K, MEHTA A and DATTA A (2000), 'Oxalate decarboxylase from *Collybia velutipes*. Molecular cloning and its overexpression to confer resistance to fungal infection in transgenic tobacco and tomato', *J Biol Chem*, 275, 7230–7238.

KNOESTER M, PIETERSE CM, BOL JF and VAN LOON LC (1999) Systemic resistance in Arabidopsis induced by rhizobacteria requires ethylene-dependent signaling at the site of application', *Mol Plant Microbe Interact*, 12,720–727.

LI R, WU N, FAN Y and SONG B (1999), 'Transgenic potato plants expressing osmotin gene inhibits fungal development in inoculated leaves', *Chin J Biotechnol*, 15, 71–75.

LIANG J, WU Y, ROSENBERGER C, HAKIMI S, CASTRO S and BERG J (1998), 'AFP genes confer disease resistance to transgenic potato and wheat plants', Abstract no. L-49. In *5th International workshop on pathogenesis-related proteins in plants; Signalling pathways and biological activities 1998*; Aussois, France.

LIMON MC, PINTOR-TORO JA and BENITEZ T (1999), 'Increased antifungal activity of *Trichoderma harzianum* transformants that overexpress a 33 kDa chitinase', *Phytopathol*, 89, 254–261.

LIN W, ANURATHA CS, DATTA K, POTRYKUS I, MUTHUKRISHNAN S and DATTA SK (1995), 'Genetic engineering of rice for resistance to sheath blight', *Bio/Technol*, 3, 686–691.

LIU D, RAGHOTHAMA KG, HASEGAWA PM and BRESSAN RA (1994), 'Osmotin overexpression in potato delays development of disease symptoms', *Proc Natl Acad Sci USA*, 91, 1888–1892.

LOGEMANN J, JACH G, TOMMERUP H, MUNDY J and SCHELL J (1992), 'Expression of a barley ribosome inactivating protein leads to increased fungal protection in transgenic tobacco plants', *Bio/Technol*, 10, 305–308.

LOMONOSSOFF GP (1995), 'Pathogen-derived resistance to plant viruses', *Annu Rev Phytopathol*, 33:323–343.

LORITO M, WOO SL, GARCIA I, COLUCCI G, HARMAN GE, PINTOR-TORO JA, FILIPPONE E, MUCCIFORA S, LAWRENCE CB, ZOINA A, TUZUN S, SCALA F and FERNANDEZ IG (1998), 'Genes from mycoparasitic fungi as a source for improving plant resistance to fungal pathogens', *Proc Natl Acad Sci USA*, 95, 7860–7865.

MALEK K and LAWTON K (1998), 'Plant strategies for resistance to pathogens', *Curr Op Biotech*, 9, 208–213.

MARTIN GB (1999), 'Functional analysis of plant disease resistance genes and their downstream effectors', *Curr Op Plant Biol*, 2, 273–279.

MELCHERS LS and STUIVER MH (2000), 'Novel genes for disease-resistance breeding', *Curr Op Plant Biol*, 3,147–152.

MITSUHARA I, MATSUFURU H, OHSHIMA M, KAKU H, NAKAJIMA Y, MURAI N, NATORI S and OHASHI Y (2000), 'Induced expression of sarcotoxin IA enhanced host resistance against both bacterial and fungal pathogens in transgenic tobacco', *Mol Plant Microbe Interact*, 13, 860–868.

MOLINA L and GARCIA-OLMEDO F (1997), 'Enhanced tolerance to bacterial pathogens caused by the transgenic expression of barley lipid transfer protein LTP2', *Plant J*, 12, 669–675.

MOURRAIN P, BECLIN C, ELMAYAN T, FEUERBACH F, GODON C, MOREL JB, JOUETTE D, LACOMBE AM, NIKIC S, PICAULT N, REMOUE K, SANIAL M, VO TA and VAUCHERET H (2000), 'Arabidopsis SGS2 and SGS3 genes are required for posttranscriptional gene silencing and natural virus resistance', *Cell*, 101, 533–542.

OERKE EC (1994), *Crop protection and crop production*, Amsterdam, Elsevier.

OGAWA T, HORI T and ISHIDA I (1996), 'Virus-induced cell death in plants expressing the mammalian $2', 5'$ oligoadenylate system', *Nat Biotechnol*, 14, 1566–1569.

OLDROYD GED and STASKAWICZ BJ (1998), 'Genetically engineered broad-spectrum disease resistance in tomato', *Proc Natl Acad Sci USA*, 95, 10300–10305.

OSUSKY M, ZHOU G, OSUSKA L, HANCOCK RE, KAY WW and MISRA S (2000), 'Transgenic plants expressing cationic peptide chimeras exhibit broad-spectrum resistance to phytopathogens', *Nat Biotechnol*, 18, 1162–1166.

PENNINCKX IA, EGGERMONT K, TERRAS FR, THOMMA BP, DE SAMBLANX GW, BUCHALA A, METRAUX JP, MANNERS JM and BROEKAERT WF (1996), 'Pathogen-induced systemic activation of a plant defensin gene in Arabidopsis follows a salicylic acid-independent pathway' *Plant Cell*, 8, 2309–2323.

PIETERSE CM, VAN WEES SC, VAN PELT JA, KNOESTER M, LAAN R, GERRITS H, WEISBEEK PJ and VAN LOON LC (1998), 'A novel signaling pathway controlling induced systemic resistance in Arabidopsis', *Plant Cell*, 10, 1571–1580.

RATCLIFF F, HARRISON BD and BAULCOMBE DC (1997), 'A similarity between viral defense and gene silencing in plants', *Science*, 276, 1558–1560.

ROMMENS CM and KISHORE GM (2000), 'Exploiting the full potential of disease resistance genes for agricultural use'. *Curr Opin Biotechnol*, 11, 120–125.

ROSSI M, GOGGIN FL, MILLIGAN SB, KALOSHIAN I, ULLMAN DE and WILLIAMSON VM (1998), 'The nematode resistance gene Mi of tomato confers resistance against the potato aphid', *Proc Natl Acad Sci USA*, 95, 9750–9754.

RYAN CA (1990), 'Protease inhibitors in plants: Genes for improving defenses against insects and pathogens', *Annu Rev Phytopathol*, 28, 425–449.

RYAN CA (2000), 'The systemin signaling pathway: differential activation of plant defensive genes', *Biochim Biophys Acta*, 7;1477, 112–121.

RYAN CA and PEARCE G (1998), 'Systemin: a polypeptide signa for plant defensive genes', *Annu RevCell Dev Biol*, 14, 1–17.

SANCHIS V and LERECLUS D (1999), 'Bacillus thuringiensis: a biotechnology model', *J Soc Biol*; 193(6): 523–530.

SCHALLER A (1999), 'Oligopeptide signalling and the action of systemin', *Plant Mol Biol*, 40, 763–769.

SHEN S, LI Q, HE SY, BARKER KR, LI D and HUNT AG (2000), 'Conversion of compatible plant-pathogen interactions into incompatible interactions by expression of the *Pseudomonas syringae* pv. syringae 61 hrmA gene in transgenic tobacco plants', *Plant J*, 23, 205–213.

SOMSSICH IE and HAHLBROCK K (1998), 'Pathogen defence in plants – a paradigm of biological complexity', *Trends Pl Sci*, 3, 86–90.

SONG EK, KOH HK, KIM JK and LEE SY (1999), 'Genetically engineered transgenic plants with the domain 1 sequence of tobacco mosaic virus 126 kDa protein gene are completely resistant to viral infection', *Mol Cells*, 9, 569–575.

SPALL VE, SHANKS M and LOMONOSSOFF GP (1997), 'Polyprotein processing as a strategy for gene expression in RNA viruses', *Sem Virol* 8, 15–23.

STAHL EA and BISHOP JG (2000), 'Plant-pathogen arm races at the molecular level', *Curr Op Plant Biol*, 3, 299–304.

STICHTER L, MAUCH-MANI BN and MÉTRAUX JP (1997), 'Systemic acquired resistance', *Annu Rev Phytopathol*, 35, 235–270.

TABASHNIK BE, ROUSH RT, EARLE ED AND SHELTON AM (2000), 'Resistance to Bt toxins', *Science*, 287: 42.

TAI TH, DAHLBECK D, CLARK ET, GAJIWALA P, PASION R, WHALEN MC, STALL RE and STASKAWICZ BJ (1999), 'Expression of the Bs2 pepper gene confers resistance to bacterial spot disease in tomato', *Proc Natl Acad Sci USA*, 96, 14153–14158.

TANG X, XIE M, KIM YJ, ZHOU J, KLESSIG DF and MARTIN GB (1999), 'Overexpression of Pto activates defense responses and confers broad resistance', *Plant Cell*, 11, 15–29.

TERRAS FR, EGGERMONT K, KOVALEVA V, RAIKHEL NV, OSBORN RW, KESTER A, REES SB, TORREKENS S, VAN LEUVEN F, VANDERLEYDEN J, et al. (1995), 'Small cysteine-rich antifungal proteins from radish: their role in host defense', *Plant Cell* 7, 573–588.

TERRAS FR, PENNINCKX IA, GODERIS IJ and BROEKAERT WF (1998), 'Evidence that the role of plant defensins in radish defense responses is independent of salicylic acid', *Planta*, 206, 117–24.

THEVISSEN K, TERRAS FR and BROEKAERT WF (1999), 'Permeabilization of fungal membranes by plant defensins inhibits fungal growth', *Appl Environ Microbiol*, 65, 5451–5458.

TOBIAS CM, OLDROYD GE, CHANG JH and STASKAWICZ BJ (1999), 'Plants expressing the *Pto* disease resistance gene confer resistance to recombinant PVX containing the avirulence gene *AvrPto*', *Plant J*, 17, 41–50.

TRUDGILL DL (1991), 'Resistance to and tolerance of plant parasitic nematodes in plants', *Annu Rev Phytopathol*, 29, 167–193.

TU J, ZHANG G, DATTA K, XU C, HE Y, ZHANG Q, KHUSH GS and DATTA SK (2000), 'Field performance of transgenic elite commercial hybrid rice expressing *Bacillus thuringiensis* δ-endotoxin', *Nat Biotechnol*, 18, 1101–1104.

TUMER NE, HUDAK K, DI R, COETZER C, WANG P and ZOUBENKO O (1999), 'Pokeweed antiviral protein and its applications', *Curr Top Microbiol Immunol*. 240, 139–158.

TUMER NE, HWANG DJ and BONNESS M (1997), 'C-terminal deletion mutant of pokeweed antiviral protein inhibits viral infection but does not depurinate host ribosomes', *Proc Natl Acad Sci USA*, 94, 3866–3871.

VAN DER VOSSEN EA, VAN DER VOORT JN, KANYUKA K, BENDAHMANE A, SANDBRINK H, BAULCOMBE DC, BAKKER J, STIEKEMA WJ and KLEIN-LANKHORST RM (1997), 'Homologues of a single resistance-gene cluster in potato confer resistance to distinct pathogens: a virus and a nematode', *Plant J*, 23, 567–576.

VAN DER VOSSEN EA, VAN DER VOORT JN, KANYUKA K, BENDAHMANE A,

SANDBRINK H, BAULCOMBE DC, BAKKER J, STIEKEMA WJ and KLEIN-LANKHORST RM (2000), 'Homologues of a single resistance-gene cluster in potato confer resistance to distinct pathogens: a virus and a nematode', *Plant J*, 23, 567–576.

VERBERNE MC, VERPOORTE R, BOL JF, MERCADO-BLANCO J and LINTHORST HJB (2000), 'Overproduction of salicilyc acid in plants by bacterial transgenes enhances pathogen resistance', *Nat Biotechnol*,18, 779–783.

VIDAL S, PONCE DE LEON I, DENECKE J and PALVA ET (1997), 'Salicylic acid and the plant pathogen *Erwinia carotovora* induce defense genes via antagonistic pathways', *Plant J*, 11, 115–123.

WANG P, ZOUBENKO O and TUMER NE (1998) Reduced toxicity and broad spectrum resistance to viral and fungal infection in transgenic plants expressing pokeweed antiviral protein II. *Plant Mol Biol*, 38, 957–964.

WATANABE Y, OGAWA T, TAKAHASHI H, ISHIDA I, TAKEUCHI Y, YAMAMOTO M and OKADA Y (1995), Resistance against multiple plant viruses in plants mediated by a double stranded-RNA specific ribonuclease', *FEBS Lett*, Sep 25; 372, 165–168.

WHITHAM SA, ANDERBERG RJ, CHISHOLM ST and CARRINGTON JC (2000), 'Arabidopsis RTM2 gene is necessary for specific restriction of tobacco etch virus and encodes an unusual small heat shock-like protein', *Plant Cell*, 12, 569–582.

WILLIAMSON VM (1998), 'Root-knot nematode resistance genes in tomato and their potential for future use', *Annu Rev Phytopathol*, 36, 277–293.

WU G, SHORTT BJ, LAWRENCE EB, LEVINE EB, FITZSIMMONS C and SHAW DM (1995), 'Disease resistance conferred by expression of a gene encoding H_2O_2-generating glucose oxidase in transgenic potato plants', *Plant Cell*, 7, 1357–1368.

YEDIDIA I, BENHAMOU N and CHET I (1999) Induction of defense responses in cucumber plants (*Cucumis sativus* L.) by the biocontrol agent trichoderma harzianum', *Appl Environ Microbiol*, 65, 1061–1070.

YU D, XIE Z, CHEN C, FAN B and CHEN Z (1999), 'Expression of tobacco class II catalase gene activates the endogenous homologous gene and is associated with disease resistance in transgenic potato plants', *Plant Mol Biol*, 39, 477–488.

ZHOU F, COLLINGE DB and THORDAL-CHRISTENSEN H (1995), 'Germin-like oxalate oxidase, a H_2O_2 producing enzyme, accumulates in barley attacked by the powdery mildew fungus', *Plant J*, 8, 139–145.

ZHU Q, MAHLER EA, MASOUD S, DIXON RA and LAMB CJ (1994), 'Enhanced protection against fungal attack by constitutive co-expression of chitinase and glucanase genes in transgenic tobacco', *Bio/Technol,* 12, 807–811

ZOUBENKO O, UCKUN F, HUR Y, CHET I and TUMER N (1997), 'Plant resistance to fungal infection induced by nontoxic pokeweed antiviral protein mutants', *Nat Biotechnol*, 15, 992–996.

5

Genes selected for their role in modifying post-harvest life

J. R. Botella, University of Queensland, Brisbane

5.1 Introduction

Once a plant is harvested it starts a process of decay that will result in the inevitable death of the organism and the deterioration of its organic matter. This process is even more accentuated if the harvested tissue is a particular organ (fruit, leaf, root, etc.) instead of the whole plant. In an ideal world this fact should not be a problem, we would grow our vegetables and fruits in our backyard and consume them fresh every day. In real life most of the food we consume has been grown hundreds or even thousands of kilometres away from the shop where we bought them. Therefore we are daily confronted with the fact that food crops, after they are grown, need to be transported to intermediate destinations, stored, transported again and distributed before finally reaching the consumer. Aside from the natural/physiological reasons stated above, sociological aspects need to be considered; the backyard self-sufficient growing strategy will not work nowadays because many of us do not have a backyard. The world is evolving from a predominantly rural population to a new demographic distribution based on high human concentration on urban areas. Areas of food production are therefore far away from areas of food consumption. We are consequently fighting a continuous battle to bring the right food to the right place with a minimum of losses. Sadly, although important advances have been made in post-harvest technology, we are still losing the battle.

Post-harvest problems can account for substantial losses, the magnitude of which depends on the crop, the country and the year. It is important to stress that post-harvest losses are one of the most significant factors limiting agricultural production in third-world and developing countries. Whereas technically advanced countries such as the USA, Japan, Australia and European countries can apply relatively sophisticated technologies to minimise losses; developing countries cannot afford them. In addition, many of the developing regions in the world are situated in the tropics, where high temperature and humidity exacerbate the problem.

Senescence, the final stage in the life of a plant (or a particular organ) was once thought to have an essentially chaotic nature in which cell components break down without a particular order. Nevertheless, recent advances in our understanding of the process have revealed that senescence is a very well programmed developmental stage, involving highly coordinated cellular events that require the sequential action of many genes. Our fundamental knowledge of senescence has greatly improved in the last few years but we are still far from having a good understanding of the underlying biochemical and molecular mechanisms. There are two kinds of senescence processes, a natural one that comes after the end of the useful life of an organ is reached, and a stress- or environmentally-induced senescence. Fruits will develop and ripen to entice predators and ensure seed dispersal. When the seeds contained in the fruit are not viable any more, a natural senescence process will eventually end up with the spoilage of the fruit. Old leaves shrivel and fall while new ones develop to keep the photosynthetic capacity of the plant. These events are just natural processes pre-programmed in the genetic code. Harvesting plant tissue for commercialisation causes a series of stresses in the detached tissue that will inexorably trigger senescence. Natural and induced senescence do not always share the same mechanisms.

The main objective of post-harvest technology is to increase the useful life of a particular foodstuff but increasing the useful life is a very ambiguous term. For some produce it means keeping the tissue turgor for longer (we all want our lettuces, broccoli and apples to have that fresh crispy feel), for other foods such as mangoes and bananas it means keeping the right degree of softness for longer without over-ripening (we like our mangoes soft but not mushy). Horticultural produce is extremely heterogeneous with a large variety of plant tissues being commercialised such as fruit, leaves, flowers, roots and tubers. Even though senescence has some common features, each tissue has specific characteristics and therefore needs to be studied separately. This is the reason why there is not a single universal post-harvest treatment useful for all horticultural crops. Leafy vegetables must be treated in a different way from fruits and even fruits have a wide variety of post-harvest problems depending on the fruit and even the commercial variety.

From a biotechnological point of view it is therefore important to establish the nature of the crop and the particular problem before establishing the approach to be attempted. It is also important to emphasise that the same problem can be tackled with very different approaches. Metabolic engineering can target internal processes but is not restricted to endogenous genes. Modern genetic engineering techniques allow us to cross species barriers (and even kingdom barriers) and therefore genes that would not normally be accessible by conventional breeding can now be incorporated into the plant species being targeted.

Post-harvest technologies such as atmosphere control (CO_2 and humidity), refrigeration, irradiation, etc., have proven useful in controlling post-harvest losses, but the implementation of many of the existing and new technologies is quite difficult in many developing countries due to the lack of infrastructure and specialised personnel in rural areas. Many crops are also grown in small farms meaning that a single grower is not able to afford the economic cost of setting up treatment plants and has no access to specialised packaging requirements. Regional centres, either governmental or private, can alleviate the situation but require a level of organisation commonly non-existent.

5.2 Biotechnological control of fruit ripening and post-harvest diseases

Most fruits ripen, deteriorate in appearance and eating quality and succumb to post-harvest diseases very rapidly after harvest. Poor post-harvest characteristics such as deficient flavour development, very short shelf life, quick softening, easy spoilage, sensitivity to low temperatures (chilling injury) and easy pathogen attack (fungi, etc.), are major constraints to profitability for the domestic market, and to the expansion of existing and new export markets. Among all fruits, tropical fruits are notorious for their poorer-than-average post-harvest qualities.

Two major obvious targets to improve the post-harvest characteristics of fruits are (i) extension of shelf life and (ii) resistance to pathogen attack. The ripening process involves a large number of biochemical pathways in the fruit that will result in marked changes in texture, taste and colour. At the molecular level there are a large number of genes involved and they are tightly regulated in order to induce the right changes at the right time in a highly coordinated process. In general, fruits are classified as climacteric or non-climacteric depending upon their patterns of respiration and ethylene synthesis during ripening. Climacteric fruits are characterised by an increased respiration rate at an early stage in the ripening process accompanied by autocatalytic ethylene production whereas non-climacteric fruits show a different respiration pattern and display a lack of autocatalytic ethylene synthesis. Many of the economically important fruit crops are climacteric; therefore a large amount of research has been devoted to studying the biochemical and molecular pathways operating during the climacteric ripening of fruits.

Most of the genetic engineering approaches attempted in order to improve the shelf life and general appearance of fruits have centred on the set of genes controlling fruit firmness (membrane and cell wall properties) and the ripening rate (ethylene production or perception). These approaches have targeted endogenous genes with vital functions in the ripening process aiming to down-regulate their activity by gene silencing.

5.2.1 Control of fruit firmness

Softening is an important contributor to losses experienced during the handling and transport of fruit. Among the genes involved in firmness, the most extensively studied is the one coding for polygalacturonase (PG) (Della-Penna *et al.*, 1986; Grierson *et al.*, 1986), a cell wall enzyme that catalyses the hydrolysis of polygalacturonic acid chains. Polygalacturonic acid is an important component of the plant cell wall that significantly contributes to the fruit firmness. Partial silencing of the PG gene has been achieved in tomato by sense and antisense techniques. Experiments using either a partial or the full length PG gene successfully reduced the levels of PG mRNA and enzyme activity (Sheehy *et al.*, 1988; Smith *et al.*, 1988; Smith *et al.*, 1990). It is important to remark that these were the first examples of the successful use of the antisense technique in plants. Different transgenic lines showed different degrees of gene silencing, indicating that the position where the transgene is inserted in the genome plays an important role in the effectiveness of the technique, as has been emphasised in Chapter 1. In tomato, PG has been extensively associated with softening because of the temporal and spatial coincidence of the increase of PG activity during the softening period of the fruit. Contrary to all expectations, the antisense PG fruits produced by Smith *et al.*, (1988) did not show any appreciable change in softening when measured by classical methods (such as compression tests). A debate was started on whether PG had any effect in internal softening of the fruits, and the widely accepted idea that PG was directly responsible for the softening process was shaken. Nevertheless, a new and more detailed study of transgenic sense and antisense PG plants revealed a number of important changes in the transgenic fruits (Schuch *et al.*, 1991). Low PG tomatoes were more resistant to cracking and splitting than regular fruit. They also had superior handling and transport characteristics showing a severely reduced degree of damage during those processes (Schuch *et al.*, 1991). How much inhibition of the PG gene is necessary to observe any changes in the fruit phenotype? This factor has not been fully answered yet because of the difficulty in regulating the exact amount of gene silencing in transgenic lines. Genetic engineering of plants has not reached the high level of sophistication needed to pre-determine or precisely regulate the level of gene silencing. Nevertheless, molecular analysis of the transgenic tomato lines showed that fruits in which PG activity had decreased to less than 1% of normal levels contain longer polygalacturonic acid chains, affecting cell adhesion and making the fruits sturdier.

Agronomically, the effect of low PG can be translated in fruits that can be left on the vine for a longer time, therefore enhancing the flavour, since the softening process has been slightly delayed. Interestingly, the main commercial use of the low PG tomato fruits has been in the processing industry. Transgenic low PG tomatoes show enhanced viscosity of the processed products and produce less waste. The new characteristics of these fruits have also allowed us to simplify the manufacturing process.

'Flavr Savr', the commercial name for a low PG tomato, marked an important milestone in plant biotechnology being the first genetically modified plant food to reach the market, commercialised by Calgene in the USA in 1994. Zeneca and associates are currently commercialising a tomato puree based on genetically modified low PG tomatoes. This product went on sale in the UK in 1996.

The results of the genetically modified low PG tomatoes have shown that although PG plays a significant role in texture changes during fruit ripening in tomato, it is not the primary factor controlling softening. There are a number of cell wall modifying enzymes that have been characterised at the biochemical level and shown to be active during fruit softening including cellulases, pectinesterases, galactanases, etc. It is also important to remark that, based on the research data available, it is likely that there is not a single softening pathway common to all fruits. Different species have been shown to have very different cell wall modifying enzymes' activity patterns during ripening, therefore it is not possible to devise a single universal strategy to control softening.

5.2.2 Control of ethylene synthesis and perception

Ethylene is one of the simplest organic molecules with biological activity and is the only gaseous hormone known to date. In climacteric fruits ethylene controls the onset and rate of ripening and therefore several strategies have been devised to interfere with either the rate of ethylene synthesis or its perception by the fruit. The elucidation of the ethylene biosynthetic pathway by Yang and coworkers (1984) (Fig. 5.1) opened the door for the isolation of the enzymes involved and the cloning of the corresponding genes.

Aminocyclopropane carboxylate (ACC) synthase and ACC oxidase are the key enzymes in this pathway controlling the last two steps in the production ethylene. Both of them are encoded by multigene families and normally only one or two members of the family are active in the fruit during ripening. In tomato, the ACC synthase gene active during ripening (LEACC2) was silenced using antisense techniques effectively reducing the production of ethylene by the ripening fruit by 99.5% (Oeller *et al.*, 1991). While control fruits begin to produce ethylene 48–50 days after pollination and immediately undergo a respiratory burst, genetically modified tomatoes produced minimal levels of ethylene and failed to produce the respiratory burst (at least during the 95-day period analysed in the report). Transgenic fruits started showing symptoms of chlorophyll degradation 10 to 20 days after the control fruits turned to yellow, and eventually developed an orange colour two months later; meanwhile control tomato fruits needed only ten days for the transition from full mature green to fully ripe red tomatoes.

The transgenic tomatoes studied by Oeller *et al.*, never turned red and soft and never developed aroma when kept in the plant or stored in an air

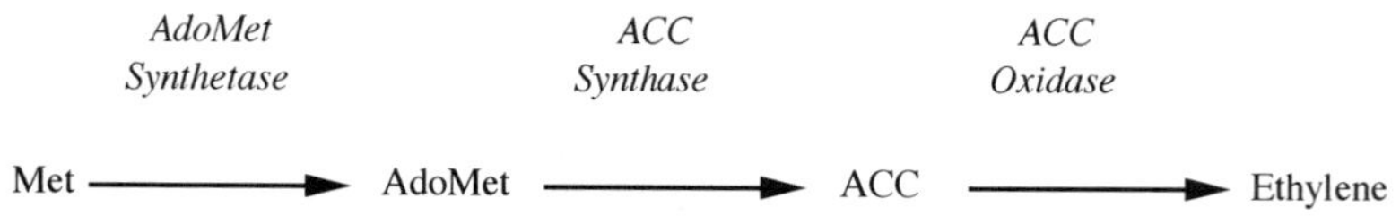

Fig. 5.1 Ethylene biosynthetic pathway.

atmosphere. Obviously, these characteristics are not desirable for a commercial fruit crop since the consumer wants a ripe product with all the attributes of colour, aroma, flavour, etc., fully developed. An obvious question arises of whether this phenotype is reversible by treatment with ethylene or the genetic change has created fruits completely unable to undergo the ripening process. When mature green transgenic fruits (49 days after pollination) were treated with ethylene, they developed a fully ripe phenotype within seven days (as opposed to two days for control fruits). The ethylene-treated genetically modified ripe fruits were indistinguishable from naturally ripened control fruits in colour, texture, aroma and compressibility. Although scientifically this work is of great importance, such extreme phenotypes may not prove useful in a commercial situation and intermediate phenotypes should be targeted. The above studies strongly suggest that ethylene is the trigger that starts the respiratory burst in climacteric fruits and controls the rate of ripening.

The ripening-related ACC oxidase gene has also been cloned in tomato and its expression inhibited by 95% (Hamilton *et al.*, 1990; Picton *et al.*, 1993). This level of inhibition did not block ripening in the transgenic plants allowing normal development of the fruits but delaying the onset of senescence, over ripening, cracking of fruits and other general over-ripening effects. Nevertheless, when mature green fruits were picked from the plant they never fully ripened. Even when exposed to ethylene, although they developed full red colour, the levels of carotenoids never reached those achieved by plant-grown fruits (Picton *et al.*, 1993).

Instead of altering the levels of enzymes controlling the biosynthesis of ethylene, two commercial companies (Monsanto and Agritope) have opted for alternative strategies aimed at depleting the intermediate substrates of the pathway. Monsanto used a bacterial enzyme (ACC deaminase) to drain the cell of the immediate precursor of ethylene (ACC). Overexpression of an ACC deaminase gene in tomato plants led to a marked depletion of the levels of ACC and therefore reduced the availability of this precursor to be converted into ethylene (Klee *et al.*, 1991). Transgenic plants overexpressing ACC deaminase were indistinguishable from controls with no differences observed during development even though there was a dramatic decrease in the levels of ethylene produced in vegetative tissues. Out of all the independent transgenic lines obtained, the best one produced fruits with ethylene levels of only 10% of the controls. When fruits were picked from the plant at the breaker stage and stored at room temperature, controls achieved fully red stage in seven days compared with 24 days for the transgenic fruits. Softening behaviour was also affected with controls showing a strong incidence of softening two weeks after picking; in contrast transgenic fruits remained firm for five months. When fruits were left on the plant to ripen, transgenic fruits remained firm for much longer than controls and did not abscise for more than 40 days. Agritope has used a bacteriophage gene encoding S-adenosyl methionine (SAM) hydrolase, in conjunction with a ripening specific promoter, to hydrolyse the first intermediate of the ethylene biosynthetic pathway (SAM) in ripening cherry tomato fruits

(Kramer *et al.*, 1997). The transgenic fruits exhibited a delayed ripening phenotype and a reduction of spoilage due to over-ripening.

Is it possible to control ripening in other fruits? All the studies previously described have been achieved in tomato. The reason for the choice of this system is clear: tomato is a very important crop with an extensive research history into the biochemistry and genetics of ripening. In addition tomato transformation is relatively easy when compared to other fruit species and the results of a transformation experiment can be evaluated in a glasshouse in a year (as opposed to an entire field and 5–7 years for fruit trees). Nevertheless there are clear indications that the approaches described earlier can be applied to other crops as is the case of melon. Ayub *et al.* (1996) used antisense techniques to inhibit ACC oxidase levels and concomitant ethylene production during ripening of cantaloupe 'Chanterais' melons. This variety has excellent eating quality but a notoriously poor storage capacity. Genetically modified plants were produced with ethylene synthesis severely impaired (less than 1% of controls). Storage capacity was extended, with transgenic fruits remaining fresh after ten days at 25ºC while control fruits had spoiled. The softening of the fruits was also affected with transgenic fruits remaining twice as firm as non-transformed controls. Exposing the transgenic fruits to external ethylene restored the ripening phenotype. A recent report by Ben-Amor *et al.* (1999) has revealed that the low-ethylene melons have considerably less sensitivity to chilling injury. This is an additional important improvement since most tropical and subtropical fruits are very sensitive to low temperatures and this fact severely impairs their transport and storage potential causing significant losses. The antisense ACC oxidase melons did not develop chilling injury when stored for up to three weeks at 2ºC while controls exhibited extensive damage.

An alternative to the control of ethylene production during ripening is to decrease the sensitivity of the fruit to the hormone. It has been established that during ripening, fruits not only increase the production of ethylene but they also become more sensitive to it (Theologis, 1994). The cloning of the ethylene receptor (*etr1*) has opened the door to the manipulation of ethylene perception instead of ethylene production (Chang *et al.*, 1993). A mutated version of *etr1* in *Arabidopsis* (*etr1-1*) confers ethylene insensitivity as a genetically inheritable dominant trait. The same mutated gene has also been introduced into tomato and petunia conferring ethylene insensitivity and producing fruits that fail to ripen or flowers with extremely delayed senescence respectively (Wilkinson *et al.*, 1997). It is clear that complete ethylene insensitivity is not a desirable trait for a fruit since it would render the fruit unable to ripen even when exposed to ethylene. On the other hand, selective, partial or induced insensitivity to ethylene could be commercially useful.

5.2.3 Non-climacteric fruits

Non-climacteric fruits do not experience a surge in ethylene production that triggers a respiratory rise. Research in non-climacteric fruit ripening has been traditionally dragging behind its climacteric counterparts and although a large body of information is being accumulated there is not yet a clear picture of the common mechanisms governing the ripening process in this large class of fruits. Numerous ripening-induced genes are being cloned encoding proteins involved in cell wall degradation, sucrose and lipid metabolism, anthocyanin synthesis, cell expansion and flavor development (Civello *et al.*, 1999; Medina-Escobar *et al.*, 1997; Moyano *et al.*, 1998; Nam *et al.*, 1999; Trainotti *et al.*, 1999). DNA microarray techniques have recently been used to identify ripening related genes with the prospect of providing a large amount of data to study the coordination of gene expression during the ripening of strawberry in particular and other non-climacteric fruit crops in general (Aharoni *et al.*, 2000).

Interestingly, even though ethylene does not play a role in the coordination of ripening, it has been known for some time that it can accelerate senescence of non-climacteric produce (including fruits and vegetables) (Kader, 1985) therefore it is important to avoid the presence of ethylene during transport and storage. Recent evidence (Wills *et al.*, 1999) shows that ethylene can affect the ripening process of 23 different kinds of produce, many of them non-climacteric, at levels much lower than previously reported. Also recently, genes encoding ACC synthase and ACC oxidase the two key enzymes in the biosynthesis of ethylene, have been cloned in pineapple, a non-climacteric fruit. It has been shown that both genes are induced during ripening in a very similar way to the induction patterns observed in climacteric fruits (Cazzonelli *et al.*, 1998).

5.2.4 Disease resistance

Resistance to post-harvest pathogens is another priority target for genetic engineers but the necessary basic knowledge on the physiology, biochemistry and genetics of the resistance mechanisms is not as advanced as in ripening. Moreover, there is not a common defence mechanism applicable to all pathogens or all crops (as is the case of ethylene in climacteric fruit ripening) which implies that resistance genes need to be found on an individual basis. Nevertheless, there are an increasing number of genes being cloned and the mechanisms underlying the resistance process are being rapidly unravelled. Aside from specific resistance genes, an interesting secondary effect of reducing the production of ethylene during ripening has been recently reported by Cooper *et al.*, (1998). An extensive study of the susceptibility of transgenic tomato plants to *Colletotrichum gloeosporioides* was reported on two groups of transgenic tomato plants. The first group contained genetically modified plants in which the levels of polygalacturonase had been reduced in transgenic fruits. The second group consisted of genetically modified tomato plants in which antisense constructs had been used to partially silence the ACC oxidase gene and therefore the fruits produced reduced levels of ethylene during ripening. The ripening characteristics of these fruits have been previously discussed in this chapter and in numerous reports (Hamilton *et al.*, 1990; Picton *et al.*, 1995;

Sheehy *et al.*, 1988; Smith *et al.*, 1988). Wild type and antisense ACO fruits were manually inoculated with *C. gloeosporioides* and the extent of the infection scored five days after inoculation showing an average infection score of 44.8% and 15.8% respectively. Wild type fruits inoculated with *C. gloeosporioides* showed a marked increase in ethylene production in response to the infection whereas in transgenic ACO fruits this response was reduced by 96%. Transgenic fruits with reduced levels of PG did not show any noticeable change in behaviour in response to infection or any resistance to fungal infection. Despite the results of this research, it is known that ethylene is an important part of the plant defence mechanism against many pathogens. Therefore impaired ethylene production or insensitivity could result in increased disease susceptibility in many cases.

5.3 Biotechnological control of vegetable ripening and post-harvest diseases

Many vegetables exhibit a very short life span after harvesting and require very elaborate measures to expand their life. Reducing the rate of senescence in these crops is not an easy task either by conventional or biotechnological methods. The main obstacle to devising new technologies is the complexity of the problem and lack of basic knowledge about the biochemical and cellular processes accompanying post-harvest induced senescence. This is accentuated by the extraordinary variety of tissue types that are commercialised. Early attempts to use genetic manipulation to alter senescence have been based on hormone physiology, either enhancing cytokinin production or blocking ethylene production or perception.

In order to extend the post-harvest life of leafy vegetables we first need to focus on the events that occur in regular leaves during senescence. It has been known for some time that cytokinins can delay leaf senescence and that during senescence there is a drop in endogenous cytokinin levels (van Staden *et al.*, 1988). Overproduction of IPT, a bacterial enzyme that catalyses the rate-limiting reaction in the biosynthesis of cytokinins under the control of the strong constitutive promoter CaMV 35S, resulted in transgenic plants with high levels of cytokinins and delayed leaf senescence. But these plants also showed many developmental abnormalities since apart from senescence cytokinins are implied in a myriad of other developmental processes (Smart *et al.*, 1991). The last example stresses the importance of the availability of adequate promoters to express the right gene in the right place at the right time. An ingenious solution to the use of cytokinins to delay senescence has been provided by Gan and Amasino (1995) who placed the IPT gene under the control of SAG12, a senescence-specific promoter. In this system, the onset of senescence activates the SAG12 promoter, leading to the production of cytokinins. The accumulation of cytokinins inhibits the emerging senescence process and consequently reduces the activity of the SAG12 promoter, therefore avoiding the accumulation of cytokinins. Transgenic tobacco plants obtained in this way contained leaves with extremely delayed senescence that maintained high levels of photosynthetic activity. It remains to be proved whether this approach can be applied to leafy vegetables.

Ethylene often has an opposite effect to cytokinins in promoting senescence but its role is not likely to be essential for the regulation of the process in vegetative plant tissues. However, blocking the production or the perception of ethylene could have a positive effect in the longevity of green tissues. Transgenic tomato plants with reduced levels of ethylene production have shown retarded leaf senescence (John *et al.*, 1995). Transgenic *Arabidopsis* plants with the *etr1-1* dominant mutation that renders them insensitive to ethylene have also shown delayed leave senescence (Grbic and Bleecker, 1995). It is important to remark that in both cases delay was observed in the onset of senescence but once the process had been started it proceeded at normal speed. From the available data it seems that senescence-related genes are activated by ethylene only if the leaf is ready to senesce; when that happens ethylene enhances the process.

In floral vegetables such as broccoli, ethylene is likely to play an important role in both the onset and the regulation of the senescence process. Ethylene has already been proven to play such a role in flowers such as carnations (Woodson *et al.*, 1992). Transgenic broccoli has been produced containing antisense copies of a tomato ACO gene (Henzi *et al.*, 1999a; Henzi *et al.*, 1999b). Analysis of respiration rates, ethylene production and ACO activity performed in several transgenic lines showed puzzling results. Transgenic lines showed a marked increase in ethylene production in the early phase of post-harvest with levels three times higher than control samples; nevertheless, 74 hours after harvest ethylene production in controls markedly increased whereas the transgenic lines showed reduced ethylene levels. Respiration rates in control and transgenic samples were comparable immediately after harvest but transgenic samples showed a linear decrease of respiration up to 98h after sampling. Paradoxically, ACC oxidase activity levels in the transgenic samples were always higher than controls. In order to evaluate and interpret these experiments further research is needed to determine the gene expression patterns of the endogenous ACO genes since the authors used a relatively low homology tomato ACO gene in their genetic constructs. Preliminary agronomic evaluation has revealed some promising transgenic lines with significant improvements over the controls (Henzi *et al.*, 2000).

In addition to the factors discussed above, some vegetables such as lettuce, broccoli, cauliflower and asparagus are harvested while they are still immature and undergoing a phase of rapid growth in the plant. In these vegetables there are very rapid changes after harvest with broccoli losing large amounts of sucrose within four hours of harvesting. Significant changes in gene expression are also observed with accumulation of different transcripts such as asparagine synthase that catalyses the synthesis of the amino acid asparagine. The metabolic parameters of immature vegetables after harvest undergo important changes with loss of proteins and lipids and accumulation of free amino acids and ammonia that ultimately lead to tissue breakdown. This data strongly resembles starvation responses and suggests that starvation might be a critical stress regulating the senescence process in harvested immature vegetables.

5.4 Future trends

Even though the results already available are very promising there is a long road in front of us for improvement. The examples discussed in this chapter are technically simple and in many cases they were providing only 'proof of principle'. Almost all of those examples were 'world firsts' demonstrating the feasibility of gene inactivation by antisense techniques, ethylene biosynthesis inhibition by gene knockout, artificial endogenous enhancement of plant hormone levels and so on (Grbic and Bleecker, 1995; Hamilton *et al.*, 1990; Oeller *et al.*, 1991). Most of those experiments were not even designed to produce commercially useful crop varieties but to perform basic research on the roles of ethylene in fruit ripening, different hydrolases in fruit softening, cytokinins in leaf senescence, etc. Therefore we have to judge them for what they are; scientific experiments. Commercial applications will need a lot of refinement that will be possible only as a result of extensive basic research.

There are several constraints to the production of commercially viable crops engineered to provide better post-harvest characteristics.

- *More basic research.* There is a lack of understanding of the fundamental biochemical, cellular and molecular processes that take place in harvested tissues. Even though our knowledge is advancing at a very rapid pace we still need more research. Companies are too eager to cash in on the new technologies and many new products are being developed without a full understanding of the metabolical processes taking place and we are running the risk of developing products with unintended but potentially adverse agronomic characteristics. This problem is accentuated by recent government attitudes, reducing funds for basic research and pushing universities and government-funded institutions towards applied research that might generate full cost recovery.
- *New genes.* As our fundamental knowledge advances we will discover new genes that will allow better and more refined control of post-harvest processes. Instead of interfering with the whole ripening process we might decide to target very specific processes, such as the kinetics of sugar accumulation or fruit colour development. New genes are urgently needed to confer resistance to different post-harvest pathogens.
- *More advanced technical tools.* There is an urgent need for new developmentally regulated promoters that will allow the precise expression of genes in very specific tissues and developmental situations. Inducible promoters are also needed with cheap and easy means of induction in order

artificially to turn genes on and off as required. More reliable gene knockout techniques such as homologous recombination need to be developed (although some promising results are already available in model systems). New, simpler and faster transformation systems are also needed.

The next generation of horticultural crops will need to target improvements that benefit consumers rather than producers. Longer shelf life is an obvious target but a complex objective. A longer-lasting fruit will need to ripen more slowly but at the same time be protected against pathogens that very effectively attack ripe fruits and will also need to maintain not only its visual appeal but the ideal levels of proteins, vitamins, sugars and aroma.

Metabolic engineering will allow us to increase the nutritional value of fruits and vegetables by adding new components that are normally lacking in the traditional varieties. A good example is the recently developed 'Golden Rice' that has been enhanced with high levels of protein A (Ye *et al.*, 2000). This new product has an enormous potential to alleviate the important problem of vitamin A deficiency in developing countries in which rice is the main component of the daily diet.

Biotechnology is emerging as a powerful tool for plant improvement. Although in its initial stages, the potential of applying biotechnology to enhance the agronomical and nutritional characteristics of crops is immense. We are seeing only the tip of the iceberg and we are bound to see huge developments in the next ten years.

5.5 Sources of further information

Due to the extremely dynamic nature of the agricultural biotechnology field the best way to keep abreast of recent developments is the World Wide Web. Most of the sources identified bellow are Internet resources.

BINAS
The Biosafety Information Network and Advisory Service (BINAS) is a service of the United Nations Industrial Development Organization (UNIDO). BINAS monitors global developments in regulatory issues in biotechnology.
http://binas.unido.org/binas/

United States Department of Agriculture, Biotechnology site
The United States Department of Agriculture (USDA) is one of three Federal Agencies, along with the Environmental Protection Agency (EPA) and the US Food and Drug Administration (FDA), primarily responsible for regulating biotechnology in the United States. Products are regulated according to their intended use, with some products being regulated under more than one agency.
http://www.aphis.usda.gov/biotechnology/

Agbioworld
Agbioworld is devoted to bringing information about technological advances in agriculture to the developing world.
http://www.agbioworld.org/

The International Centre for Genetic Engineering and Biotechnology
An international organisation dedicated to advanced research and training in molecular biology and biotechnology, with special regard to the needs of the developing world.
http://www.icgeb.trieste.it/

Agbioforum
A magazine devoted to the economics and management of agro-biotechnology
http://www.agbioforum.missouri.edu/

Access Excellence, The National Health Museum (USA)
Access Excellence, launched in 1993, is a national educational program that provides high school biology and life science teachers access to their colleagues, scientists, and critical sources of new scientific information via the World Wide Web. The program was originally developed and launched by Genentech Inc., a leading biotechnology company.
http://www.accessexcellence.org/index.html

Nature debates
Benefits and risks of genetic modification in agriculture. Mike Wilkinson, moderator for this debate, surveys the terrains on which conflicting interests will do battle.
http://helix.nature.com/debates/gmfoods/gmfoods_frameset.html

REDENBAUGH, K., HIATT, W., MARTINEAU, B., KRAMER, M., SHEEHY, R., SANDERS, R., HOUCK, C. and EMLAY, D. (1992). *Safety Assessment of Genetically Engineered Fruits and Vegetables: A Case Study of the FLAVR SAVR Tomato*. Boca Raton: CRC Press.

5.6 References

AHARONI, A., KEIZER, L.C.P., BOUWMEESTER, H.J., SUN, Z., ALVAREZ-HUERTA, M., VERHOEVEN, H.A., BLAAS, J., VAN HOUWELINGEN, A.M.M.L., DE VOS, R.C.H., VAN DER VOET, H., JANSEN, R.C., GUIS, M., MOL, J., DAVIS, R.W., SCHENA, M., VAN TUNEN, A.J. and O'CONNELL, A.P. (2000). Identification of the *SAAT* Gene Involved in Strawberry Flavor Biogenesis by Use of DNA Microarrays. *Plant Cell*. **12**, 647–661.

AYUB, R., GUIS, M., AMOR, M.B., GILLOT, L., ROUSTAN, J.P., LATCHE, A., BOUZAYEN, M. and PECH, J.C. (1996). Expression of ACC oxidase antisense gene inhibits ripening of cantaloupe melon fruits. *Nature Biotech.* **14**, 862–866.

BEN-AMOR, M., FLORES, B., LATCHE, A., BOUZAYEN, M., PECH, J.C. and ROMOJARO, F. (1999). Inhibition of ethylene biosynthesis by antisense ACC oxidase RNA prevents chilling injury in Charentais cantaloupe melons. *Plant Cell Environ.* **22**, 1579–1586.

CAZZONELLI, C.I., CAVALLARO, A.S. and BOTELLA, J.R. (1998). Cloning and characterization of ripening-induced ethylene biosynthetic genes from non-climacteric pineapple (*Ananas comosus*) fruits. *Aust. J. Plant Physiol.* **25**, 513–518.

CHANG, C., KWOK, S.F., BLEECKER, A.B. and MEYEROWITZ, E.M. (1993). *Arabidopsis* ethylene-response gene ETR1: Similarity of product to two-component regulators. *Science* **262**, 539–544.

CIVELLO, P.M., POWELL, A.L.T., SABEHAT, A. and BENNETT, A.B. (1999). An expansin gene expressed in ripening strawberry fruit. *Plant Physiol.* **121**, 1273–1279.

COOPER, W., BOUZAYEN, M., HAMILTON, A., BARRY, C., ROSSALL, S. and GRIERSON, D. (1998). Use of transgenic plants to study the role of ethylene and polygalacturonase during infection of tomato fruit by *Colletotrichum gloeosporioides*. *Plant Path.* **47**, 308–316.

DELLA-PENNA, D., ALEXANDER, D.C. and BENNETT, A.B. (1986). Molecular cloning of tomato (*Lycopersicon esculentum* cultivar Castlemart) fruit polygalacturonase: Analysis of polygalacturonase messenger RNA levels during ripening. *Proc. Natl. Acad. Sci. USA* **83**, 6420–6424.

GAN, S. and AMASINO, R.M. (1995). Inhibition of leaf senescence by autoregulated production of cytokinin. *Science*. **270**, 1986–1988.

GRBIC, V. and BLEECKER, A.B. (1995). Ethylene regulates the timing of leaf senescence in Arabidopsis. *Plant J.* **8**, 595–602.

GRIERSON, D., MAUNDERS, M.J., SLATER, A., RAY, J., BIRD, C.R., SCHUCH, W., HOLDWORTH, M.J., TUCKER, G.A. and KNAPP, J.E. (1986). Gene expression during tomato ripening. *Phil. Trans. R. Soc. Lond.* **B314**, 399–410.

HAMILTON, A., LYCETT, G.W. and GRIERSON, D. (1990). Antisense gene that inhibits synthesis of the hormone ethylene in transgenic plants. *Nature*. **346**, 284–287.

HENZI, M.X., CHRISTEY, M.C. and MCNEIL, D.L. (2000). Morphological characterisation and agronomic evaluation of transgenic broccoli (*Brassica oleracea* L. var. *italica*) containing an antisense ACC oxidase gene. *Euphytica*. **113**, 9–18.

HENZI, M.X., CHRISTEY, M.C., MCNEIL, D.L. and DAVIES, K.M. (1999a). *Agrobacterium rhizogenes*-mediated transformation of broccoli (*Brassica oleracea* L. var. *italica*) with an antisense 1-aminocyclopropane-1-carboxylic acid oxidase gene. *Plant Sci.* **143**, 55–62.

HENZI, M.X., MCNEIL, D.L., CHRISTEY, M.C. and LILL, R.E. (1999b). A tomato antisense 1-aminocyclopropane-1-carboxylic acid oxidase gene causes reduced ethylene production in transgenic broccoli. *Aust. J. Plant Physiol.* **26**, 179–183.

JOHN, I., DRAKE, R., FARRELL, A., COOPER, W., LEE, P., HORTON, P. and GRIERSON, D. (1995). Delayed leaf senescence in ethylene-deficient ACC-oxidase antisense tomato plants: Molecular and physiological analysis. *Plant J.* **7**, 483–490.

KADER, A.A. (1985). Ethylene-induced senescence and physiological disorders in harvested horticultural crops. *HortSci.* **20**, 54–57.

KLEE, H.J., HAYFORD, M.B., KRETZMER, K.A., BARRY, G.F. and KISHORE, G.M. (1991). Control of ethylene synthesis by expression of a bacterial enzyme in transgenic tomato plants. *Plant Cell.* **3**, 1187–1194.

KRAMER, M.G., KELLOG, J., WAGONER, W., MATUMURA, W., GOOD, X., PETERS, S., CLOUGH, G. and BESTWICK, R.K. (1997) Reduced ethylene synthesis and ripening control in tomatos expressing S-adenosyl-methionine hydrolase. In *Biology and Biotechnology of the Plant Hormaone Ethylene* (Kanellis, A.K., Chang, C., Kende, H. and Grierson, D., eds). Dordrecht.: Kluwer Academic Publishers.

MEDINA-ESCOBAR, N., J. CARDENAS, E. MOYANO, J.L. CABALLERO and B.J. MUNOZ. (1997). Cloning, molecular characterization and expression pattern of a strawberry ripening-specific cDNA with sequence homology to pectate lyase from higher plants. *Plant Molecular Biology* **34**, 867–877.

MOYANO, E., PORTERO, R.I., MEDINA-ESCOBAR, N.,N., VALPUESTA, V., MUNOZ, B.J. and CABALLERO, J.L. (1998). A fruit-specific putative dihydroflavonol 4-reductase gene is differentially expressed in strawberry during the ripening process. *Plant Physiol.* **117**, 711–716.

NAM, Y.W., TICHIT, L., LEPERLIER, M., CUERQ, B., MARTY, I. and LELIEVRE, J.M. (1999). Isolation and characterization of mRNAs differentially expressed during ripening of wild strawberry (*Fragaria vesca* L.) fruits. *Plant Mol. Biol.* **39**, 629–636.

OELLER, P.W., MIN WONG, L., TAYLOR, L.P., PIKE, D.A. and THEOLOGIS, A. (1991). Reversible inhibition of tomato fruit senescence by antisense RNA. *Science* **254**, 437–439.

PICTON, S., BARTON, S.L., BOUZAYEN, M., HAMILTON, A.J. and GRIERSON, D. (1993). Altered fruit ripening and leaf senescence in tomatoes expressing an antisense ethylene-forming enzyme transgene. *Plant J.* **3**, 469–481.

PICTON, S., GRAY, J.E. AND GRIERSON, D. (1995). The manipulation and modification of tomato fruit ripening by expression of antisense RNA in transgenic plants. *Euphytica* **85**, 193–202.

SCHUCH, W., KANCZLER, J., ROBERTSON, D., HOBSON, G., TUCKER, G., GRIERSON, D., BRIGHT, S. and BIRD, C. (1991). Fruit quality characteristics of transgenic tomato fruit with altered polygalacturonase activity. *Hort. Sci.* **26**, 1517–1520.

SHEEHY, R.E., KRAMER, M. and HIATT, W.R. (1988). Reduction of polygalacturonase activity in tomato fruit by antisense RNA. *Proc. Natl. Acad. Sci. USA* **85**, 8805–8809.

SMART, C.M., SCOFIELD, S.R., BEVAN, M.W. and DYER, T.A. (1991). Delayed leaf senescence in tobacco plants transformed with tmr, a gene for cytokinin production in *Agrobacterium*. *Plant Cell.* **3**, 647–656.

SMITH, C.J.S., WATSON, C.F., BIRD, C.R., RAY, J., SCHUCH, W. and GRIERSON, D. (1990). Expression of a truncated tomato polygalacturonase gene inhibits expression of the endogenous gene in transgenic plants. *Mol. Gen. Genet.* **224**, 477–481.

SMITH, C.J.S., WATSON, C., RAY, J., BIRD, C.R., MORRIS, P.C., SCHUCH, W. and GRIERSON, D. (1988). Antisense RNA inhibition of polygalacturonase gen expression in transgenic tomatoes. *Nature*. **334**, 724–726.

THEOLOGIS, A. (1994). Control of ripening. *Curr. Opin. Biotech.* **5**, 152–157.

TRAINOTTI, L., SPOLAORE, S., PAVANELLO, A., BALDAN, B. and CASADORO, G. (1999). A novel E-type endo-beta-1,4-glucanase with a putative cellulose-binding domain is highly expressed in ripening strawberry fruits. *Plant Mol. Biol.* **40**, 323–332.

VAN STADEN, J., COOK, E.L. and NOODEN, L.D. (1988). Cytokinins and senescence. In *Senescence and aging in plants* (Nooden, L.D. and Leopold, A.C., eds). San Diego: Academic Press, pp. 281–328.

WILKINSON, J.Q., LANAHAN, M.B., CLARK, D.G., BLEECKER, A.B., CHANG, C., MEYEROWITZ, E.M. and KLEE, H.J. (1997). A dominant mutant receptor from *Arabidopsis* confers ethylene insensitivity in heterologous plants. *Nature Biotech.* **15**, 444–447.

WILLS, R.B.H., KU, V.V.V., SHOHET, D. and KIM, G.H. (1999). Importance of low ethylene levels to delay senescence of non-climacteric fruit and vegetables. *Aust. J. Exp. Ag.* **39**, 221–224.

WOODSON, W.R., PARK, K.Y., DRORY, A., LARSEN, P.B. and WANG, H. (1992). Expression of ethylene biosynthetic pathway transcripts in senescing carnation flowers. *Plant Physiol.* **99**, 526–532.

YANG, S.F. and HOFFMAN, N.E. (1984). Ethylene bioshynthesis and its regulation in higher plants. *Ann. Rev. Plant Phys.* **35**, 155–189.

YE, X.D., AL-BABILI, S., KLOTI, A., ZHANG, J., LUCCA, P., BEYER, P. and POTRYKUS, I. (2000). Engineering the provitamin A (beta-carotene) biosynthetic pathway into (carotenoid-free) rice endosperm. *Science*. **287**, 303–305.

6

The use of molecular genetics to improve food properties

I. Amaya, M. A. Botella and V. Valpuesta, Universidad de Málaga

6.1 Introduction

Many of the initial studies in plant biotechnology were focused on developing new plant varieties with better yield rather than changing the properties of the plant-derived fruits. Thus, genes conferring resistance to several biotic and abiotic stresses were incorporated in cultivated plants. As a consequence, since 1995 a number of crops with improved agronomic traits have been available for farmers in some parts of the world. Biotechnology has also the ability to create new varieties focused on product quality and output traits rather than agronomic traits. Fruit commercialisation of a genetically engineered canola plant with modified oil content occurred in 1996 (Yuang and Knauf, 1997). It was a milestone in the long path followed by many research groups towards genetically modified plants with improved properties for human consumption. This has been partially the result of combating the attitudes of consumers to genetically modified foods. These aspects constitute the topics reviewed in the present chapter. In addition, it has recently become clear that the use of transgenic plants as living reactors constitutes an advantage for the inexpensive production of some proteins and metabolites that are economically important. This new trend in the application of plant biotechnology is also reviewed in this chapter.

6.2 Changing the nutritional value of foods

6.2.1 Amino acid content of proteins

An early application of biotechnology for improving the nutritional value of foods has involved changing the amino acid composition of some common

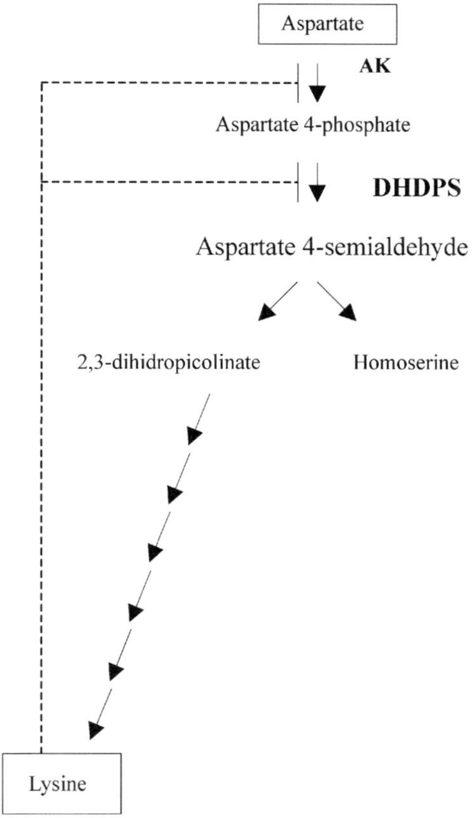

Fig. 6.1 Biochemical regulatory mechanism proposed to regulate the synthesis of the amino acid Lysine, in higher plants, derived from the amino acid Aspartate. Enzyme activities, aspartate kinase (AK) and dihidrodipicolinate synthase (DHDPS) are indicated. Broken lines indicate the inhibition of these enzymes by Lysine.

proteins of the human diet. It has been long known that humans cannot live on a protein-free diet. The reason is that we are incapable of synthesising half of the 20 standard amino acids present in proteins. These are known as essential amino acids and must be provided in the diet. Consequently, the nutritive value of a protein-based diet is directly related to the content of these essential amino acids. In general, cereals have proteins with a low content of the essential amino acids lysine, threonine and tryptophan, while legume proteins have a deficiency of cystein, methionine and triptophane. Among the most valuable sources of vegetable protein are the grain legumes. In one of these, soybean, genetic engineering has been applied to increase the content of the essential amino acid lysine in the seed proteins. The rationale was that increasing the synthesis of lysine in the seeds of soybean would increase the synthesis of proteins with high lysine content. Lysine synthesis in this species is finely regulated by a feed-back mechanism, i.e., when the lysine content is high there is an inhibition of two of the enzymes involved in the metabolic pathway leading to the synthesis of this amino acid (Fig. 6.1).

The strategy consisted of the integration in the genome of soybean of genes from other species encoding for enzymes without the feed-back mechanism. The transformation of soybean with the gene *lysCM4* from *Escherichia coli* (encoding AK) and the *dapA* gene from *Corynebacterium* (encoding DHDPS), both insensitive to the feed-back inhibition by lysine, resulted in a transgenic soybean plant with a duplicated pathway of lysine biosynthesis, one sensitive and the other insensitive to lysine. As a result, the lysine content of the transgenic soybean plants was over 100-fold the value of the untransformed plants. Other plant species like corn, wheat and canola have been subjected to the same genetic manipulation to increase their lysine content with similar results to those obtained in soybean.

6.2.2 Fatty acid composition of triacylglycerols

Lipids are also main components of the human diet. The consumer preference for plant-derived oils is increasing to the detriment of animal fats. Annual plant oil production is increasing worldwide and most of it is used for human consumption as margarines, oils and food ingredients. The triacylglycerols are the most important components of plant seed oils. Interestingly, the physical and chemical properties of an edible oil are related to the chemical structure of the fatty acids esterifying the glycerol (Table 6.1). Properties such as melting point, colour, flavour, mouthfeel, spreadability, stability, and effects on human health are determined by the fatty acid composition of the triacylglycerols. Most efforts in developing changes in the lipid composition of plant oils have been directed to change the proportion among the fatty acids of the triacylglycerols.

Common fatty acids in the commercial seed oils are lauric, myristic, palmitic, stearic, oleic, linoleic and linolenic. As is apparent, their differences occur in the length of the carbon skeleton (C12 to C20) as well as in the presence of double bonds (unsaturations). Long chain fatty acids containing two or more double

Table 6.1 Nomenclature and representative examples of naturally occurring fatty acids

Systematic nomenclature	m:n $\Delta^{aZbZ...}$ m: carbon atoms n: double bonds superscript: positions of the double bonds a, b, ...: carbon numbered from the carboxyl end Z: configuration *cis* of the double bond	
Examples:		
Saturated	18:0	Stearic acid
Unsaturated	18:1 Δ^{9Z}	Oleic acid
Polyunsaturated (PUFA)	20:5 $\Delta^{5Z,8Z,11Z,14Z,17Z}$	Eicosapentaenoic acid

bonds are named polyunsaturated fatty acids (PUFA). Different studies on the effect of dietary fatty acids consumption on human health have noticed the trend of consumers towards a reduction of saturated acids in the diet and, accordingly, an increase in unsaturated acids. Epidemiological studies have shown that intake of monounsaturated acids was associated with a low incidence of coronary artery disease (Keys *et al.*, 1986), which has been explained by its reduction in the low density lipoproteins (LDL) levels and their oxidation (Mata *et al.*, 1997). Therefore the unsaturation of fatty acids have been the target for modification by genetic engineering studies. Although the metabolic pathways leading to the synthesis of these compounds are not simple, some genes have been adequately selected to be modified. Oleic acid, the major monounsaturated acid of the diet, reduces cholesterol and LDL in the serum. Transgenic plants overexpressing the desaturase gene, that encodes for the enzyme catalysing the conversion of the saturated precursor stearic acid C18 into oleic, have been obtained (Fig. 6.2). This has determined that oleic content of soybean has been raised to values up to 80% of the total fatty acids content of the seeds (Kinney, 1996). Although the relationship between PUFA and disease remains contentious, there is a consensus among the health organisations that PUFA should form 8–23% of the total lipid intake in the human diet (Gill and Valivety, 1997). However, the production of high PUFA oil plants is not straightforward (Fig. 6.2), and no report on transgenic plants with high PUFA content is known.

Interestingly, there have been cases where saturation of the fatty acids has been the purpose of the plant genetic modification. Saturation of fatty acids determines properties such as melting temperature and viscosity that may be important to a commercial product. Margarine, for example, needs to be easily

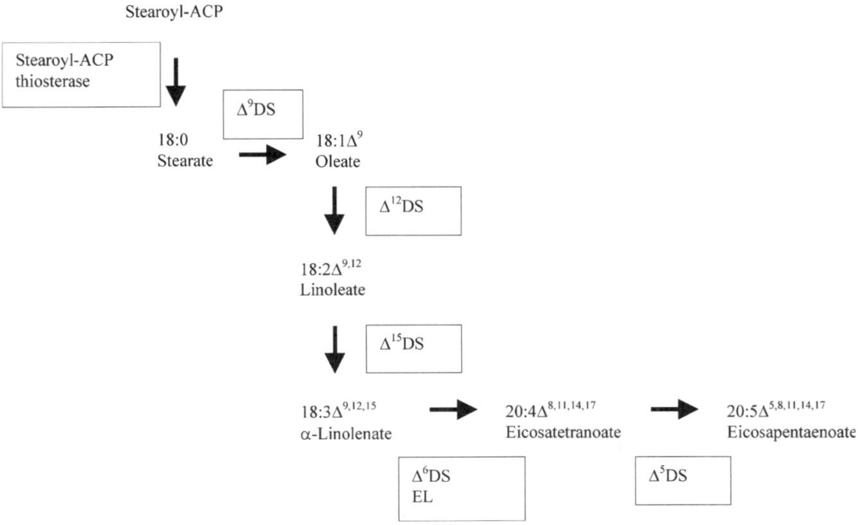

Fig. 6.2 Interconversions of the fatty acids indicated in Table 6.1 as examples, Enzymes: Δ^nDS: n-desaturase: EL: elongase.

spreadable within a range of temperatures. In addition, saturation may also be beneficial for oil stability since it is known that unsaturated acids are more readily oxidised, resulting in an increased tendency to rancidity and off odours. Finally, vegetable oils used for frying require partial saturation by hydrogenation in order to give adequate characteristics of stability and melting temperature to these oils. The chemical hydrogenation has been proven to induce also a change in the configuration of the double bonds of the fatty acids, from the naturally occurring *cis* to *trans*. The presence of the *trans* unsaturated fatty acids has been correlated to a risk of coronary heart disease. Therefore, plants with a high content of natural oil and with a high level of saturation have been engineered. High stearate content of the oil in a *Brassica* plant has been achieved by two methods. One method transformed this plant species with the antisense construct of the gene encoding the stearoyl-ACP desaturase, the enzyme that catalyses the transformation of stearic acid into oleic acid (Fig. 6.2, Δ^9DS) (Knutzon *et al.* 1992). The silencing of the endogenous gene produced the accumulation of stearic acid up to 40% of the total fatty acids content. The second method transformed the *Brassica* plant with a gene encoding the stearoyl -ACP thiosterase specific for the synthesis of stearic acid (Fig. 6.2). Using this approach, the transgenic plants yielded up to 68% of this fatty acid.

6.2.3 Vitamins and diet enrichment compounds

Vitamins are essential compounds for humans and other vertebrates and they must be obtained from the diet. In addition, some vitamins are used as functional additives in food products. Ascorbic acid (vitamin C) is used to prevent oxidation in apples, peaches, apricots, potatoes, peanut butter, potato chips, beer, fat and oils. The carotenoids (vitamin A precursors) are used as colorants in margarine, cheese, ice cream, pasta, juices and beverages (Giese, 1995). It is generally believed that people in Western countries have adequate vitamin intake. There are, however, susceptible groups within the general population who may have inadequate vitamin intake. Such groups include dieters, people on medication, pregnant women, alcoholics, adolescents, and people with diabetes and other chronic ailments. In developing countries very little information exists concerning their nutritional status.

The edible part of rice grains, the endosperm, lacks an essential nutrient as vitamin A. A diet mostly based on rice consumption may eventually cause vitamin A deficiency. It is estimated that improved vitamin A nutrition could prevent worldwide 1–2 million deaths annually among children. A very promising achievement has been the introduction of genes into rice that enabled the biosynthesis in the endosperm of β-carotene, the precursor of vitamin A (Ye *et al.*, 2000). β-carotene is synthesised from the precursor geranylgeranyl bisphosphate which is converted to the colourless phytoene by the enzyme phytoene synthase (Fig. 6.3). The phytoene undergoes four desaturations to form lycopene, which is red and gives colour to ripened tomato fruits. Further cyclisation of lycopene results in the formation of β-carotene. Immature rice endosperm is capable of synthesising geranylgeranyl biphosphate which can be used to produce phytoene by expression of a phytoene synthase gene. The introduction of three genes in rice via *Agrobacterium* allowed the expression of the entire β-carotene pathway into the endosperm. These genes were a phytoene synthase and a lycopene β-ciclase from daffodil, and a bacterial phytoene desaturase from *Erwinia uredovora* (Figure 6.3). The grain of the transgenic rice had a yellow-golden colour and by itself contained sufficient β-carotene for human vitamin A requirements. In rapeseed, transformation with a phytoene synthase gene also increased the level of vitamin A precursor (Kishore and Shewmaker, 1999). In tomato the transformation with a bacterial phytoene desaturase increased up to twofold the β-carotene content in fruits (Römer *et al.*, 2000).

Another lipid-soluble vitamin whose function is linked to an antioxidant role is vitamin E (α-tocopherol). Daily intake of this vitamin in excess of a recommended minimum is associated with decreased incidence of several diseases. Plant oils are the main source of dietary vitamin E and they generally have a high content of the vitamin E precursor γ-tocopherol. Overexpression of

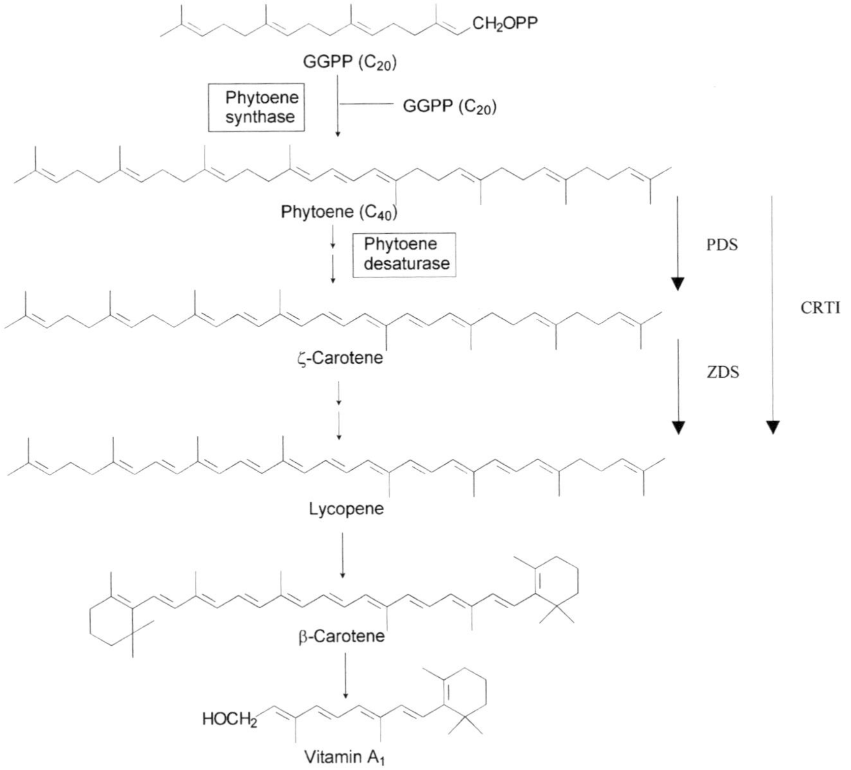

Fig. 6.3 Biosynthetic pathway of vitamin A from geranylgeranyl pyrophosphate (GGPP). Vertical arrows indicate steps catalysed by phytoene desaturase (PDS), carotene desaturase (ZDS), and *Erwinia uredovora* phytoene desaturase (CRTI).

γ-tocopherol methyl transferase greatly increased the seed level of α-tocopherol in the model plant *Arabidopsis thaliana* (Shintani and DellaPenna, 1998). This process seems ready to be eventually applied to some commercial crops in the future.

Flavonols are another group of secondary metabolites whose inclusion in the human diet may give protection against cardiovascular diseases. The biosynthetic pathway leading to the synthesis of these compounds has been known for a long time. However, recent information regarding the pathway has allowed the design of specific strategies to increase the content of selected bioactive compounds. Thus, the transformation of tomato with a gene from Petunia encoding a chalcone isomerase has produced tomato fruits with a 78-fold increase in the content of flavonols in the peel (Muir *et al.* 2001). What is more important, 65% of the flanonols were retained in the paste obtained after processing the transgenic fruits.

A high risk of iron deficiency has been reported when vegetables are the major components in the diet. Although some plants are rich in this element, its availability is limited by the fact that the same plants contain oxalic acid and phytate-like substances that may complex this element. Some studies have shown that oral administration of ferritin, a protein used by plants and animals to store iron, can provide the iron needed to treat anaemia in rats. With this information, rice has been transformed with a soybean gene encoding ferritin, under the control of a seed-specific promoter (Goto *et al.* 1999). Transgenic rice plants accumulated ferritin in the endosperm tissue and up to three-fold levels of iron in comparison to normal seeds. Interestingly, plants overexpressing ferritin have been reported to be tolerant to oxidative damage and pathogens. This seems to be an additional agronomic trait for these transgenic plants (Deák *et al.* 1999).

6.3 Modification of fruit colour and sweetness

It is becoming widely accepted that plant biotechnology is entering a second phase of development that looks for differentiated crop development (Kishore and Shewmaker, 1999). This phase is characterised for being more focused on output traits than input traits. Some of these output traits have been described in the previous section, but still there are others that are related to fruit quality for human consumption. In general, market trends predict growth in the consumption of fruit if quality can be improved. These traits remain a challenge for plant biotechnology developers. However, there are two aspects that are present in the efforts of various research laboratories. They focus on fruit colour and sweetness, and they may be on the market in the near future.

Carotenoids are a group of natural pigments that determine the colour of many flowers, fruits and vegetables. In fact, the golden-yellow colour of high β-carotene transgenic rice has previously been indicated to be a side effect of rice plant transformation. In some cases, the goal has been specifically to increase the colour of the transgenic product. This trait is particularly important in fruits that are going to be processed to produce jams, marmalades, pastes, and even wine. Recently, the carotenoid biosynthesis pathway has been modified in tobacco plants using the *CrtO* gene encoding the β-carotene ketolase, from the alga *Haematococcus pluvialis* (Mann *et al.*, 2000). The transgenic plants accumulated several ketocarotenoids that changed the colour of the nectary from yellow to red. The researchers claim that plant transformation with this gene may be used in the future to change the colour of fruit.

Flavonoids are a major class of secondary plant products well known for the colouration they provide in blue, red and purple flowers, fruits and leaves. Flavonoids are derived from simple metabolites as phenylalanine and acetyl-CoA in a highly branched pathway that leads to flavonols, flavanones, isoflavonoids and anthoycianins. Each type may undergo further modifications that result in a great diversity of colours. It is known that this complex pathway is regulated at the level of transcription of structural genes thus giving a main role to transcriptional factors, i.e. proteins that control the expression of some other genes at transcriptional level. This has been applied to the model plant *Arabidopsis* and tobacco, which were transformed with the maize *R* and *C1* genes. These are regulators that produced an activation of genes encoding for enzymes of the anthocyanine production pathway in the transgenic plants. In general, changes at the level of intermediate enzymes of the flavonoids pathway may change the final balance of these coloured compounds and eventually the colour of a given plant organ, for example, the production of yellow flowers in acyanic lines of Petunia after its transformation with a chalcone reductase gene from *Medicago sativa* (Davies *et al.*, 1998). Only redirection of the flavonoid biosynthesis pathway could explain the reported effect since chalcone reductase activity is not naturally present in Petunia as it is not the product of the reaction, which is further transformed in a coloured compound. This illustrates the complex equilibrium of the complete pathway and the difficulty of predictable effects after plant transformation with heterologous genes.

A critical component of fruit palatability is sweetness. There are two goals associated with this food property to be achieved. The first is that many fruits are not sweet enough to make them appetising to consumers. The second is that the most common natural sweetener is sucrose, whose caloric content has caused a change in consumer preference for other sweeteners. Both problems have been addressed by plant biotechnologists.

It was known that a protein from the plant *Dioscorephyllum cumminsii*, named monellin, was about 10^5 sweeter than sucrose on a molar basis. This property is common among some plant proteins of the thaumatin-like class. Tomato plants overexpressing the monellin gene under the control of a fruit-specific promoter (E8) have been produced (Peñarrubia *et al.*, 1992). Although no further reports on fruit properties have been reported, the experience indicates a possible way to increase fruit sweetness, at the time that protein content is enhanced.

Other researchers have also looked for substitutes for the natural sweetener sucrose (Sévenier *et al.*, 1998). They identified the low molecular weight fructans, polymers of fructose, as an adequate replacement. These compounds resemble sucrose in their organoleptic properties, but are indigestible by humans. In addition, they cannot be used as a carbon source by cariogenic bacteria. To obtain high fructan plants, the gene encoding 1-sucrose:sucrose fructosyl transferase from *Helianthus tuberosus* was introduced into sugar beet. Most of the stored sucrose in the root was converted to low molecular weight fructans. The experience is ready to be extended to other commercial crops whose sugar content make them unattractive to consumers.

6.4 Modification of food-processing properties of fruit

Wheat is a cereal commonly used to make bread and a number of end-products such as cakes, cookies or pastry. The adequacy of the wheat grains used for each purpose is related to two main properties of the grain that have been consequently targeted as susceptible to being modified by genetic engineering. One is the gluten protein composition that may eventually determine the viscoelasticity (elasticity and extensibility) of the dough. The other is the grain hardness that has a direct effect on the milling and baking properties of the grain.

What is known as gluten is in fact a complex mixture of up to 50 proteins most of them classified as prolamins. Among these, a group is assembled in high molecular weight (HMW) polymers whose subunit composition determines some properties of the dough, including its elasticity. The entrapment of CO_2 in the formed network favours a porous structure that determines the viscoelasticity of the dough. Since wheat is hexaploid, up to six possible genes for the HMW subunits may be expected. However, it has been demonstrated that the modification of one or two genes may be sufficient to change the properties of the dough prepared from the grain. Thus, wheat transformed with one or two HMW subunits produced grains that caused a stepwise increase in dough elasticity (Barro *et al.*, 1997).

Recent studies have demonstrated that two wheat proteins, named puroindolines A and B, acting together in a 1:1 ratio, control to some extent wheat grain softness. In cereals with a hard texture, such as corn and rice, these two proteins are not present in the grain. To genetically modify the texture of these cereals, transgenic plants expressing the wheat puroindoline genes were obtained. The result was that rice grain softness has been increased significantly (Krishnamurthy and Giroux, 2001).

Potato constitutes a basic food in the diet of many Western countries where it is commonly consumed as potato chips. When harvested, potato tubers are frequently stored in the cold to prevent sprouting. Under this condition part of the starch of the tuber is converted into hexoses. The increased hexoses induce the sweetening of the tuber and have an adverse effect on the quality of the processed chips. The reason is that an excess of hexoses react with the amino acids during frying causing an undesirable browning of the chips. Hexoses increase in the cold stored tubers results from an imbalance between the rate of their production from starch and the rate of their degradation in the glycolytic pathway. The accumulated hexoses may be metabolically converted into sucrose that is eventually split into glucose and fructose by the action of the enzyme invertase. Several strategies have been attempted with variable success to keep the hexose level low in the cold stored tubers. The latest is based on a gene encoding a protein that inhibits the invertase enzyme (Greiner *et al.*, 1999). Thus, in the presence of this protein, hexoses would not be produced from the hydrolysis of sucrose. Transgenic potato plants overexpressing this invertase inhibitor gene reduced by 70% the hexose accumulation after the cold treatment and, as expected, browning of the chips was also prevented. Most importantly, this occurred without any change in the starch quality or quantity.

Around 30% of the starch produced in plants is used for direct human and animal consumption. The ratio between the two main components of this polymer, linear amylose and branched amylopectin, determines its applicability. The reason is that the amylose:amylopectin ratio influences the physicochemical properties of the starch. This ratio fluctuates among different crops, between 20–30% for amylose to 70–80% for amylopectin. There has been recently a successful report on the variation of this ratio in a transgenic potato plant (Schwall *et al.*, 2000). The species was transformed with the antisense of two genes encoding two isoforms of a starch branching enzyme. The result was that amylopectin was practically absent as a component of the starch.

6.5 Molecular farming and therapeutic food

The production of plant-derived biopharmaceutical products is sometimes named as molecular farming. The word biopharmaceutical is applied to a naturally occurring or modified polypeptide, protein, DNA or RNA product that is to be used for therapeutic, prophylactic or *in vivo* diagnostic use in humans. The main categories of biopharmaceutical products are recombinant proteins, therapeutic monoclonal antibodies, polyclonal antibodies, non-recombinant proteins and antisense oligonucleotides. Between 1995 and 1999, 15 recombinant proteins, six monoclonal antibodies, two polyclonal antibodies, two non-recombinant proteins and one antisense oligonucleotide were approved by the FDA. By mid-2000 there were between 80–90 biopharmaceuticals in general medical use, and around 500 more were undergoing clinical trials. Major targets of these compounds include cancer, cardiovascular diseases, and infectious diseases. Most of these products have been produced in cultured mammalian cells, bacteria and fungi. Now, the use of plants as alternative production systems is being evaluated since plants are potentially a cheap source of recombinant products (Table 6.2).

One possible disadvantage of using plants as bioreactors for biopharmaceuticals is that post-translational modification of synthesised proteins may differ from mammals. However, these modifications are few compared with the differences between mammals and microorganisms that have been commonly

Table 6.2 Some examples of proteins whose production in various plant species have been reported (Summarised from Giddings *et al.*, 2000)

	Potential application	Plant	Protein
Vaccines			
	Hepatitis B	Tobacco	Hepatitis B surface antigen
	Cholera	Potato	V. cholerae toxin Ctoxa and Ctox B subunits
	HIV	Tobacco	HIV epitope (gp120)
	Malaria	Tobacco	Malarial B-cell epitope
Antibodies (single chain Fv fragments)			
	Production of protein in tubers	Potato	Phytochrome binding scFv
	Treatment of non Hodgkin's lymphoma	Tobacco	scFv of IgG mouse B-cell lymphoma
	Production of tumour associated marker antigen	Cereals	scFv against carcinoembryogenic antigen
Biopharmaceuticals			
	Anticoagulant	Tobacco	Human protein C
	Anaemia	Tobacco	Human erythopoietin
	Provitamin A deficiency	Rice	Daffodil phytoene synthase
	Hypertension	Tobacco	Angiotensin-1-converting enzyme

used as a source of biopharmaceuticals. There is also the risk of impurities in the plant-derived biopharmaceuticals that may include secondary plant metabolites, pesticides and herbicides. Such impurities could have a direct toxic effect, could affect product stability, or could even have immunogenicity leading to allergic reactions. However, biopharmaceuticals derived from transgenic plants could be safer than those derived from human cells that could be contaminated by human pathogens.

Transgenic plants have already been developed to produce proteins such as enkephalines, α-interferon, human serum albumin, glucocerobrosidase and granulocyte-macrophage colony-stimulating factor. These last two are among the most expensive drugs before their production in plants (Giddings *et al.* 2000). Rice plants have been engineered to produce α-1-antitrypsin and the product is presently under trial (Giddings *et al.*, 2000). In one case the human somatotropin has been reported to be produced in the chloroplast of a non-food crop such as tobacco (Staub *et al.*, 2000) claiming the additional advantage of biological containment in the field cultivation of this plant.

Among possible proteins to be produced in transgenic plants there have been great efforts to produce antibodies whose therapeutic potential has been largely recognised. Functional antibodies have already been produced in plants and sometimes have been named as plantibodies. Plants have been successfully used to generate complex secretory antibodies that would be of particular benefit for topical immunotherapy in mucoses. This is the case with the production of humanised monoclonal antibodies for immunoprotection against genital herpes (Zeitlin *et al.*, 1998). Moreover, the use of edible plant parts as a source of antibodies opens the door to the possibility of treatment of the mucoses of the gastrointestinal tract.

The development of plants expressing vaccine antigens is another relatively new potential application of plant biotechnology. Vaccines consisting of macromolecules with a protective immune response has a limited use in developing countries, mainly owing to their high cost and low stability. The low requirements of growing plants make their production much cheaper. The expression of vaccines in plant tissues also eliminates the risk of contamination with animal pathogens, make them more stable, and may allow oral delivery if expressed in edible parts of the plants. The first clinical trial with a plant-derived vaccine in 1997 demonstrated the induction of a mucose immune response (Tacket *et al.*, 1998). Potatoes genetically modified to produce a cholera toxin B subunit has been shown to induce antibody production in humans after oral administration (Arakawa *et al.*, 1998). Also, in preclinical animal trials, mice fed with transgenic potato expressing hepatitis B surface antigen results in a primary immune response (Richter *et al.*, 2000). More recently, three plant synthesised antigens of cholera, rotavirus and enterotoxigenic *E. coli* were expressed in potato and showed a strong immune response in potato-fed mice (Yu and Langridge, 2001). Since it is well documented that delivery of plant-derived vaccine to a mucosal site induces both local and systemic immune responses, the list of plant-derived vaccinogens continues to grow, and includes viral, bacterial, enteric and non-enteric antigens.

6.6 Future trends

Experts in the field indicate that it is most likely that Biotechnology will play a significant role in the 21st century (Cantor, 2000). At present, most of the commercial applications of transgenic plants have been for crops with improved agronomic benefits such as resistance to pests or herbicides. This trend will continue in developing countries since there is still the possibility of increasing the yield ceiling. However, it is likely that the focus for development in the coming years will be in multiple gene introductions to increase output traits such as increased nutritional value, vitamin content, or improved flavour components. Specifically, in the area covered by this chapter, the main trend pointed out by experts is in the production of nutraceuticals. This term was coined in 1979 for 'foods, or parts of foods, that provide medical or health benefits, including the prevention and treatment of disease' (Brower, 1998). The term includes some other definitions such as dietary supplements, functional foods and medical foods. It was estimated that by 1998, 47% of the population of Japan consumed nutraceuticals. Although there is no agreement among experts on the size of the world market for these products, it seems that the trend to a higher consumption is expected (Brower, 1998). Nutraceuticals include a broad range of products from high-tech foods, such as engineered canola with increased levels of antioxidants, to low-tech candy bars supplemented with vitamins. In plants, the number of possibilities for increased demand from consumers can only be estimated at present. The abundance of basic information, the improvement of plant tranformation and protein delivery methods, the development of computational biology, bioinformatics, and the trend of individualised medicine will mark the direction of this area in the future.

6.7 Sources of further information and advice

FDA's guidance concerning demonstration of comparability of human biologic products, including therapeutic biotechnology-derived products:
http://www.fda.gov/gdlns/comptest.pdf

European Medicines Evaluation Agency's note for guidance on comparability of medicinal products containing biotechnology-derived proteins as active substances:
http://www.emea.eu.int/pdfs/human/bwp/320700en.pdf

European Parliament and Council of the European Union (2001) Directive 2001/18/EC of the European Parliament and of the Council of 12 March on the deliberate release into the environment of genetically modified organisms and repealing Council Directive 90/220/EEC. *Off. J. Eur. Commun.* L106/1. On line magazine of Biotechnology Industry analysis:
http://www.recap.com/mainweb.nsf

Biotech Rumor Mill, a bulletin board run by Biofind on which people swap stories about what is being sold to whom and why, what companies are not what they used to be, what other people are earning, etc.
http://www.biofind.com/rumor

Site to search for GM crops by trait, by producer and by test site, offering a link to a wide number of useful sources in the debate, in the legal documents and the patent applications:
http://www.genewatch.com

6.8 References

ARAKAWA T., CHONG D.K.X. and LANGRIDGE W.H.R. (1998) Efficacy of a food plant-based oral cholera toxin B subunit vaccine. *Nature Biotechnol.* 16: 292–297.

BARRO R., ROOKE L., BÉKÉS F., GRAS P., TATHAM A.S., FIDO R., LAZZERI P.A., SHEWRY P.R. and BARCELÓ P. (1997) Transformation of wheat with high molecular weight subunit genes results in improved functional properties. *Nature Biotech.* 15: 1295–1299.

BROWER V. (1998) Nutraceuticals: Poised for a healthy slice of the healthcare market? *Nature Biotechnol.* 16: 728–731.

CANTOR C.R. (2000) Biotechnology in the 21st century. *Trends Biotechnol.* 18: 6–7.

DAVIES K.M., BLOOR S.J., SPILLER G.B. and DEROLES S.C. (1998) Production of yellow colour in flowers: redirection of flavonoid biosynthesis in Petunia. *Plant J.* 13: 259–266.

DEÁK M., HORVÁTH G.V., DAVLETOVA S., TÖRÖK K., SASS L., VASS I., BARNA B., KIRÁLY Z and DUDITS D. (1999) Plants ectopically expressing the iron-binding protein, ferritin, are tolerant to oxidative damage and pathogens. *Nature Biotechnol.* 17: 192–196.

GIDDINGS G., ALLISON G., BROOKS D. and CARTER A. (2000) Transgenic plants as factories for biopharmaceuticals. *Trends Biotechnol.* 18: 1151–1155.

GIESE J. (1995) Vitamins and mineral fortifications of foods. *Food Technol.* 49: 110–122.

GILL I. and VALIVETY R. (1997) Polyunsaturated fatty acids, Part I: occurrence, biological activities and applications. *Trends Biotechnol.* 15: 401–409.

GOTO F., YOSHIHARA T., SHIGEMOTO N., TOKI S. and TAKAIWA F. (1999) Iron fortification of rice seed by the soybean ferritin gene. *Nature Biotechnol.* 17: 282–286.

GREINER S., RAUSCH T., SONNEWALD U. and HERBERS K. (1999) Ectopic expression of a tobacco invertase inhibitor homolog prevents cold-induced sweetening of potato tubers. *Nature Biotech.* 17: 708–711.

KEYS A., MENOTTI A., KARVONEN M.J., ARAVANIS C., BLACKBURN H., BUZINA R., DJORDJEVIC B.S., DONTAS A.S., FIDANZA F., KEYS M.H., KROMHOUT D., NEDELJKOVIC S., PUNSAR S., SECCARECCIA F. and TOSHIMA H. (1986) *Am. J. Epidemiol.* 124: 903–915.

KINNEY A.J. (1996) Development of genetically engineered soybean oil for food applications. *J. Food Lipids* 3: 273.

KISHORE G.M. and SHEWMAKER C. (1999) Biotechnology: Enhancing human nutrition in developing and developed worlds. *Proc. Natl. Acad. Sci.* 96: 5968–5972.

KNUTZON D.S., THOMPSON G.A., RADKE S.E., JOHNSON W.B., KNAUF V.C. and KRIDL O.C. (1992) Modification of Brassica seed oil by antisense expression of a stearoyl-acyl carrier protein desaturase gene. *Proc. Natl. Acad. Sci.* 89: 2624–2628.

KRISHNAMURTHY K. and GIROUX M.J. (2001) Expression of wheat puroindoline genes in transgenic rice enhances grain softness. *Nature Biotechnol.* 19: 162–166.

MANN V., HARKER M., PECKER I. and HIRSCHBERG J. (2000) Metabolic engineering of astaxanthin production in tobacco flowers. *Nature Biotech.* 18: 888–892.

MATA P., VARELA O., ALONSO R., LAHOS C., OYA M. and BADIMÓN L. (1997) Monounsaturated and polyunsaturated n-6 fatty acid-enriched diets modify LDL oxidation and decrease human coronary smooth muscle cell DNA synthesis. *Arterioscler. Thromb. Vasc. Biol.* 17: 2088–2095.

MUIR S.R., COLLINS G.J., ROBINSON S., HUGHES S., BOVY A., RIC DE VOS C.H., VAN TUNEN A.J. and VERHOEYEN M.E. (2001) Overexpression of petunia chalcone isomerase in tomato results in fruit containing increased levels of flavonols. *Nature Biotechnol.* 19: 470–474.

PEÑARRUBIA L., KIM R., GIOVANNONI J., KIM S.-H. and FISCHER R.L. (1992) Production of the sweet protein monellin in transgenic plants. *Bio/ Technology* 10: 561–564.

RICHTER L.J., THANAVALA Y., ARNTZEN C.J. and MASON H.S. (2000) Production of hepatitis B surface antigen in transgenic plants for oral immunization. *Nature Biotechnol.* 18: 1167–1171.

RÖMER S., FRASER P.D., KIANO J.W., SHIPTON C.A., MISAWA N., SCHUCH W. and BRAMLEY P.M. (2000) Elevation of the provitamin A content of transgenic tomato fruits. *Nature Biotechnol.* 18: 666–669.

SCHWALL G.P., SAFFORD R., WESTCOTT R.J., JEFFCOAT R., TAYAL A., SHI Y-C., GIDLEY M.J. and JOBLING S.A. (2000) Production of very-high-amylose potato by inhibition of SBE A and B. *Nature Biotechnol.* 18: 551–554.

SÉVENIER R., HALL R.D., VAN DER MEER I.M., HAKKERT H.J.C., VAN TUNEN A.J. and KOOPS A.J. (1998) High level fructan accumulation in a transgenic sugar beet. *Nature Biotechnol.* 16: 843–846.

SHINTANI D. and DELLAPENNA D. (1998) Elevating the vitamin E content of plants through metabolic engineering. *Science* 282: 2098–2100.

STAUB J.M., GARCÍA B., GRAVES J., HAJDUKIEWICZ P.T.J., HUNTER P., NEHRA N., PARADKAR V., SCHLITTLER M., CARROLL J.A., SPATOLA L., WARD D., YE G. and RUSSELL D.A. (2000) High-yield production of a human therapeutic protein in tobacco chloroplasts. *Nature Biotechnol.* 18: 333–338.

TACKET C.O., MASON H.S., LOSONSKY G., CLEMENTS J.D., LEVINE M.M., and ARNTZEN C.J. (1998) Immunogenicity in humans of a recombinant bacterial antigen delivered in a transgenic potato. *Nat. Med.* 4: 607–609.

YE X., AL-BABILI S., KLÖTI A., ZHANG J., LUCCA P., BEYER P. and POTRYKUS I. (2000) Engineering the provitamin A (β-carotene) biosynthetic patwaya into (carotenoid-free) rice endosperm. *Science* 287: 303–305.

YU J. and LANGRIDGE W.H.R. (2001) A plant-based multicomponent vaccine protects mice from enteric diseases. *Nature Biotechnol.* 19: 548–552.

YUAN L. and KNAUF V.C. (1997) Modification of plant components. *Current Opin. Biotechnol.* 8: 227–233.

ZEITLIN L., OLMSTED S.A., MOENCH T.R., CO M.S., MARTINELL B.J., PARADKAR V.M., RUSSELL D.R., QUEEN C., CONE R.A. and WHALEY K.J. (1998) A humanized monoclonal antibody produced in transgenic plants for immunoprotection of the vagina against genital herpes. *Nature Biotechnol.* 16: 1361–1364.

7

Nutritional enhancement of plant foods

D. G Lindsay, CEBAS-CSIC, Murcia

7.1 Introduction

The question of why it is necessary to improve the nutritional value of plant foods is one that at first hand might seem difficult to justify. What evidence is there that this is a problem? In the developed world there are no overt signs of malnutrition even amongst strict vegans. The reasons for this are that many processed plant foods are fortified with essential nutrients. Fortification is utilised to replace nutrients lost in the heat processing of foods and through oxidation. Few vegetarians are dependent on a single plant source to provide their basic nutritional needs. In addition, vegetarians frequently consume vitamins as supplements and the growth in this industry has been rapid. The fact that people are resorting to the consumption of vitamins as supplements is a reflection of their belief that more of a good thing will result in an improvement in their health. This is a very dubious argument. Nonetheless, it is important to recognise that the recommended intakes of nutrients, that have been determined by expert groups of nutritionists, are based on the evidence that a specific intake level for a nutrient is required to ensure healthy growth and development. They do not reflect the growing body of evidence that suggests different, and often higher, intakes of these same nutrients are required to optimise health and lead to an active life through the prevention of chronic degenerative diseases associated with ageing.[1,2,3] The critical issue is to determine what intakes are required to optimise health rather than to compromise it.

7.2 The nutritional importance of plants

Plants are the staple food for the vast majority of the world's population. It is known that many staple plant foods are deficient in essential nutrients and, consequently, malnutrition is widespread. It has been estimated that over 100 million children worldwide are vitamin A deficient and improving the vitamin A content of their food could prevent as many as two million deaths annually in young children.[4] This is apart from the deficiencies in iodine intake, resulting in goitre, and in iron-deficient anaemia which are estimated to affect millions in the developing world. There is also an important need to improve the amino acid content of legume proteins that are deficient in essential sulphur amino acids. Nutritional deficiencies can lead to a reduction in immune responsiveness, rather than a specific attributable disorder, making it difficult clearly to establish how many people are suffering from malnutrition.[5]

In the developed world all public health authorities are urging consumers to consume more plant-based foods as part of a healthy diet. There is a significant body of evidence to suggest that the traditional Mediterranean diet, rich in plant foods, reduces the risk of many age-related diseases. Epidemiological studies show a strong and consistent inverse relationship between fruit and vegetable intake and the risk of cardiovascular diseases and some cancers.[2] An explosion of interest in trying to define what are the factors in fruit and vegetables which might be responsible for these observations has not yet led to a clear set of explanations although many theories abound.

Plants contain 17 mineral nutrients, 13 vitamins and numerous phytochemicals that have been shown to have potentially beneficial effects on health especially against the initiation or progression of degenerative diseases. Almost all human nutrients can be obtained from plant foods, the exceptions are vitamins B_{12} and D. However, the adequacy of a plant diet in delivering a health benefit from a specific component will depend on the amount ingested, and its bioavailability. Many beneficial plant compounds that are associated with the plant cell wall are not easily bioavailable. Any way in which overall levels can be increased will help overcome this difficulty.

7.3 Strategies for nutritional enhancement

There is no single approach to the improvement of the nutritional quality of plant foods since this is affected by a wide variety of factors. Amongst these are:

- the application of traditional breeding methods to select for varieties with an increased level of the bioactive compound
- a reduction in the content of antinutritional factors
- the use of genetic manipulation to introduce new traits in plants
- improvements in handling, storage and food processing technologies.

Each of these approaches has a role to play but genetic manipulation provides a mechanism for the improvement of nutrional quality that overcomes the problem of the absence of a specific biochemical pathway in a staple crop

7.3.1 Application of 'traditional' breeding methods

Plant varieties have not been selected to date on the basis of nutritional qualities but there are wide natural variations that can be found in the gene pool of crop plants. Examples of where significant variations in the nutrient content of genotypes have been documented include a:

- 2-fold variation in calcium concentration in beans[6]
- 4-fold variation in β-carotene concentrations in broccoli[7]
- 4-fold variation in folates in beetroot[8]
- 2–3 fold variation in iron and zinc levels in maize[9]

In the case of the pro-vitamin A carotenoids, plants provide highly variable amounts depending on their colour. Varieties of sweet potato may contain levels varying from 0.13 mg to 11.3 mg g^{-1} dry weight of β-carotene.[10] Similar variations in levels can be found in carrots and cassava. In the case of the tomato, genes have been identified that are associated with high and low lycopene content. Incorporation of genes that increase lycopene content and or elimination of genes that decrease the lycopene content, can be achieved by pedigree selection and backcross programmes. Such techniques have produced hybrids with a three- or four-fold content of lycopene in tomato fruits.[11]

7.3.2 Reduction in antinutritional factors

The interest in reducing antinutritional factors in plants has been predominantly focused around improving the nutritional value of feedstuffs. Phytates are present in many plant seeds and limit phosphorous uptake as well as other elements. The potential for introducing a phytase gene into feedstuffs has been explored.[12] However, there are other strategies that seem to be of greater overall value in human nutrition. Thioredoxin is thought to be an activator of the germination process in seeds.[13] It is able to activate proteins to degradation by proteolysis and results in improved digestibility.[14] It also has the potential advantage of being able to reduce allergenicity, presumably because of its capacity to break disulphide bonds by the action of the reduced thiol groups in the molecule and ensure the tertiary structure of the protein is accessible to degradation by proteases.[14] The insertion of the wheat thioredoxin gene into barley has produced a transgenic plant where thioredoxin accounts for 7% of the total protein content in the barley and is a good source of sulphur amino acids.[15]

7.3.3 The application of genetic manipulation

Genetic engineering is being applied to enhance levels of functional compounds in food crops. Indeed for some purposes it will be the only approach feasible especially where there are widespread deficiency diseases and the population is dependent on staple crops which are not sources of the nutrient required. There are many examples where technology has been applied with success although there are no products which have yet reached the marketing stage where nutritional benefits have been the main focus.

Potential strategies for the enhancement of specific metabolites could target on:

1. over expression of enzymes that control the final steps in the biosynthesis of a metabolite
2. over expression of rate-limiting enzymes
3. silencing of genes whose expression causes the metabolite to be degraded
4. increased expression of genes that are not subject to metabolic feedback control
5. increasing the number of plastids in a plant
6. increasing metabolic flux into the pathway of interest
7. expression in storage organs using site-specific promoters.

The strategy that has had the greatest success at present is the first one, especially in conjunction with the last strategy. In practice if a substantial increase in the concentration of a metabolite is required, the use of specific promoters directing the synthesis to a particular organelle normally used for storage purposes, or where the plant normally synthesises the metabolite, is essential. Failure to use these could cause toxicity in the plant by interfering with the production or function of other essential metabolites. However, this strategy presupposes the metabolite of interest is the final one in a particular pathway.

Few strategies have yet been applied where multiple gene insertions are necessary to produce the metabolite, although these are progressing rapidly, and none where plastid numbers have been increased. However, the accumulation of sequence data of both chromosomal DNA and expressed sequence tags of plants and other species is providing rapid advances in knowledge of the genetic make-up and functions of several plants and it is expected that these other possibilities will soon be feasible.

7.4 The priorities for nutritional enhancement

7.4.1 For the developed world

Although it is known that the distribution and processing of food can lead to a significant loss in nutritional quality, there are few instances where present evidence suggests there is a need to change current practices. There is very little evidence for nutritional deficiencies. In those cases where public health authorities have thought there is a potential problem, food supplementation with nutrients is a commonly adopted policy. The use of nutritional supplements is widespread. Whilst the focus of current interest is on the need to consider nutrients and other phytochemicals as protective against the development of disease in later life, the levels of intake that may be necessary to optimise protection are far from resolved at the present time.

The only plant-derived food product on the market where nutritional health benefits are claimed (as opposed to implied) is the enrichment of margarines

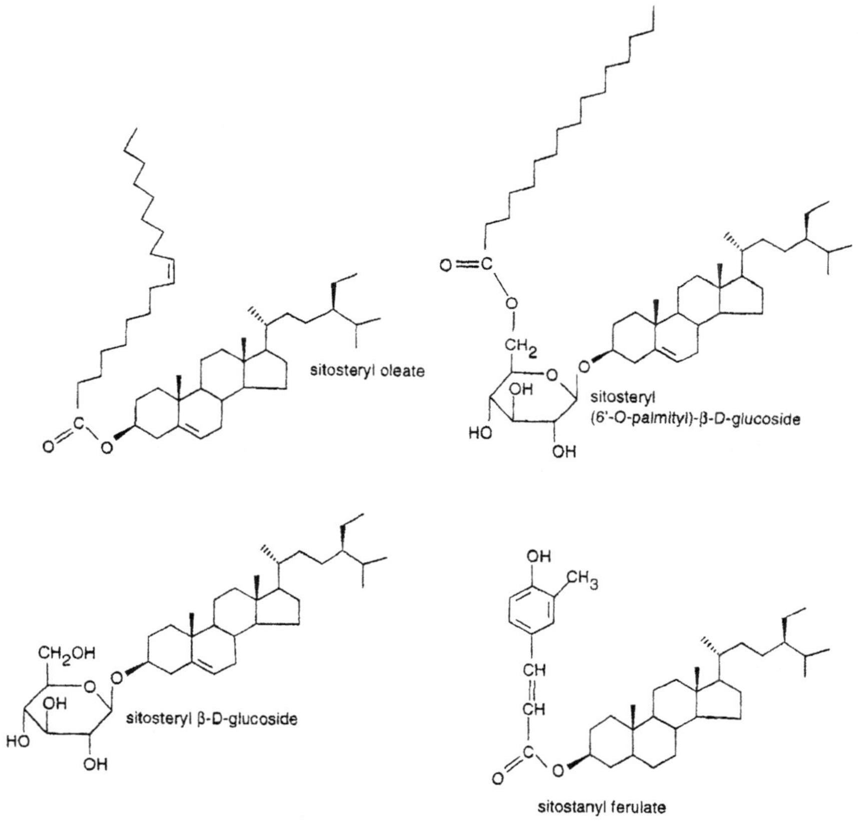

Fig. 7.1 Structure of plant sterol and stanol esters.

with plant sterol and stanol esters for the reduction in plasma cholesterol levels (Fig. 7.1). These products do not require the development of specifically bred plants since it is possible to extract stanols and sterols from existing plants (albeit in the case of the stanols from the bark of a tree) for use in their manufacture.

Experiments with plant stanol esters were shown to lower serum cholesterol consistently by about 10–15% and LDL-cholesterol by about 20% in patients with high serum cholesterol levels as well as in normal individuals.[16, 17] Similar effects have been seen with plant sterol esters but at least 1g/day of plant sterols need to be consumed.[18] Consequently they require extraction and addition to foods.

Plant sterols can be in the free form or predominantly esterified with long chain fatty acids or with phenolic acids as in rice-bran oil (ferulate) and shea butter (cinnamates). Sterol esters are better absorbed than the free sterols and most sterol esters are hydrolysed to the free sterols in the intestine.

As campesterol esters are better absorbed than sitosterol esters, serum levels of campesterol could rise to those levels that are found in the very few people who suffer toxic symptoms from phytosterolemia. Thus there may a benefit in increasing the sitosterol to campesterol ratio in plants.

The ideal situation would be for sufficient sterols to be present in our diets to ensure that plasma cholesterol levels are kept reasonably low without the need to buy a specific functional food, and that they would be in a fat soluble form for effective uptake. The evidence favours in increasing order of preference the use of:

1. plant sterol esters with low campesterol contents
2. sterol esters from tall oil (derived from pine wood) which have a higher stanol content than edible oils
3. plant stanol esters.

A vegetable oil rich in plant stanols, especially in sitostanol esterified with polyunsaturated fatty acids, would also have the benefit of being less susceptible to oxidation at frying temperatures than the sterols. The potential health benefits of this class of bioactive compounds are unlikely to be met by the use of classical plant breeding methods but genetic engineering could make these targets feasible.

7.4.2 For the developing world

The world-wide deficiency of vitamin A is being tackled both through conventional plant breeding and by genetic manipulation. However, the use of conventional plant breeding to deliver adequate intakes is dependent on availability of carotenoid-rich staple foods. Often these are available for very restricted times of the year in some societies. In those countries where rice is a dietary staple the problem is particularly severe and the deficiency is likely to be corrected only by the introduction of rice that has been genetically manipulated to produce β-carotene. However, yellow rice is produced and this may give rise to problems of acceptability to consumers used to white rice.

Manipulation of the carotenoid pathway in rice

The nature of the challenges faced in manipulating plant secondary metabolites is well illustrated through the attempts that have been made to produce carotenoids in rice plants. A simplified version of the pathways leading to the synthesis of the carotenoids principally found in food plants is shown in Fig. 7.2.

Immature rice endosperm is capable of synthesising the early pathway intermediate geranylgeranyl diphosphate (GGDP). Four plant genes corresponding to the enzymes phytoene synthase (psy) (1), phytoene desaturase (2), zeta carotene desaturase (3) and lycopene cyclase (crt) (4) are required. Enzyme (1) was obtained from the daffodil (Narcissus pseudonarcissus), (2) from a bacterium Erwinia uredovora – which is capable of achieving steps (2) and (3) from the single enzyme, and (4) from the daffodil.

The genes need to be expressed in a tissue-specific manner through the insertion of specific promoters. This has been achieved in rice through the use of

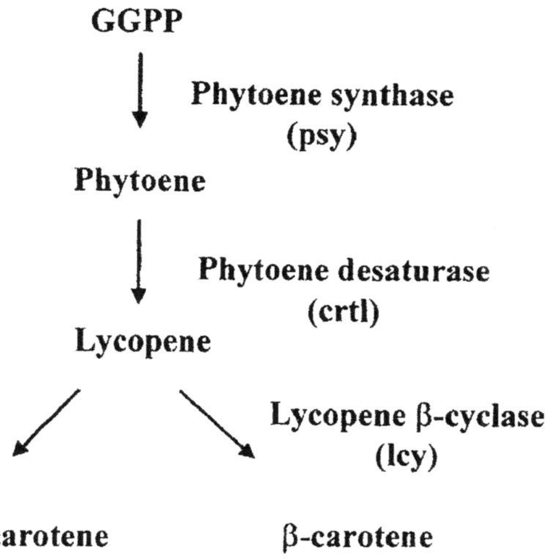

Fig. 7.2 The carotenoid biosynthetic pathway (simplified).

the daffodil psy gene.[19] In rice the daffodil psy cDNA insertion is under the control of an endosperm-specific promoter. The choice of promoter will very much affect the timing and tissue-specific expression of a gene.

Surprisingly, seeds that expressed psy and crt did not accumulate lycopene. Instead they contained β-carotene and other xanthophylls. Thus it would seem that the enzymes required to make these metabolites are either normally expressed in rice endosperm or are induced if lycopene is formed. The maximum level of carotenoids in the endosperm of plants that were heterozygous for the transgenes was 1.6 mg kg^{-1} which is likely to help to meet the nutritional needs of people consuming rice as a staple. Interestingly, good progress is being made in adding a gene coding for ferritin – the iron storage protein found in mammals and plants – to rice.[20] It is likely that this would also help improve the iron deficiency also seen in these communities if it is shown to be bioavailable.

The controversy over the use of advanced technologies for producing sustainable food in the developing world has been addressed by the developers of modified rice. They have in effect waived all intellectual property rights for exploitation of the technologies in the developing world, and are actively involved in assisting the International Rice Research Institute to breed stable and agronomically successful lines for use in vitamin A-deficient areas.

7.5 Relationship of structure to nutritional quality (bioavailability)

The overall content of a given nutrient in a food is not always a useful indicator of its nutritional value as not all of the nutrient present is absorbed. Nutritionists must concern themselves with understanding the proportion of an available nutrient that is digested, absorbed, and ultimately utilised. In the case of nutrients or phytochemicals, whose beneficial effects are directed towards inhibiting degenerative diseases, it is important to know whether or not the nutrient is reaching the particular target organ and in a form which is active. Otherwise the claims for the health benefits of that chemical would not be justified, especially as it is difficult to demonstrate benefits from long-term human studies.

Diet plays an important role in the uptake of specific nutrients and phytochemicals. Those that are lipophilic are absorbed much more readily from a lipid-rich diet. Frying tomatoes in oil dramatically improves the uptake of lycopene compared with the consumption of fresh tomatoes.[21] Raw carrots, which have high levels of pro-vitamin A carotenoids, are poorer sources of β-carotene than gently cooked carrot.[22] The bioavailability of certain trace elements is increased on cooking or processing; for example, the bioavailability of iron is increased in canned spinach.[23]

The chemical form of the phytochemical present in food is very important in determining uptake through the gastro-intestinal tract. Quercitin-β-glucoside is more easily absorbed than the aglycone quercitin. Isorhamnetin-β-glucoside, which is chemically similar to quercitin, differing only by a single methoxyl group, is more readily absorbed. Flavonoid rutinosides (rhamnosyl 1–6 glucosides) are less easily absorbed.[24] Thus, while some phenols might be better antioxidants than others when tested in *in vitro* systems, this is of little significance in terms of health relevance. What matters is whether the compounds are easily absorbed, are not quickly degraded in tissues, and are able to reach the target sites. Flavonoids that are not absorbed undergo extensive degradation by gut microorganisms, and may play only a limited role in preventing oxidative damage in the colon.

7.6 Nutritional enhancement versus food fortification

The importance of enhancing the levels of a natural protective constituent in plant foods is well illustrated in the case of the folates. There is a good chance that folate status even in affluent countries is not optimal.[25, 26] The most important sources of folates in the diet are liver, products derived from yeast, eggs, green vegetables, legumes and certain fruits. Plant foods (vegetables, fruits and potatoes) are by far the single largest contributor to the overall folate intake of adults.[27, 28, 29] Some 40% of the total folate intake is from fruit and vegetable consumption in these countries even when the average consumption is not very high.

Folates have the effect of reducing the levels of plasma homocysteine which is a sensitive biomarker of folate status. A variety of studies have suggested that increased plasma homocysteine levels are a risk factor for cardiovascular disease and stroke.[30] Human studies have shown that if individuals consume a supplement of 100 μg/day of folic acid their plasma homocysteine is reduced to a level of about 7.0 μM/l. Increasing the intake of folic acid beyond that level has no further effect. However, the bulk of the population have homocysteine levels in excess of 7.0 μM/l.

Folic acid is not the natural form of folate that is found in plants where natural folate consists of ten different polyglutamate complexes. Folic acid is however the form of folate that is used in the fortification of food as it is more stable. It is also found to be more bioavailable. Natural folates show only 50% or less of the bioavailability of folic acid.[31,32]

There is good evidence that to achieve the ideal level of plasma homocysteine dietary levels of folate (as opposed to folic acid) would have to increase from the current average of 200 μg/day to 600 μg/day.[33] This increased intake is also likely to have an important impact on the reduction of Neural Tube Defects (NTDs). Women with a low folate status (about 150 μg/l red cell folate) have a 0.7% risk of NTDs in their offspring, whereas supplementation with folic acid at doses of between 100–200 μg/day, resulted in red blood cell folate levels that have been associated with an optimal reduction in NTD incidence. Since average intakes of natural folates are about 100 μg/day from the diet it would require at least 500 μg of natural folate to be consumed (preferably 600 μg/day) to ensure that the incidence of NTD in the population was kept to a minimum.

The fact that supplemental folic acid can achieve these same effects whilst being more stable and bioavailable would imply that there was little purpose in supplementing natural levels of folate. This ignores the intrinsic difference between the cellular metabolism of synthetic pro-vitamin folic acid compared to the natural folates. The mucosa converts all of the natural forms of folate into 5-methylenetetrahydrofolate monoglutamate. This reaction also occurs when folic acid is consumed but the difference is that for folic acid the process can be saturated at around 300 μg. Intakes in excess of this cause un-metabolised folic acid to enter the circulation.[34] The control of how much natural folate is taken up and retained by cells is regulated by the enzyme methionine synthase which acts on 5-methyltetrahydrofolate to conjugate it into a polyglutamate which is then retained in the cell. Conversely folic acid does not pass through the methionine synthase pathway and can be conjugated directly, retained and metabolised.

The ability of folic acid to bypass an important regulatory step is that excess cellular levels cause DNA biosynthesis in vitamin B_{12}-deficient cells in cases of pernicious anaemia via the DNA cycle. This causes a haematological response with the risk that the anaemic state is masked and the associated neuropathy is not avoided. Natural folates on the other hand will be poorly metabolised by vitamin B_{12}-deficient cells enabling the anaemia to be detected at an earlier stage. Another concern that has been raised against increasing population levels of folate is that the increased capacity to cause DNA biosynthesis could promote tumour growth. This would be expected to be more of a problem with folic acid than natural folates because of folic acid's less controlled uptake into cells.

7.7 Constraints on innovation

The potential fully to exploit GM technologies is severely limited by constraints on the use of the technology itself, as well as in satisfying the legislation that exists on the pre-market approval of foods that have been produced by the technology, or are in some way novel. These constraints are so severe in Europe that in very few cases will any producer see a return on their investment if nutritional improvement is their goal. This market is also affected by the widespread addition of specific nutrients as additives to certain processed foods. Enhancement of any component considered to be beneficial is likely to be of market value only if positive claims can be made. Whenever possible 'conventional' plant breeding will be used.

7.7.1 Genetic manipulation

There has been a *de facto* European Union moratorium on the approval of GMO products since October 1998. Eighteen products have already been approved under the general EU Directive (90/220/CEE) whilst 14 are pending approval. Five Member States have temporarily banned already approved GM products, which is permitted under the Directive. Two new EU labelling regulations have been drafted but have not been implemented because of a lack of testing methodologies, certifying labels and inspection procedures. The ultimate intention is to ensure that products can be labelled GM free to enable consumers to make an informed choice. It has been argued that products labelled as containing products derived from GM will convey negative messages to consumers. This is likely to be so in the absence of benefits that are clearly seen by consumers. This will occur if plants are used as factories for the production of vaccines and pharmaceutical products. However, it is unclear at present where the benefits will lie in the nutritional field other than for the developing world.

The lack of public confidence in the European food safety system is already causing harm to markets in the US and in developing nations where the technology is already embraced. This is likely to lead to major problems in international trade unless it is resolved.

7.7.2 Safety

No scientific development in food can ignore the very strict regulatory controls that exist before any new or 'novel' product or process can be applied in its production. Food plants produced by 'conventional' plant breeding techniques in general are not subject to any regulatory controls. In some countries voluntary codes of practice have been developed within the plant breeding sector when it was discovered that varieties of potatoes, with good agronomic characteristics, were found to contain high levels of toxic glycoalkaloids.[35]

At the present time, genetically–modified (GM) foods are regulated applying the concept of 'substantial equivalence'.[36] This concept is applied as the basis from which to determine the extent of the requirements for food safety assessment. If a genetically modified food can be characterised as substantially equivalent, it can be assumed to pose no new health risks over its conventional counterpart and can be marketed without the need to undertake extensive toxicological and nutritional studies to determine its safety-in-use.

The principle of substantial equivalence was adopted into the EU Regulation on Novel Foods and Novel Food Ingredients.[36] The Regulation excludes from its controls foods and food ingredients obtained through traditional propagating or breeding practices and which have a history of safe use. GM plants are considered as 'novel' under the terms of the Regulation. However, the detailed safety evaluation provisions of the Regulation do not apply to foods produced by genetic manipulation 'if on the basis of the scientific evidence available they are substantially equivalent to existing foods with regard to their composition, nutritional value, metabolism, intended use, and the level of undesirable substances present'. The Regulation regards food as 'novel' if the characteristics of the food differ from the conventional food regarding the accepted limits of natural variation of such characteristics. It is clear that most nutritionally enhanced plants would be caught under the definition of a 'novel' food.

The principle of substantial equivalence is vague and difficult to define in many cases. Consequently the whole issue of regulation of GM foods is under intensive debate. Meanwhile the EU has applied a *de facto* moratorium on GM plant introductions. The US attitude to regulation has so far been to regard safety as an issue that relates to the characteristic of the food and not to the process(es) that lead to it. Novel food products, of which products produced by GM are included in the definition, are not subject to any specific approval on safety grounds if the constituents of the food are the same, or substantially similar, to substances currently found in other foods.

It is clear that it is never going to be possible to argue that a GM plant is safe any more than to be able to argue that a plant produced by conventional plant breeding is safe. The very concept can be addressed only in the context of a history of safe use as a human food. Clearly, the overwhelming evidence supports the view that health benefits arise as a consequence of the regular consumption of a variety of fruits and vegetables, few if any of which have any close compositional relationship to the wild types from which they were bred. Similarly their production, storage and distribution has depended on the use of a wide range of chemical fertilisers and pesticides. These chemicals are extensively tested for safety before approval is given for their marketing and use but this has not removed the widely held view amongst consumers that 'organic products' are better for your health. There is no evidence to support this view and any adverse health effects that there might be as a consequence of the use of pesticides appear to be outweighed by the beneficial effects from the consumption of fruit and vegetables. What determines 'safety' is the overall effect of consumption over a period not the effects of a specific chemical that might be present.

The issue of 'safety' in the context of the ability to market foods which are 'novel' is emotionally charged and without a solid scientific base. Consequently it is unlikely that any industry would want to take on these issues unless they had a product with a potentially large market.

7.8 Future trends

It is clear that the developing world will adopt whatever approach is technically feasible for them to meet the food and nutritional needs of their populations. Genetically modified crops will be used if there are clear benefits. In terms of resistance to disease and adaptation to harsh environments the technology has clear potential. Improvements in nutritional quality can be added to the list of benefits.

In Europe and other developed countries the impetus for improving the nutritional value of foods will occur only if there are clear health benefits in doing so. As there is growing evidence that nutritional needs will vary according to age and genetic susceptibility, it will be hard to convey a consistent message since intakes that benefit one sector of society might not benefit another. The priority is to demonstrate clearly what are the functional effects of nutrients, or beneficial phytochemicals, at the physiological level. The information is generally rudimentary. In those cases where the function is clearer the relationships between dose and effect are not known. When it comes to marketing foods that have been genetically manipulated the benefits will have to be very great indeed if current consumer resistance to their use is to be overcome.

7.9 Further information

Sources of further information about the potential for nutritional enhancement can be found in reviews by Willis, Lencki and Marangoni (1998),[38] Grusak and DellaPenna (1999),[39] Dixon *et al.*, (1996),[40] Yamauchi and Minamikawa (1998).[41] An overview of the subject is contained in Lindsay, D.G. (2000) 'Maximising the functional benefits of plant foods'. In: *Functional Foods.* Chapt. 8. Ed. by Williams, C. M. and Gibson, G.R.. Woodhead Publishing Ltd. Cambridge, England. pp. 183–208. ISBN 1 85573 503 2.

An account of the issues to be addressed in tackling the nutritional enhancement of plants (including increasing levels of bioactive secondary metabolites) can be found in published reviews commissioned under the EU's concerted action project 'The Nutritional Enhancement of Plant Foods in European Trade (NEODIET)', 2000, *J Sci Food & Agric.* **80** (7): 793–1137. Some of the issues raised in relation to the use of plant biotechnology in food and feed production are discussed in papers contained in a special edition of *Science* (Plant Biotechnology: Food & Feed, 1999, *Science* 285: 367–389.)

7.10 References

1. BURING JE and SMITH CV, (1997), Antioxidant vitamins and cardiovascular disease *Nutrn Rev,* **55(1)**, S53–S58.
2. STEINMETZ KA and POTTER JD, Vegetables, fruit and cancer prevention: A review (1996). *J Am Diet Assoc*, **96**, 1027–1039.
3. AIFR WORLD CANCER RESEARCH FUND, (1997), *Food, Nutrition and the Prevention of Cancer – A Global Perspective*, Washington, DC.
4. WORLD HEALTH ORGANISATION, (1995), Global prevalence of vitamin A deficiencies' Micronutrient deficiency information systems, Working Paper No 2, Geneva, WHO.
5. CALDER PC and JACKSON AA, (2000), Undernutrition, infection and immune function, *Nutrn Res Rev*, **13**, 3–29.
6. QUINTANA JM, HARRISON HC, NIENHUIS J, PALTA JP and GRUSAK MA, (1996), Variation in calcium concentration among sixty S1 families and four cultivars of snap bean (Phaseolus vulgaris L.), *J Amer Soc Hort Sci*, **121**, 789–793.
7. SCHONHOF I and KRUMBEIN A, (1996), Gehalt an wertgebenden Inhaltstoffen verschiedener Brokkolitypen (Brassica oleracea var italica Plenck). *Gartenbauwissenschaft*, **61**, 281–288.
8. WANG M and GOLDMAN IL, (1996), Phenotypic variation in free folic acid content among F1 hybrids and open-pollenated cultivars of red beets, *J Am Soc Hort Sci*, **121**, 1040–1042.
9. IFPRI, (1999), International Food Policy Research Institute, *Agricultural strategies for micronutrients*. http.//www.cgair.org/ifpri/themes/grp06.htm
10. SOLOMONS NW and BULUX J, (1997), Identification of local carotene-rich foods to combat vitamin A malnutrition, *Eur J Clin Nutrn*, **51**, (Suppl.), S39-S45.
11. AMITOM, (1999) *Role and control of antioxidants in the tomato processing industry.* EU FAIR Project (FAIR CT97–3233). http//:www.tomato.org/ Antioxidantnetwork.html
12. PEN J, VERWOERD TC, VAN PARIDON PA, BEUDEKER RF, VAN DEN ELZEN PJM, GEERSE K, VAN DER KILS JD, VERSTEEG HAJ, VAN OOYEN AJJ and HOEKEMA A, (1993). Phytase-containing transgenic seeds as a novel feed additive for improved phosphorous utilisation, *Biotechnology*, **11**, 811–814.
13. LOZANO RM, WONG JH, YEE BC, PETERS A, KOBREHEL K, BUCHANAN BB, (1996) New evidence for a role for thioredoxin h in seedling development and germination, *Planta*, **200 (1)**, 100–106.
14. BUCHANAN BB ADAMIDI C, LOZANO RM, YEE BC, MOMMA M, KOBREHEL K, ERMEL R and FRICK OL, (1997), Thioredoxin-linked mitigation of wheat allergies, *Proc Natln Acad Sci USA* **94**, 5372–5377.
15. BUCHANAN BB. Unpublished data.
16. GYLLING H, RADHAKRISHNAN R and MIETTINEN TA, (1997), Reduction of serum cholesterol in postmenopausal women with previous myocardial infarction and cholesterol malabsorption induced by dietary sitostanol

ester margarine: women and dietary sitostanol, *Circulation*, **96**, 4226–4231.

17. GYLLING H and MIETTINEN TA, (1999), Phytosterols, analytical and nutritional aspects, In, *Functional foods – a new challenge for the food chemist.* Eds Lasztity R, Pfannhauser W, Simon-Sarkadi L and Tômôskôzi S, Publishing Company of TUB, Budapest. *Proceedings of the Euro Food Chem X.* Vol. 1, 109.
18. HENDRICKS HFJ, WESTRATE JA, VAN VLIET T and MEIJER GW, (1999), Spreads enriched with three different levels of vegetable oil sterols and the degree of cholesterol lowering in normocholesterolemic subjects, *Eur J Clin Nutrn*, **53**, 319–327.
19. BURKHARDT PK, BEYER P, WUNN J, KLOTI A, ARMSTRONG GA, SCHLEDZ M, VONLINTIG J and POTRYKUS I, (1997), Transgenic rice (Oryza sativa) endosperm expressing daffodil (Narcissus pseudonarcissus) phytoene synthase, accumulates phytoene a key intermediate in pro-vitamin A synthesis, *Plant J*, **11**, 1071–1078.
20. SHINTANI D and DELLAPENNA D, (1998), 'Elevating the vitamin E content of plants through metabolic engineering' *Science*, **282**, 2098–2100.
21. GÄRTNER C, STAHL W and SIES H, (1997), Increased lycopene bioavailability from tomato paste as compared to fresh tomatoes, *Am J Clin Nutrn*, **66**, 116–122.
22. ROCK CL, LOVALVO JL, EMENHISER C, RUFFIN MT, FLATT SW and SCHWARTZ SJ, (1998), Bioavailability of beta-carotene is lower in raw than in processed carrots and spinach in women, *J Nutr*, **128(5),** 913–916.
23. LEE K and CLYDESDALE FM, (1981) Effect of thermal processing on endogenous and added iron in canned spinach, *J Food Sci*, **46**, 1064–1067.
24. AZIZ AA, EDWARDS CA, LEAN M EJ and CROZIER A, (1998), Absorbtion and excretion of conjugated flavonols, including quercitin-4′-O-β-glucoside and isorhamnetin-4′-O-β-glucoside by human volunteers after the consumption of onions. *Free Rad Res*, **29**, 257–269.
25. WARD M, MCNULTY H, MCPARTLIN J, STRAIN JJ, WEIR DG and SCOTT JM, (1997), Plasma homocysteine, a risk factor for cardiovascular disease can be effectively reduced by physiological amounts of folic acid, *Quart J Med*, **90**, 519–524.
26. SELHUB J, JACQUES P, WILSON PWF, RUSH D and ROSENBERG IH, (1993), Vitamin status and intake as primary determinants of homocysteinemia in an elderly population, *J.A.M.A.*, **270**, 2693–2698.
27. BAUSCH-GOLDBOHM RA, HULSHOF, KFAM, BRANTS HAM, VAN DEN BERG H and BOUMAN, M, (1995), TNO Report V95.84 Zeist, The Netherlands, TNO Nutrition and Food Research Institute.
28. GREGORY J, FORSTER K, TYLER H and WISEMAN, M, (1990). *The dietary and nutritional survey of British adults*, Office of population and surveys, London HMSO.
29. LEE P and CUNNINGHAM K, (1990), *Irish Nutrition Survey.* Dublin: The Irish Nutrition and Dietetic Institute.

30. BOUSHEY CJ, BERESFORD SAA, OMENN GS and MOTULSKY AG, (1995), A quantitative assessment of plasma homocysteine as a risk factor for vascular disease: probable benefits of increasing folic acid intake, *J.A.M.A.*, **274**, 1049–1057
31. GREGORY JF, (1995), The bioavailability of folate, In: *Folate in Health and Disease* Ed. Bailey LB, New York, Marcel Dekker, pp. 195–235
32. CUSKELLY GJ MCNULTY H and SCOTT JM, (1996), Effect of increasing dietary folate on red-cell folate: implication for prevention of neural tube defects, *Lancet*, **347**, 657–659.
33. SCOTT J REBEILLE F and FLETCHER J, (2000), Folic acid and folates: the feasibility for nutritional enhancement in plant foods, *J Sci Fd Agric*, **80**, 795–824.
34. KELLY P, MCPARTLIN J, GOGGINS M, WEIR DG and SCOTT JM, (1997), Unmetabolized folic acid in serum: acute studies in subjects consuming fortified food and supplements, *Amer J Clin Nutrn*, **65**, 1790–1795.
35. HELLENAS K-E, BRANZELL C, JOHNSSON H, SLANINA P, (1995), High levels of glycoalkaloids in Swedish potato variety Magnum-Bonum, *J Sci Fd Agric*, **68**, 249–255.
36. OECD. *Safety evaluation of foods derived by modern biotechnology.* OECD, Paris, 1993.
37. EU (1997). Regulation No 258/97 of the European Parliament and the Council concerning novel foods and novel food ingredients, *Off J European Communities*, **43**, 1–7.
38. WILLIS WM LENCKI RW and MARANGON AG, (1998), Lipid modification strategies in the production of nutritionally functional fats and oils, *Crit Rev in Food Sci & Nutrn*, **38**, 639–674.
39. GRUSAK MA and DELLAPENNA D, (1999), Improving the nutrient composition of plants to enhance human nutrition & health, *Annu Rev Plant Physiol, Plant Mol Biol*, **50**, 133–161.
40. DIXON RA LAMB CJ MASOUD S SEWALT VJH and PAIVA NL, (1996), Metabolic engineering: prospects for crop improvement through the genetic manipulation of phenylpropanoid biosynthesis and defense responses – a review, *Gene*, **179**, 61–71.
41. YAMAUCHI D and MINAMIKAWA T (1998). Improvement of the nutritional quality of legume seed storage proteins by molecular breeding, *J. Plant Res*, **111**, 1–6.

Part II

Case studies

8

Tomato

A.L.T. Powell and A.B. Bennett, University of California, Davis

8.1. Introduction

Tomatoes originated in the Andean region of South America under extremely variable climatic conditions. Wild relatives of tomato grow from sea level to sub-alpine elevations with some ecotypes adapted to flooded conditions and others to extreme drought. Domestication of tomato led to its cultivation as a crop on all continents and traits have been selected to promote abundant production of fruit. Selective breeding from the narrow genetic base of domesticated tomato as well as the introduction of exotic germplasm from the numerous wild relatives of tomato have developed tomato plants producing high-quality fruit for fresh consumption as well as for processed, prepared and stored products valued at approximately US$5 billion annually. Advances in agricultural biotechnology recently have provided the opportunity to expand the genetic resources available for tomato improvement. The goals of tomato genetic engineering have been to protect the tomato crop from environmental and biological assaults, and to improve the quality of tomato fruit in order to deliver greater value in processed tomato products or more healthful and attractive fresh fruit.

Tomato fruit are a significant source of nutrition for substantial portions of the world's human population because this vegetable crop is widely cultivated and consumed extensively as both a fresh vegetable and concentrated processed products. Tomatoes are rich sources of vitamins, especially ascorbic acid and β-carotene, and antioxidants such as lycopene. A single small tomato is sufficient to supply about a quarter of the vitamins A and C recommended for humans to consume daily (Hamner and Maynard, 1942; Beecher, 1998). Most of the nutritional components in tomato fruit are stabilized by the acid pH of the fruit tissue and many of the human nutrients are conserved during the relatively short and mild processing used in the preparation of most tomato food products. Tomatoes are grown in industrial quantities in many temperate locations, but the stability of the concentrated processed product has made it possible to transport tomato products widely and to prolong the storage of tomato products.

Tomato was one of the first plants to be transformed by *Agrobacterium tumefaciens* and regenerated into fertile, productive plants (Fillatti *et al.*, 1987). The success of early work to obtain transgenic plants allowed for the first commercial release of a transgenic food product, the Flavr Savr tomato, with extended shelf life of the ripe fruit. The transformation of a large number of tomato varieties has been reported, suggesting that essentially any variety is amenable to genetic transformation. For example, fresh market varieties (Moneymaker, Better Boy), greenhouse varieties (Ailsa Craig), small-fruited fresh varieties (VFNT Cherry) and processing varieties (UC82b) as well as wild tomato relatives (*L. chilense* (Agharbaoui *et al.*, 1995), *L. peruvanium* (Rudas *et al.*, 1997) and *L. hirsutum* (Smith *et al.*, 1996)) have all been transformed in academic and commercial research laboratories. Most of the successful transformation protocols for tomato utilize *Agrobacteria* to deliver transgenes to the hypocotyl sections of newly germinated seedlings, but biolistic approaches also have been utilized. The success of the floral dip methods used in *Arabidopsis* has not been reported for tomato. Antibiotic resistance of transformed tissues is frequently the preferred method of selection of transgenic tissues, because of its historical success. However, public concerns about the contents of genetically modified food products will undoubtedly lead to the utilization of new selection methods, including positive selection for growth on selective media (Haldrup *et al.*, 1998).

Because fruit are the economically significant crop from tomato plants, many transgenic modifications have targeted the fruit ripening processes to develop products that better withstand harvest, handling, transportation and storage practices utilized in commercial distribution. To reduce processing costs and effectively increase processing yield, transgenic fruit have been developed with increased solids content. To provide novel value-added products, tomato fruit have also been engineered to produce increased components of nutritional value and to produce pharmaceutical compounds. To enhance production efficiency and yield, tomato plants have been engineered for resistance to herbicides, extreme temperatures and pathogens by the transgenic expression of foreign genes not accessible by classical breeding methods. A comprehensive listing of transgenic tomato modifications that have successfully altered aspects of plant growth, morphology and cultivation are summarized in Table 8.1.

8.2 Modifications targeting fruit

Ripening is a genetically regulated developmental process that triggers metabolic changes enhancing the flavor, texture and aroma of fruit but that simultaneously initiates fruit senescence and deterioration. A major goal of tomato genetic engineering has been to manipulate the ripening process in order

Table 8.1 Transgenic tomato modifications

Trait	Gene	Regulation	Expression	Reference
Fruit ripening				
Ethylene reduction	Bacterial ACC deaminase	Constitutive	Expression	(Klee *et al.*, 1991)
Ethylene reduction	Phage SAMase	Fruit specific	Expression	(Good *et al.*, 1994)
Ethylene reduction	Tomato ACC synthase	Constitutive	Antisense suppression	(Oeller *et al.*, 1991)
Ethylene reduction	Tomato ACC synthase	Constitutive	Sense suppression	(Lee *et al.*, 1997)
Ethylene reduction	Tomato ACC oxidase	Constitutive	Antisense suppression	(Picton *et al.*, 1995)
Fruit softening	Tomato fruit PG	Constitutive	Antisense suppression	(Sheehy *et al.*, 1988)
Fruit softening	Tomato fruit PG	Constitutive	Sense suppression	(Smith *et al.*, 1990)
Fruit softening	Tomato fruit PME	Constitutive	Antisense suppression	(Tieman *et al.*, 1992)
Fruit softening	Tomato fruit PG and PE	Constitutive	Antisense suppression	(Seymour *et al.*, 1993)
Fruit softening	Tomato fruit expansin	Constitutive	Sense suppression	(Brummell *et al.*, 1999)
Fruit abscission	Tomato fruit Cel1 and Cel2	Constitutive	Antisense suppression	(Brummell *et al.*, 1999; Lashbrook *et al.*, 1998)
Fruit composition				
Sucrose accumulation, solids content	Tomato fruit invertase	Constitutive	Antisense suppression	(Klann *et al.*, 1996)
Solids content	Bacterial *ipt*	Constitutive	Expression	(Martineau *et al.*, 1995)
Starch accumulation	*Arabidopsis sucrose synthase*	Constitutive	Expression	(Dai *et al.*, 1999)
Fatty acid and flavor content	Yeast $\Delta 9$ desaturase	Constitutive	Expression	(Wang *et al.*, 1996)
Color	Tomato phytoene synthase	Constitutive	Antisense suppression	(Fraser *et al.*, 1995)
Parthenocarpic	Bacterial tryptophan monoxygenase	Constitutive	Expression	(Martineau *et al.*, 1995)

Table 8.1 Continued

Trait	Gene	Regulation	Expression	Reference
Seeds				
Increased dormancy	Tomato NCED	Constitutive	Expression	(Thompson *et al.*, 2000)
Decreased dormancy	*Arabidopsis abi-1*	Constitutive	Expression	(Carrera and Prat, 1998)
Pathogen and pest resistance and tolerance				
TMV	TMV N	Constitutive	Expression	(Whitham *et al.*, 1996)
CMV	Cucumber mosaic virus CP	Constitutive	Expression	(Fuchs *et al.*, 1996)
TSWV	Tomato spotted wilt virus N	Constitutive	Expression	(Kim *et al.*, 1994; Ultzen *et al.*, 1995)
PhMV	Physalis mottle tymovirus CP	Constitutive	Expression	(Sree Vidya *et al.*, 2000)
Pseudomonas syringae pv tomato	Tomato *Pto*	Constitutive	Expression	(Chandra *et al.*, 1996)
Xanthomonas campestris pv. Vesicatoria	Pepper *Bs2*	Constitutive	Expression	(Tai *et al.*, 1999)
Cladosporium fulvum	Tomato *Cf9*	Constitutive	Expression	(Hammond-Kosack *et al.*, 1998)
Verticillium dahliae	Tomato chitinase	Constitutive	Expression	(Tabaeizadeh *et al.*, 1999)
Fusarium oxysporum f.sp. lycopersici	Tobacco chitinase and β1,3- glucanase	Constitutive	Expression	(Jongedijk *et al.*, 1995)
Trichoderma hamatum	Rubber tree hevein	Constitutive	Expression	(Lee and Raikhel, 1995)
Xanthomonas campestris pv. vesicatoria	Tomato *LeETR4*	Constitutive	Expression	(Ciardi *et al.*, 2000)
Sclerotinia sclerotiorum	*Collybia velutipes* oxalate decarboxylase	Constitutive	Expression	(Kesarwani *et al.*, 2000)
Botrytis cinerea	Pear fruit PGIP	Constitutive	Expression	(Powell *et al.*, 2000)
Phytophthora infestans	Grape reserveratrol	Constitutive	Expression	(Thomzik *et al.*, 1997)
Manduca Sexta	Tomato prosystemin	Insect induced	Expression	(McGurl *et al.*, 1994)
Insect resistance	*Bt* toxins	Constitutive	Expression	(Rhim, 1998; Van Der Salm *et al.*, 1994)

Insect resistance	*Bt cry1Ac*	Constitutive	Expression	(Mandaokar *et al.*, 2000)
Nematode resistance	Rice cystatin *Oc-I*	Constitutive	Expression	(Atkinson *et al.*, 1996)
Root knot nematode, aphid, viral resistance	Tomato *Mi*	Root-specific	Expression	(Brommonschenkel *et al.*, 2000; Williamson, 1998)
Plant defense responses				
Extracellular responses	*Agrobacterium ipt*	Constitutive	Expression	(Storti *et al.*, 1994)
Environmental stresses				
Salt stress	Yeast *HAL2*	Constitutive	Expression	(Arrillaga *et al.*, 1998)
Drought	*Arabidopsis ABI-1*	Constitutive	Expression	(Carrera and Prat, 1998)
Chilling and oxidative stress sensitivity	Tomato catalase	Constitutive	Antisense suppression	(Kerdnaimongkol and Woodson, 1999)
Heavy metal tolerance	Bacterial ACC deaminase	Root-specific or stress induced	Expression	(Grichko *et al.*, 2000)
Herbicide resistance				
Thiazopyr resistance	Rabbit liver esterase	Constitutive	Expression	(Feng *et al.*, 1997)
Quinclorac resistance	Tomato ACC synthase	Constitutive	Antisense suppression	(Grossman and Schmuelling, 1995)
Fenthion (insecticide) sensitivity	Tomato *prf*	Constitutive	Antisense suppression	(Martin *et al.*, 1994)
Metabolic modifications				
Increased sucrose unloading unloading	Sucrose phosphate synthase	Root-specific or fruit-specific	Expression	(Micallef *et al.*, 1995; Nguyen-Quoc *et al.*, 1999)
Foliage coloration				
Increased anthcyanin	*Antirrhinum del*	Constitutive	Expression	(Mooney *et al.*, 1995)
Floral patterns				
Indeterminate flowering	Tomato *agamous*	Constitutive	Expression	(Pnueli *et al.*, 1994)
Precocious termination	Tomato *agamous*	Constitutive	Antisense suppression	(Pnueli *et al.*, 1994)

to delay fruit senescence and deterioration while retaining the beneficial metabolic attributes of the ripening process. Because of the importance of this developmental process, several approaches have been used to manipulate tomato fruit ripening (Grierson and Fray, 1994). The most general approach has been to modify the expression of regulators of suites of genes that control fruit development or ripening and over-ripening. Ethylene gas is produced by ripening tomato fruit and is the natural hormonal regulator of the ripening process. The amount of ethylene produced has been specifically modified in several lines of transgenic plants as a means to regulate the ripening process (Theologis *et al.*, 1992; 1993). A more specific approach to the regulation of aspects of ripening has been to modify the expression of genes whose products encode enzymes or proteins that are instrumental in a targeted component of the process of ripening. In this regard, modification of the expression of various cell wall hydrolases has been attempted several times to suppress cell wall disassembly and fruit softening while allowing other aspects of ripening to proceed normally (Gray *et al.*, 1992; 1994; Tucker *et al.*, 1999).

8.2.1 Regulation of ripening and senescence

Tomato is a climacteric fruit and ripening is naturally regulated by ethylene produced by the fruit at the onset of ripening. Two transgenic approaches have been used to reduce endogenous ethylene production in ripening fruit in order to delay the onset and rate of fruit ripening. The pathway of ethylene biosynthesis is now well known (Kende, 1993; Barry *et al.*, 2000) and the final two steps in the pathway, conversion of S-adenosyl methionine (SAM) to 1-aminocyclopropane-1-carboxylic acid (ACC) and its oxidization to ethylene have been targeted for modification in transgenic plants. One approach has been to metabolize either SAM or ACC to an inactive product and the second approach has been to specifically suppress the expression of the two ethylene biosynthetic enzymes required to catalyze the final steps in the pathway. Both approaches resulted in fruit with significantly reduced ethylene content and greatly delayed ripening.

To metabolize the ethylene precursors, SAM and ACC, microbial genes have been expressed in tomato. Metabolic inactivation of SAM was carried out by transgenic expression of a bacteriophage T3 S-adenosylmethionine hydrolase (SAMase) gene in tomato (Good *et al.*, 1994). The SAMase enzyme converts S-adenosyl methionine, to methylthioadenosine and homoserine rather than ACC, and serves as a means to divert SAM from the ethylene biosynthetic pathway. The transgenic tomato plants were engineered to express SAMase only in ripening fruit after the breaker stage by linking the T3 SAMase coding sequence to the *E8* promoter (Deikman *et al.*, 1992), a tomato fruit-specific and ripening-regulated promoter. Consequently, the effects of reduced ethylene were observed only in ripening fruit and these fruit exhibited delayed ripening and enhanced firmness (Good *et al.*, 1994). An alternative strategy was to express a microbial gene encoding ACC deaminase, an enzyme that converts ACC to α-ketoglutarate in transgenic tomato. The ACC deaminase enzyme was identified in a strain of *Pseudomonas* that utilized ACC as a nitrogen source, and the corresponding gene was introduced into tomato as a means to divert ACC from the ethylene biosynthetic pathway. The transgenic fruit exhibited reduced ethylene production and the fruit ripened at a slower rate and remained firm (Klee *et al.*, 1991). No effects were observed in the vegetative tissues of these transgenic plants and, surprisingly, fruit ripening was delayed only in fruit detached from the plant but not in fruit allowed to ripen on the plant (Klee, 1993). Interestingly, transgenic tomato lines expressing a bacterial ACC deaminase with the root-specific *RolB* promoter or the pathogen inducible tobacco promoter (*PRB-1b*) were able to grow in the presence of heavy metals such as Cd, Co, Cu, Ni, Pb and Zn (Grichko *et al.*, 2000).

Transgenic tomato plants with reduced ethylene levels also were developed by suppressed expression of endogenous genes encoding the ultimate and penultimate steps in the biosynthetic pathway. Constitutive expression of a tomato fruit-specific ACC synthase gene in its sense orientation resulted in two phenotypically different groups of plants, those over-expressing the ACC synthase gene, and those in which ACC synthase gene expression was reduced by co-suppression of the endogenous gene (Lee *et al.*, 1997). The transgenic lines with reduced ACC synthase gene expression exhibited reduced ethylene production and reduced ripening.

Another transgenic approach to reduce ethylene in ripening fruit relied on expression of antisense genes to suppress the expression of the endogenous ACC oxidase or ACC synthase genes. This approach was first explored as a means to deduce the function of a tomato gene (*pTOM13*) by antisense suppression of its expression (Grierson *et al.*, 1990). Fruit from these plants produced considerably reduced amounts of ethylene and ripened more slowly. Subsequently, *pTOM13* was shown to encode the ethylene-forming enzyme of tomato, ACC oxidase (Hamilton *et al.*, 1991; Spanu *et al.*, 1991). This initial finding was later expanded to demonstrate that constitutive expression of the tomato antisense ACC oxidase gene caused delayed fruit ripening (Picton *et al.*, 1995) and delayed leaf senescence by 10 to 14 days (John *et al.*, 1995). A cDNA encoding apple fruit ACC oxidase has also been expressed in its antisense orientation in tomato, resulting in greater than 95% reduction in the endogenous tomato ACC oxidase mRNA accumulation, a reduction in the ethylene production, and delayed ripening (Bolitho *et al.*, 1997). Theologis and colleagues also demonstrated that constitutive expression of an antisense ACC synthase gene significantly reduced endogenous ethylene production and delayed fruit ripening (Oeller *et al.*, 1991). Using these transgenic plants, the requirement for ethylene to initiate and maintain the progression of ripening and senescence was demonstrated.

Altering the plants' perception of ethylene is another approach that has been taken to modify the role of ethylene in transgenic tomato plants. The primary ethylene receptor is encoded by the family of *ETR* genes, which in tomato consists of five members (*LeETR1, LeETR2, Nr, LeETR4* and *LeETR5*) that collectively control ethylene sensitivity throughout the plant. These genes encode histidine kinase sensors homologous to two-component signaling proteins found in bacteria and in *Arabidopsis*. *LeETR4*, *Nr* and *LeETR5* are expressed in ripening fruit and their expression is stimulated by ethylene, suggesting that the *ETR* genes are important in fruit for both the competence to respond to ethylene and the specific tissue responses to ethylene. Expression of a mutant form of *Nr* in tomato renders the plants insensitive to ethylene (Wilkinson *et al.*, 1995). Fruit ripening was also delayed in transgenic tomato plants with antisense suppressed expression of *Nr* (Tieman *et al.*, 2000), although in this case suppression of *Nr* is compensated, at least in part, by an increase in *LeEXP4* expression. Paradoxically, suppression of *LeETR4* expression in transgenic tomato results in increased ethylene sensitivity and premature flower senescence and more rapid fruit ripening. Ethylene regulation of *ETR* gene expression is complex, but future transgenic lines of tomato with specifically designed expression or suppression of multiple members of the *ETR* family may provide the basis to produce plants and tissues that exhibit precise responses to both endogenous and exogenous ethylene.

8.2.2 Fruit texture during ripening

Endo-Polygalacturonases

Fruit ripening is accompanied by disassembly of several cell wall polymers, including pectin and hemicellulose, which are primarily responsible for ripening-associated changes in fruit texture. Extensive studies on ripening-associated pectin disassembly and the expression of the endo-polygalacturonase (PG) gene family, suggested that tomato fruit texture could be modified by transgenic modification of PG gene expression. The expression of both antisense and sense constructs of the tomato fruit PG catalytic subunit (PG2) gene resulted in greater than 95% reduction in PG activity (Sheehy *et al.*, 1988; Smith *et al.*, 1998; 1990a). Fruit with reduced expression of PG were analyzed for alterations in the expression of other cell wall hydrolases and none were detected (Smith *et al.*, 1990b). Fruit with reduced PG activity provided the basis for testing the significance of this PG during softening and ripening as well as the basis for the commercial introduction of fresh and processed tomato fruit whose texture was modified by this genetic modification. Analysis of cell wall polymers of these fruit demonstrated that diminished PG expression contributed to reduced depolymerization of the chelator-solubilized pectins and increased viscosity of processed tomato products but did not reduce fruit softening (Taylor *et al.*, 1991; Carington *et al.*, 1993; Fenwick *et al.*, 1996; Brummell and Labavitch, 1997; Porretta and Poli, 1997; Porretta *et al.*, 1998). The effect of antisense suppression of a single fruit PG on fruit softening may be partially offset by expression of other tomato PG genes in ripening fruit (Sitrit and Bennett, 1998). Interestingly, transgenic plants with reduced expression of fruit PG did not exhibit changes in leaf abscission, suggesting that PGs involved in abscission are distinct from those that participate in fruit ripening (Taylor *et al.*, 1991). The tomato fruit PG gene has also been inactivated by transposition and stabilization of a maize transposon, *DS*, within the PG gene (Cooley and Yoder, 1998).

Suppression of the non-catalytic β subunit of the PG1 isozyme complex in transgenic tomato plants by expression of an antisense gene construct also reduced pectin metabolism during fruit ripening (Watson *et al.*, 1994). Specifically, the reduced expression of the PG1 β subunit, a regulatory subunit, reduced cell wall pectin solubilization and depolymerization, suggesting that the dynamics of pectin associations and structure in the cell wall are determined by several factors, perhaps some acting cooperatively (Watson *et al.*, 1994).

Several other fruit characteristrics have been measured in tomato fruit with suppressed PG gene expression. Transgenic tomato fruit were evaluated for sensory characteristics and their color and flavor outperformed a similar variety that was heterozygous for the *rin* (ripening inhibited) locus, a variety that had been bred for long shelf life (Sozzi Quiroga and Fraschina, 1997). The tomatine content of transgenic fruit was unaffected by antisense suppression of PG (Furui *et al.*, 1998). Furthermore, PG antisense fruit generally had improved integrity and were less susceptible to cracking and pathogen attack specifically at the over-ripe stage (Kramer *et al.*, 1992; Hadfield and Bennett, 1998). However, the susceptibility of PG suppressed transgenic tomato fruit to *Colletotrichum gloeosporioides* was not measurably different than in wild-type fruit (Cooper *et al.*, 1998).

Pectin methylesterases

Because methylation of pectins affects their structural properties in the cell wall as well as their susceptibility to pectinases, expression of pectin methylesterases (PMEs) has been altered to modify pectin metabolism of tomato fruit (Tieman *et al.*, 1992; Tieman and Handa, 1994; Tieman *et al.*, 1995). Suppression of the expression of a single PME in tomato by the transgenic introduction of a truncated sense tomato PME gene resulted in significantly higher molecular weight pectins isolated from the fruit cell walls (Thakur *et al.*, 1996a). Processed tomato products made from the PME suppressed transgenic lines also exhibited increased serum viscosity and reduced serum separation (Thakur *et al.*, 1996b; Errington *et al.*, 1998). Analysis of the PME mRNA in the transgenic plants suggested that the reduction in the endogenous PME mRNA resulted from interference by the transgenic mRNA with post-transcriptional processing of the endogenous PME mRNA (Mishra and Handa, 1998). Transgenic expression of the sequence encoding 71 amino acids of tomato fruit PG linked to the tomato PME sequence under control of the 35S CaMV promoter resulted in suppression of both PG and PE simultaneously (Seymour *et al.*, 1993).

Expansins

Expansins are cell wall proteins that have been proposed to participate in disruption of hydrogen bonding between hemicellulose and cellulose polymers at the surface of the cellulose microfibril. These polymeric associations are particularly important for cell expansion and other developmental events in which cell wall disassembly occurs, such as tissue softening during fruit ripening (Rose *et al.*, 1997). Several expansin genes are expressed during tomato fruit development and ripening (Brummell *et al.*, 1999b). The over-expression and suppression of one expansin gene, *LeExp1*, was examined in tomato plants expressing a sense full-length or truncated *LeExp1* gene (Brummell *et al.*, 1999c). Transgenic fruit from plants over-expressing *LeExp1* were significantly less firm at the mature green and breaker stages and analysis of the cell wall material demonstrated that precocious depolymerization of the hemicellulose structure of the wall correlated with constitutive over-expression of *LeExp1*. However, expression in tomato of a cucumber hypocotyl expansin, *CsExp1*, did not result in phenotypic alterations of transgenic tomato fruit, suggesting that divergent expansin proteins may have distinct functions or substrates *in vivo* (Rochange and McQueen-Mason, 2000). Suppression of *LeExp1* in transgenic tomato resulted in increased fruit firmness, especially at the early (e.g. Breaker) stage of ripening (Brummell *et al.*, 1999c). Surprisingly, in the *LeExp1* suppressed fruit, polyuronide depolymerization but not hemicellulose depolymerization was reduced in the later stages of ripening in these fruit.

β1, 4 endo-glucanases

Because hemicellulose, and xyloglucans specifically, are disassembled in ripening fruit substantial research has focused on endo-β-1, 4-glucanases as a class of enzymes because they have the capacity to cleave the β-1,4-glucan linkages of xyloglucan. At least two endo-β-1,4-glucanases are expressed in ripening tomato fruit, and one of them, *Cel1*, is also expressed in fruit abscission zones (Lashbrook *et al.*, 1994). Analysis of expression patterns demonstrated that the mRNA corresponding to a second endo-β-1,4-glucanase, *Cel2*, is more abundant in ripening fruit. However, the expression of both *Cel1* and *Cel2* during ripening suggested that the two endo-β-1,4-glucanases could act synergistically on their substrates in the softening cell wall. Transgenic plants in which *Cel1* expression was suppressed by an antisense gene construct produced fruit that softened normally, and abscission was partially reduced (Lashbrook *et al.*, 1998). Transgenic plants engineered for suppression of the *Cel2* gene also exhibited no changes in fruit ripening or softening but also were altered in abscission zones, requiring greater force for the abscission zone breakage (Brummell *et al.*, 1999a).

Galactosidases

A recent report indicated that antisense suppression of galactosidase gene expression in ripening tomato fruit reduced softening by approximately 40% (Gross and Smith, USDA report). At least four of seven tomato fruit β-galactosidases are expressed during ripening (Smith and Gross, 2000), and the release of galactosyl residues is the most dynamic cell wall change during ripening. As breakdown of galactose-containing polymers in the fruit cell wall is abundant during ripening, reduction of galactosidases, perhaps in concert with other cell wall hydrolases and expansin proteins may provide the basis to regulate the softening process of ripening tomato fruit.

8.2.3 Fruit composition

Soluble solids content of tomato fruit is a major determinant of fruit quality, particularly for processing tomatoes, and the soluble solids are comprised predominantly of soluble sugars. Thus, approaches to altering the composition of ripe fruit using transgenic strategies have focused on carbohydrate composition. One of the first attempts in tomato was to suppress the expression of acid invertase in ripe fruit in order to increase sucrose levels. Other transgenic tomato lines have been designed to alter the ratio of the monosaccharides, glucose and fructose, in the fruit.

Acid invertase and sucrose synthase

Although tomatoes transport sucrose in the phloem, tomato fruit typically have very low levels of sucrose and approximately equal ratios of the hexose sugars, glucose and fructose. Interestingly, some wild relatives of tomato accumulate primarily sucrose in their fruit and these fruit have very high levels of total soluble sugars. Introgression of the locus controlling sucrose accumulation (*sucr*) from the wild relative, *L. chmielewskii*, resulted in smaller fruit with increased soluble sugar levels (Chetelat *et al.*, 1995). Because the *sucr* locus from *L. chmielewskii* was determined to encode an inactive allele of acid invertase, it was reasoned that the same trait could be produced by transgenic suppression of invertase expression. Constitutive expression of an antisense gene encoding tomato soluble acid invertase resulted in tomato fruit with an increased concentration of sucrose and decreased concentrations of the hexoses, fructose and glucose (Klann *et al.*, 1996). Fruit from the sucrose accumulating transgenic plants were approximately 30% smaller, presumably due to the osmotic effects of sucrose accumulation as compared to the hexose-accumulating non-transgenic control plants. Many of the characteristics of the transgenic plants with reduced invertase expression were similar to sucrose accumulating lines of tomato that had been derived by introgression *sucr* locus from the *L. chmielewskii*. However, transgenic plants engineered using the *E8* promoter (Deikman *et al.*, 1992) to suppress invertase gene expression only in ripening fruit, remained hexose accumulators. This suggested that expression of the invertase gene regulates sucrose to hexose conversion early in fruit development, before the developmental timing of expression specified by the *E8* promoter.

Sucrose synthase has also been a target for genetic modification with the goal of enhancing sink 'strength' by increasing the capacity to metabolize imported sucrose. However, the fruit-specific antisense suppression of sucrose synthase did not produce any changes in the accumulation of starch or sugars in the fruit tissues even though the transient increase in sucrose synthase expression normally observed in early in fruit development was suppressed (Chengappa *et al.*, 1999). Similar transgenic plants with suppressed expression of a fruit-specific sucrose synthase also exhibited no change in hexose or starch accumulation (D'Aoust *et al.*, 1999b). However, fruit set (e.g. number of fruit) on these plants was diminished and the sucrose unloading capacity of young fruit was significantly reduced.

Additional transgenic strategies have been employed to specifically alter source-sink relations in tomato by ectopic expression of sucrose phosphate synthase. When the *Zea maize* sucrose phosphate synthase gene regulated by the rubisco promoter was expressed in transgenic tomato foliar tissue sucrose partitioning was increased and this reduced limitations of photosynthesis (Micallef *et al.*, 1995). Sucrose unloading in fruit also was increased by the over-expression of sucrose phosphate synthase (Nguyen-Quoc *et al.*, 1999).

Hexokinase and fructokinases

Over-expression of hexokinase in transgenic tomato demonstrated the regulatory role of this enzyme in photosynthetic tissues, particularly affecting senescence. However, in fruit from plants over-expressing an *Arabidopsis* hexokinase gene the quantity of starch in young fruit and of hexose in ripe fruit was reduced (Dai *et al.*, 1999).

Because fructose is almost twice as sweet as glucose, modification of fructokinase expression in ripening tomato fruit has been attempted in an effort to increase the ratio of fructose and glucose and, thus, enhance fruit 'sweetness'. Two genes encoding fructokinase are expressed in tomato fruit, *Frk1* and *Frk2* (Kanayama *et al.*, 1998). *Frk2* expression is correlated with periods of starch accumulation, and *Frk1* is expressed ubiquitously, although less abundantly. The potentially complex regulation of both the expression and activity of each of these fructokinase isoforms may confound attempts to increase fructose concentrations in fruit from transgenic plants with antisense genes for either or both fructokinases.

8.2.4 Fruit flavors and aromas

Fruit flavor is a complex trait determined by the mixture and balance of sugars, acids and a large number of aldehyde, ketone and alcohol volatiles (Buttery, 1993; Baldwin *et al.*, 2000). Lipoxygenases are key enzymes in fatty acid metabolism, producing the hexanal aldehyde and alcohols that contribute substantially to the volatile tomato aromas and flavors. At least three lipoxygenase genes (*LOX*) are expressed during tomato fruit ripening (Ferrie *et al.*, 1994; Kausch and Handa, 1997; Griffiths *et al.*, 1999a). The expression of the *LOX* genes in ripening tomato fruit is regulated by both ethylene and developmental factors. In order to modify the C6 aldehyde and alcohol composition of tomatoes, antisense transgenic tomatoes were developed using a conserved region of lipoxygenases to potentially suppress expression of all LOX activities (Griffiths *et al.*, 1999b). In ripe fruit of these transgenic plants, the expression of *tomloxA* and *tomloxB*, but not *tomloxC*, was substantially reduced. However, no changes in the flavor and aroma constituents resulted, suggesting that regulation of their biosynthesis may be complex or that *tomloxC* expression was sufficient for these pathways to proceed unimpeded.

Another approach to altering the flavor composition of tomato fruit was engineered by the expression of a yeast $\Delta 9$ desaturase gene (Wang *et al.*, 1996), resulting in changes in the fatty acid content of fruit. Concentrations of palmitoleic acid, 9,12-hexadienoic acid, and linoleic acid increased and concentrations of palmitic acid and stearic acid were reduced. Concentrations of flavor compounds derived from these fatty acids were changed and resistance to powdery mildew was enhanced (Wang *et al.*, 1998). Another approach to changing the composition of flavor aldehydes in tomato fruit was taken by altering the expression of alcohol dehydrogenase 2 (*adhII*) (Speirs *et al.*, 1998; Prestage *et al.*, 1999). The amount of Z-3-hexanol decreased and the amount of 3-methylbutanal increased in plants with reduced expression of *adhII*. Fruit from plants with increased expression of *adhII* had a more intense 'ripe fruit' flavor and had increased amounts of hexanol and Z-3-hexanol.

8.2.5 Fruit color and vitamin A

Tomato fruit color is primarily determined by the concentration of the red carotenoid, lycopene, and its precursors. Because these carotenoid pigments are the source of provitamin A as well as of the other major nutritional antioxidants in tomato, their levels have significant consequences for both the appearance and nutritional value of fresh and processed tomatoes. Modification of the carotenoid composition of tomato fruit has been achieved by transgenic modification of the activity of a key enzyme in the carotenoid biosynthetic pathway, phytoene synthase (Fraser *et al.*, 1995) that converts phytoene to lycopene. Suppression of phytoene synthase gene (*Psy1*) expression significantly reduced carotene and xanthophyll as well as abscisic acid (ABA) in ripe fruit compared to control fruit. The resulting yellow fruit were very similar to naturally occurring phytoene synthase mutant fruit (Fray and Grierson, 1993). Plants constitutively over-expressing *Psy1* are dwarf, apparently because gibberellin biosynthesis is reduced, and have less chlorophyll in their leaves (Fray *et al.*, 1995). A second phytoene synthase gene (*Psy2*) also is expressed in ripening fruit but apparently is not involved in carotenoid synthesis in this tissue (Fraser *et al.*, 1999).

Recently, in an effort to increase the vitamin A content of tomato, the expression of a bacterial phytoene desaturase in tomato has been reported (Romer *et al.*, 2000). The *Erwinia uredovora crtI* gene was expressed constitutively in tomato plastids. The total carotene composition of these plants was unaltered although the proportion of β-carotene increased moderately. The lycopene produced in the transgenic plants was cyclized by the induction of two endogenous lycopene cyclases to increase the β-carotene content about two-fold in these orange fruit. Unlike retinol or vitamin A, β-carotene or provitamin A is non-toxic and can be stored by after human ingestion. Thus, this transgenic modification provides provides an effective nutritional supplement of about 40% of the recommended daily consumption of vitamin A in a single fruit.

8.2.6 Other fruit characteristics

Fruit development is promoted by auxin produced by seeds developing within the ovary walls. Parthenocarpic fruit or fruit which develop in the absence of ovule fertilization, have the obvious advantage of being seedless and in tomato also have been reported to have elevated soluble solids content. Parthenocarpic tomato fruit were generated by the transgenic expression of a microbial gene (*iaaM*) that encodes a tryptophan monoxygenase used in synthesizing an auxin precursor (Ficcadenti *et al.*, 1999). The *iaaM* gene from *Pseudomonas savastanoi* was linked to the *DefH9* promoter from *Anthirrhinum majus* that specifies expression in placental and ovule tissues, resulting in auxin production targeted to the ovary tissues destined for fruit development. As in other plants, the additional IAA in the developing fruit promoted fruit formation in the absence of pollination or ovule fertilization. The composition, size and abundance of the fruit from these plants are indistinguishable from control fruit.

Similar strategies have been used to alter the levels of other plant hormones in developing tomato fruit. The fruit-specific expression of a bacterial *ipt* gene specifying elevated cytokinin biosynthesis in ovary tissues of tomato resulted in increased total and soluble solids (Martineau *et al.*, 1995). The fruit from these plants had islands of green pericarp tissue and excess cytokinin from fruit was exported to the leaves causing increased accumulation of genes, such as PR-1 and chitinase, induced by cytokinin (Martineau *et al.*, 1994).

Transgenic approaches to alter fruit size and shape have not been reported. However, the identification of the QTL locus, *fw2.2*, in the introgressed population between *L. esculentum* and *L. pimpinellifolium* (Frary *et al.*, 2000), may suggest candidate genes that could be introduced transgenically into tomato varieties which will influence tomato fruit size.

8.3 Modifications targeting seeds and germination

Modification of the seed germination process may be possible as gene expression patterns during germination have been recently reported (Bradford *et al.*, 2000). The expression of mannanase in the seed endosperm and a PG expressed at the time of radicle emergence were candidate marker genes for early germination (Still and Bradford, 1997; Still *et al.*, 1997; Sitrit *et al.*, 1999). However, the expression of both of these genes was most abundant in the later stages of germination and not early enough to serve as a marker of the initial stages of germination. The expression of a second mannanase (*LeMAN2*) gene in the emerging micropylar endosperm appears to be more directly related to the earlier stages of germination (Nonogaki *et al.*, 2000). The identification of expansins in germinating seeds has provided the clearest opportunity for marking the germination process in individual seeds (Chen and Bradford, 2000). *LeEXP4* is expressed in the micropylar region and *LeEXP8* is expressed in the radicle tip and their expression is precisely correlated with germination. These seed germination-specific promoter elements may provide the means to monitor germination of individual seeds or to enhance or accelerate seed germination as a means to enhance stand establishment. Seed specific expression of a GUS reporter gene in tomato also has been noted using portions of the maize *Sh1* promoter (D'Aoust *et al.*, 1999a).

Seed dormancy can be promoted by increased amounts of ABA. One of the outcomes of over-expressing a tomato 9-cis-epoxycarotenoid dioxygenase (NCED) in tomato was increased seed dormancy as well as increased stomatal conductance resulting from elevated ABA levels (Thompson *et al.*, 2000). Some plants from this same population exhibited co-suppression of the endogenous NCED gene and had a wilted phenotype.

8.4 Modifications targeting biotic and abiotic stress tolerance

8.4.1 Pathogen resistance

Viral resistance

Tomato plants engineered for increased resistance to viral pathogens have been developed by expression of specific components of viral genes in transgenic plants. For example, transgenic expression of the TMV coat protein (CP) in tomato increased resistance to viral infection without altering the nutritional and biochemical constituents of the fruit (Anan *et al.*, 1996). Expression of a truncated form of the replicase gene of the tomato yellow leaf curl geminivirus (TYLCV) in tomato provided resistance to this virus but not to another geminivirus, tomato leaf curl virus (ToLCV-Au) (Brunetti *et al.*, 1997). Expression of the cucumber mosaic virus CP in transgenic tomato provided durable resistance to CMV under field conditions (Fuchs *et al.*, 1996), and expression of CP from more than one subgroup of CMV in other transgenic tomato lines has provided broader resistance to strains of CMV even in epidemic conditions in the field (Xue *et al.*, 1994; Provvidenti and Gonsalves, 1995; Murphy *et al.*, 1998; Kaniewski *et al.*, 1999; Tomassoli *et al.*, 1999). Expression of CMV satellite RNA in tomato also increased CMV resistance (McGarvey *et al.*, 1994; Stommel *et al.*, 1998; Monti *et al.*, 1999). Resistance to the tomato spotted wilt virus (TSWV) has been achieved by expression of the nucleocapsid protein (N) in transgenic tomato lines (Kim *et al.*, 1994; Ultzen *et al.*, 1995). Partial resistance and delayed symptom development to Physalis mottle tymovirus (PhMV) has been obtained by expressing the coat protein of this virus in tomato (Sree Vidya *et al.*, 2000).

Endogenous plant genes that confer virus resistance also have been employed to engineer for virus resistance in tomato. For example, expression of the tobacco race-specific *N* gene in tomato also conferred resistance to the appropriate strain of TMV (Whitham *et al.*, 1996). This demonstrated that the tobacco *N* gene functions in tomato and can be used as a source of genetic resistance.

Bacterial resistance

Resistance to *Pseudomonas syringae pv tomato* is specified by the products of a host-specific plant resistance gene (*Pto*) and a bacterial avirulence gene (*avrPto*). The sufficiency of the *Pto* gene to confer resistance to the bacterial strains producing *avrPto* was demonstrated by the transgenic expression of *Pto* in tomato (Chandra *et al.*, 1996; Tang *et al.*, 1999). *Pto* encodes a kinase and autophosphorylation is required for the resistant hypersensitive response, including the *avrPto* specific oxidative burst (Sessa *et al.*, 2000). *Prf*, a gene related to *Pto* in the resistance cluster, when expressed in transgenic plants, confers sensitivity to the insecticide fenthion (Martin *et al.*, 1994).

Resistance to bacterial speck disease is observed in transgenic tomato plants in which the pepper *Bs2* resistance gene is expressed (Tai *et al.*, 1999). This disease is caused by *Xanthomonas campestris pv. vesicatoria* strains expressing the *avrBs2* gene.

Fungal resistance

Resistance to fungal pathogens has been improved in tomato by the transgenic expression of race-specific resistance genes, targeting initially particular pathogens, and by the over-expression of proteins that augment or activate defense responses to a broader group of fungal pathogens. The approach of expressing resistance factors that recognize specific microbial avirulence factors has demonstrated that these factors are sufficient for gene-for-gene resistance, and several examples have shown that the expression of these factors also can provide some resistance against other pathogens. The over-expression of recognition factors and genes that activate defense responses against pathogens has incrementally improved resistance to fungal pathogens.

Expression of the *Cf9* gene, the extracellular and membrane anchored leucine-rich repeat resistance factor against *Cladosporium fulvum* race 9 in tomato, demonstrated that this gene is sufficient to confer resistance in an otherwise susceptible variety of tomato (Hammond-Kosack *et al.*, 1998). Furthermore, the expression of *Cf9* has allowed details of the interaction between *C. fulvum* and resistant tomato varieties to be examined. Expression of the bacterial gene, *avr9*, in a line of tomato to be crossed with a *Cf9* tomato, demonstrated that expression of the *avr* gene in plant cells triggers the hypersensitive response (Honee *et al.*, 1995). A K^+ channel of the transgenic tobacco leaf guard cells is apparently involved in the recognition of *avr9* by the host and the interaction involves at least one phosphorylation step (Blatt *et al.*, 1999; Romeis *et al.*, 2000).

Insect and nematode resistance

Expression of *Bacillus thuringiensis* (*Bt*) toxins has been used in many plant species to control insect pests on transgenic plants (Van Der Salm *et al.*, 1994; Rhim, 1998). The expression in transgenic tomato of the synthetic Bt *cry1Ac* gene coding for an insecticidal crystal protein (ICP) provided a high level of resistance to the larval stage of the fruit borer, *Helicoverpa armigera* (Mandaokar *et al.*, 2000).

Systemin, a systemic signal molecule that triggers endogenous defense mechanisms in tomato, is produced by proteolysis of a precursor protein, prosystemin (Ryan and Pearce, 1998; Ryan, 2000). The transgenic expression of prosystemin in tomato yielded tomato plants with improved resistance to the tomato hornworm pest *Manduca sexta* (Orozco-Cardenas *et al.*, 1993). As a result of the constitutive presence of systemin, these plants produced two proteinase inhibitor proteins and a polyphenol oxidase that participate in endogenous tomato defense mechanisms. Mutants of the transgenic tomato plants constitutively expressing systemin have been used to identify a group of genes that are required for wound and system induced defense gene expression (Howe and Ryan, 1999). Thus, engineering systemin expression has provided additional understanding about the regulation and function of components of defense pathways elicited in tomato vegetative tissues by this mobile signal molecule (McGurl *et al.*, 1992; 1994; Constabel *et al.*, 1995).

The expression of cystatin, a cysteine protease inhibitor or papain inhibitor, is induced in plants over-expressing prosystemin and pest-induced expression of cystatin provides an inducible mechanism for insect resistance (Jacinto *et al.*, 1998). Expression of the rice cystatin Oc-I in tomato improved the resistance of roots to the nematode pathogens *Meloidogyne incognita* and *Globodera pallida*, reducing the size of the female nematodes and frequency with which they fed on the transgenic roots (Atkinson *et al.*, 1996). Expression of the PHI-*ipt* cytokinin biosynthesis gene in tobacco increases the toxicity of foliar tissues to *Manduca sexta* and *Myzus persicae*, but expression of the same gene in tomato did not reduce the feeding capacity of these pests (Smigocki *et al.*, 2000).

In tomato, a well-known source of nematode resistance (*Mi* gene) was originally identified in the wild tomato relative, *L. peruvianum*, and has since been introgressed into most commercial tomato cultivars. The recent cloning of the tomato *Mi* gene has provided an opportunity to improve resistance to root knot nematode species (*Meloidogyne* sp.) as well as aphid species (*Macrosiphum euphorbiae*) (Rossi *et al.*, 1998; Vos *et al.*, 1998; Williamson, 1998). Root-specific expression of the *Mi* gene has been possible using *A. rhizogenes* transformation and identification of the functional roles of some portions of the protein have been clarified by transgenic expression of modified forms of *Mi* (Hwang *et al.*, 2000). A homologue of *Mi*, *Sw-5*, provides resistance to topovirus species (Brommonschenkel *et al.*, 2000), suggesting that viral and nematode resistance may share portions of a defense signalling pathway. The transgenic expression of *Mi*, *Sw-5* or other related variants that include the domains responsible for the virus and nematode resistance, provides an unprecedented opportunity to engineer effective resistance in tomato against viral, nematode and insect pests simultaneously.

General defense

Transgenic expression of genes encoding proteins involved in plant defense against many pathogens or in the regulation of defense responses has been used to elucidate the significance of specific components of plant defenses or as a strategy to reduce susceptibility to groups of pathogens. Expression of the *Agrobacterium ipt* gene resulted in increased cytokinin synthesis in tomato and the constitutive activation of extracellular defense responses such as *PR-1* and acidic chitinase gene expression (Martineau *et al.*, 1994) that normally occur in response to elicitors, including *Fusarium oxysporum* cell wall components (Storti *et al.*, 1994; Bettini *et al.*, 1998). Cell cultures established from transgenic tomato lines expressing either cytokinin or auxin biosynthetic genes from *A. tumefaciens* and were assessed for the effects on *F. oxysporum* growth. Cell cultures from a transgenic susceptible tomato variety became more resistant when cytokinin was increased as a result of *ipt* expression, and a transgenic resistant variety expressing the auxin biosynthesis gene reduced *F. oxysporum* growth (Storti *et al.*, 1994). Expression of *A. tumefaciens ipt* in tomato also increased resistance to *Manduca sexta* and *Myzus persicae* (Smigocki *et al.*, 2000).

Over-expression of a *L. chilense* chitinase in *L. esculentum* improved foliar resistance to *Verticillium dahliae* (Tabaeizadeh *et al.*, 1999). Simultaneously expressing tobacco chitinase and β1,3- glucanase in tomato reduced symptoms of *Fusarium oxysporum f.sp. lycopersici* by about 50%, suggesting that these proteins can act synergistically to defend against a fungal pathogen (Jongedijk *et al.*, 1995).

Expression of the precursor of the chitin binding protein, hevein, in tomato retarded the growth of *Trichoderma hamatum*, but did not eliminate infections. The hevein prepeptide was not cleaved efficiently in the transgenic plants to form mature hevein (Lee and Raikhel, 1995).

Antisense suppression of an anionic peroxidase in tomato was attempted in order to define the role of this enzyme in plant defenses (Sherf *et al.*, 1993). Although peroxidase expression was reduced, no influence on suberin and cell wall phenolics was ascertained, but the plants exhibited some resistance to immature stages of the insects, *Helicoverpa zea* and *Manduca sexta* (Dowd *et al.*, 1998). Over-expression of the tobacco anionic peroxidase in tomato increased lignin composition, but did not provide increased resistance to pathogens (Lagrimini *et al.*, 1993). Over-expression of the tomato basic peroxidase, *tpx1*, resulted in increased lignin in the transgenic plants but no effect on pathogen susceptibility was reported although some changes in stress responses were observed (El Mansouri *et al.*, 1999).

As ethylene synthesis and perception are signaling components for the activation of many defense responses, alterations in ethylene and its recognition had significant effects on pathogen susceptibility. However, the outcomes of ethylene modifications vary depending on the tissues, stage of development, and pathogen. Many pathogens induce ethylene synthesis and the role of ethylene receptors in *Xanthomonas campestris* pv. *vesicatoria* infections of tomato has been explored by examining the *X. campestris* susceptibility of transgenic lines expressing the ethylene receptors, *NR* or *LeETR4*. Both lines of transgenic plants respond to the pathogen with reduced necrosis and reduced ethylene sensitivity because *NR*, like *LeETR4* is a negative regulator of ethylene sensitivity. In non-transgenic tomato, the induction of ethylene receptors by pathogens may reduce ethylene sensitivity and, hence, the expansion of necrosis (Ciardi *et al.*, 2000).

Inactivating products or eliminating functions that pathogens use in virulence have been used to engineer resistance in tomato. An improvement in resistance to *Sclerotinia sclerotiorum* has been accomplished by expressing an oxalate decarboxylase from *Collybia velutipes* in tomato (Kesarwani *et al.*, 2000). As *S. sclerotiorum* requires oxalate for infection, the removal of oxalate by this transgenically expressed enzyme represents a novel approach for limiting the growth of the pathogen on plant tissues. Transgenic expression in tomato of a plant inhibitor (PGIP) of a fungal cell wall hydrolase (PG) has proven to be a means of reducing the spread of *Botrytis cinerea* infections on leaves and fruit (Powell *et al.*, 2000). *B. cinerea* uses PG, along with several other cell wall hydrolases, to break down plant tissues during establishment of decomposing lesions. Transgenic synthesis of the phytoalexin reserveratrol in tomato enhanced resistance to *Phytophthora infestans* but not to *B. cinerea* or *Alternaria solani* (Thomzik *et al.*, 1997).

8.4.2 Salt, water and temperature stress

Tolerance of growth in high salt conditions is an attribute that has been amenable to transgenic manipulation in tomato. Expression of the yeast *HAL2* gene in tomato allows hypocotyl growth and root formation on high salt media (Arrillaga *et al.*, 1998). Plants expressing the yeast *HAL1* gene probably accommodate high salt conditions by increasing the intracellular K^+ concentration (Gisbert *et al.*, 2000). Overexpressing the *A. thaliana* vacuolar Na^+/H^+ antiport in tomato allows growth in up to 200 mH NaCL (Zhang and Blumwald, 2001).

Drought tolerance can be influenced by the ABA concentration in tomato. Engineering plants with an *Arabidopsis* wild type *ABI-1* or a mutant *abi-1* gene has demonstrated that transgenic modifications affecting ABA concentration modulate responses to drought. Plants expressing *abi-1* are wilty and have reduced seed dormancy. These plants are affected as well in their responses to wounding (Carrera and Prat, 1998).

Decreased tolerance to chilling and oxidative stress was conferred by the antisense suppression of catalase that resulted in an increase in the H_2O_2 concentration in the plants (Kerdnaimongkol and Woodson, 1999). Transgenic tomato plants expressing an *E. coli* glutathione reductase were equally sensitive to chilling as wild-type plants, suggesting that ascorbate peroxidase pathway in the chloroplast is not limiting for the chilling sensitivity of photosynthesis in tomato (Brueggemann *et al.*, 1999).

8.4.3 Herbicide tolerance

Resistance to the herbicide thiazopyr has been engineered by the transgenic expression of a rabbit liver esterase (Feng *et al.*, 1997). Seedlings deactivated the herbicide in proportion to the expression of the transgene. Transgenic tomato plants with suppressed expression of the ACC synthase gene do not display deleterious effects of the herbicide quinclorac (Grossman and Schmuelling, 1995). Treatment of wild-type tomato with ethylene inhibitors similarly alleviates the effects of this herbicide

8.5 Modifications targeting vegetative tissues and flowers

Increased pigmentation of tomato vegetative tissue has been achieved by the transgenic expression of the delila (*del*) gene from *Antirrhinum* (Mooney *et al.*,

1995). The anthocyanin pigment is regulated by the expression of the *del* gene and linking the maize AC transposable element has resulted in variegated plants, suggesting that the *del* gene acts cell autonomously. The results of expressing *del* in *Arabidopsis* and in tobacco differ from the results of expressing the gene in tomato.

The responses of tomato plants to light exposure have been modified by the over-expression of avena phytochrome A (*PhyA*). The responses of seedlings to 'end of the day' exposures to light occurred largely at low Pfr/P (3–61%) for the wild-type and at high Pfr/P (61–87%) for transgenic seedlings (Casal *et al.*, 1995). Tomato plants with reduced ACC oxidase expression because of the transgenic expression of *pTOM13* demonstrated reduced yellowing in response to long day length light exposure (Jensen and Veierskov, 1998).

Alterations in floral morphology and cell fate can be achieved by the over-expression or suppression of the tomato *AGAMOUS* homologue, *TAG1* (Pnueli *et al.*, 1994). Indeterminate flowering can be achieved by the over-expression of *TAG1* and precocious termination and differentiation by the antisense suppression of *TAG1* expression.

8.6 Expression of novel proteins in tomato

The expression of pharmaceutically relevant proteins in tomato has been pursued as a possible approach to facilitate the production and delivery of compounds that enhance human and animal health. The rabies glycoprotein, G-protein, was expressed in transgenic tomato. The protein was glycosylated uniquely probably because of its expression in tomato, but the expression of the rabies virus G-protein provided an opportunity to attempt to produce an orally accessible vaccine in plants (McGarvey *et al.*, 1995). A potential oral vaccine to the respiratory syncytial virus (RSV) has been developed by expressing the RSV fusion (F) protein in ripening tomato fruit using the *E8* promoter (Sandhu *et al.*, 2000). Durable serum and mucosal RSV-F antibodies were generated in mice who had consumed the ripe tomato fruit from these transgenic plants. Other pharmaceutical compounds, including human β interferon have been reported to be produced in non-traditional varieties of tomato (Rudas *et al.*, 1997).

8.7 Regulation of transgenic gene expression in tomato

As the technology for expression of transgenes in plants becomes more sophisticated, tissue specific, developmentally timed, and inducible expression of specific genes will be required. One way to ensure appropriate expression is through the choice of promoter regulating gene expression. Several promoters have been evaluated in tomato for the specificity of their expression. Table 8.2 summarizes the specificity of promoters that have been used in transgenic tomato plants. Fruit-specific expression has been important for genes whose products influence fruit performance or content.

Table 8.2 Promoters used in transgenic tomato

Specificity	Promoter	Regulation	Reference
Fruit-specific	Tomato *2A11*		(Van Haaren and Houck, 1993)
	Tomato PG2	Outer pericarp	(Bird *et al.*, 1988)
	Tomato *E8*	Ethylene and ripening induced	(Deikman *et al.*, 1998)
	Tomato *E4*	Ethylene induced	(Montgomery *et al.*, 1993)
	Tomato *LoxA*	Outer pericarp	(Beaudoin and Rothstein, 1997)
	Apple ACC oxidase	Ripening	(Atkinson *et al.*, 1998)
	Apple PG	Ripening	(Atkinson *et al.*, 1998)
	Pepper capsanthin/ capsorubin synthase	Drought and ethylene induced	(Kuntz *et al.*, 1998)
	Pepper fibrillin	Drought and ethylene induced	(Kuntz *et al.*, 1998)
	Tomato HMG2	Coincides with lycopene synthesis	(Daraselia *et al.*, 1996)
	Tomato RBCS	Locular specific in developing fruit	(Meier *et al.*, 1995)
Sucrose sink tissues	Potato ADP-glucose pyrophosphorylase	Sucrose regulated	(Du Jardin *et al.*, 1997)
Leaf abscission zones	Soybean cellulase	Ethylene induced	(Koehler *et al.*, 1996)
	Tomato abscission zone PG		(Hong *et al.*, 2000)
Flower-specific	Tomato *LAP*	Methyl jasmonate induced and induced in wounded leaves	(Ruiz-Rivero and Prat, 1998)
	Tomato *LAT52* and *LAT 59*	Pollen and late anther specific	(Eyal *et al.*, 1995)
Germinating seeds	Tomato *LoxC*		(Beaudoin and Rothstein, 1997)
	Tomato *LeEXP8*	Micropylar region specific	Chen and Bradford, pers. comm.
Vegetative tissue	Maize *Sh-1*	Not expressed in fruit	(D'Aoust *et al.*, 1999)
Circadian expression	Tomato Lhc	Clock controlled	(Piechulla *et al.*, 1998)
Pathogen induced	Tomato prosystemin	Jasmonate induced, insect feeding induced	(Jacinto *et al.*, 1997)
Drought induced	Tomato H1-S histone	ABA induced in all tissues	(Scippa *et al.*, 2000)
Metal regulated	Tomato metallothionein	Expressed more in leaves than in roots	(Whitelaw *et al.*, 1997)

The *E8* promoter specifies ethylene-regulated fruit-specific expression, as does the *E4* gene promoter (Deikman *et al.*, 1992; 1998; Montgomery *et al.*, 1993a; Xu *et al.*, 1996). The ca. 2kb promoters for ACC oxidase and PG from apple also function in fruit to specify expression during ripening (Atkinson *et al.*, 1998). The tomato fruit PG promoter (1.4 kb) gene directs expression during fruit ripening (Bird *et al.*, 1988). Expression specified by the PG promoter is confined to the outer pericarp tissue of the fruit and is not ethylene regulated (Montgomery *et al.*, 1993b; Nicholass *et al.*, 1995). Fruit-specific expression also can be achieved using the tomato 2A11 promoter (Van Haaren and Houck, 1993). The tomato lipoxygenase genes A and B (*tomloxA* and *B*) are expressed in the pre-ripening stages of fruit development and a third tomato lipoxygenase gene (*tomloxC*) is expressed in germinating seed. The promoter for *tomloxA* is able to direct gene expression in the outer pericarp from 5–20 days postanthesis (Beaudoin and Rothstein, 1997). Gene expression is confined to tomato sink tissues and stimulated by sucrose when regulated by the potato ADP-glucose pyrophosphorylase promoter (Du Jardin *et al.*, 1997). Expression in ripening tomato fruit also can be specified by the promoters from pepper capsanthin/capsorubin synthase and fibrillin genes, genes whose expression is ripening regulated in the non-climacteric pepper (Kuntz *et al.*, 1998). The expression from these promoters is ethylene regulated in tomato but also responds to other stimuli such as drought. The expression of the ethylene inducible genes in tomato leaf abscission zones and adjacent petioles can be specified by the soybean cellulase promoter (Koehler *et al.*, 1996) and by the promoter for the tomato abscission zone specific PG (Hong *et al.*, 2000). The abscission zone PG promoter does not cause expression inducible by ethylene in fruit tissues. Expression of genes with a circadian expression pattern can be specified by the light harvest complex protein promoter (Piechulla *et al.*, 1998). Gene expression in tomato flowers can be achieved using the promoter for the tomato leucine aminopeptidase (*LAP*) (Ruiz-Rivero and Prat, 1998). The promoter for systemin has been linked to the insecticidal proteins because the promoter is activated by wounding or insect attack (Jacinto *et al.*, 1997). Gene expression from this promoter is induced by methyl jasmonate as well. Gene expression in response to ABA or drought in tomato can be specified by the H1-S histone protein promoter (Scippa *et al.*, 2000). Pollen-specific expression in tobacco has been achieved using a tomato *lat52* promoter, but the application to tomato has not been reported (Wilkinson *et al.*, 1998). Development and wound regulated expression of the promoter for the tomato anionic peroxidase promoter has been reported in tobacco but not in tomato (Mohan *et al.*, 1993).

8.8 Conclusions

With the exception of *Arabidopsis*, tomato is perhaps the best genetically characterized dicotyledonous plant. Because of its rich genetic history, the availability of numerous mutants and genetic stocks and its relatively easy transformation by *Agrobacterium*, tomato has been used as a model crop to test the effects of numerous transgenes. Although tomato was the first genetically engineered food to be commercialized with the release of the Flavr Savr tomato in the early part of the 1990s, today there are no commercial transgenic tomato cultivars. The commercial production of transgenic corn, soybeans and cotton most likely reflects the relatively higher value of the major agronomic crops and lack of commercial research attention to the majority of horticultural crops, including tomato.

Because tomatoes play a central role in the diet of many cultures, this fruit provides a unique vehicle for the delivery of vitamins and other healthful constituents to the human diet. Transgenic approaches will provide the essential genetic tools to enhance or redirect metabolism towards constituents that provide new human health benefits and to provide a high quality product for consumers. As the agricultural biotechnology industry matures and public acceptance of genetically modified foods grows, we can anticipate that the extensive research base in tomato will lead to the commercial development of a large number of transgenic tomato cultivars.

Other information about the cultivation and availability of transgenic tomato varieties is available from academic centers with farm extension programs. At the University of California, Davis, information about the genetic resources in tomato is available from the Charlie Rick Tomato Genetic Resource Center (http://tgrc.ucdavis.edu), and the Vegetable Crops Research and Information Center (http://vric.ucdavis.edu/). Information about tomato production in California is available at http://www.tomato.org/. Information about US government policy on issues related to plant biotechnology is available at http://www.state.gov/www/issues/economic/biotech/eb_biotech_index.html

8.9 References

AGHARBAOUI Z, GREER A F and TABAEIZADEH Z (1995), 'Transformation of the wild tomato *Lycopersicon chilense* Dun. by *Agrobacterium tumefaciens*'. *Plant Cell Reports* 15, 102–105.

ANAN T, ITO H and MONMA S (1996), 'Chemical contents in fruits of transgenic tomato carrying the TMV coat protein gene, nontransgenic tomato, and other Lycopersicon species'. *Journal of the Japanese Society for Horticultural Science* 65, 635–644.

ARRILLAGA I, GIL-MASCARELL R, GISBERT C, SALES E, MONTESINOS C, SERRANO R and MORENO V (1998), 'Expression of the yeast *HAL2* gene in tomato increases the *in vitro* salt tolerance of transgenic progenies'. *Plant Science* 136, 219–226.

ATKINSON H J, URWIN P E, CLARKE M C and MCPHERSON M J (1996), 'Image analysis of the growth of *Globodera pallida* and *Meloidogyne incognita* on transgenic tomato roots expressing cystatins'. *Journal of Nematology* 28, 209–215.

ATKINSON R G, BOLITHO K M, WRIGHT M A, ITURRIAGAGOITIA-BUENO T, REID S J and ROSS G S (1998), 'Apple ACC-oxidase and polygalacturonase: Ripening-specific gene expression and promoter analysis in transgenic tomato'. *Plant Molecular Biology* 38, 449–460.

BALDWIN E A, SCOTT J W, SHEWMAKER C K and SCHUCH W (2000), 'Flavor trivia and tomato aroma: biochemistry and possible mechanisms for control of important aroma components'. *HortScience* 35, 1013–1022.

BARRY C S, LLOP-TOUS M I and GRIERSON D (2000), 'The regulation of 1-aminocyclopropane-1-carboxylic acid synthase gene expression during the transition from system-1 to system-2 ethylene synthesis in tomato'. *Plant Physiology* 123, 979–986.

BEAUDOIN N and ROTHSTEIN S J (1997), 'Developmental regulation of two tomato lipoxygenase promoters in transgenic tobacco and tomato'. *Plant Molecular Biology* 33, 835–846.

BEECHER G (1998), 'Nutrient content of tomatoes and tomato products'. *Proceedings of the Society for Experimental Biology and Medicine* 218, 98–100.

BETTINI P, COSI E, PELLEGRINI M G, TURBANTI L, VENDRAMIN G G and BUIATTI M (1998), 'Modification of competence for *in vitro* response to *Fusarium oxysporum* in tomato cells. III. PR-protein gene expression and ethylene evolution in tomato cell lines transgenic for phytohormone-related bacterial genes'. *Theoretical and Applied Genetics* 97, 575–583.

BIRD C R, SMITH C J S, RAY J A, MOUREAU P, BEVAN M W, BIRD A S, HUGHES S, MORRIS P C, GRIERSON D and SCHUCH W (1988), 'The tomato polygalacturonase gene and ripening-specific expression in transgenic plants'. *Plant Molecular Biology* 11, 651–662.

BLATT M R, GRABOV A, BREARLEY J, HAMMOND-KOSACK K and JONES J D G (1999), 'K+ channels of *Cf-9* transgenic tobacco guard cells as targets for *Cladosporium fulvum Avr9* elicitor-dependent signal transduction'. *Plant Journal* 19, 453–462.

BOLITHO K M, LAY-YEE M, KNIGHTON M L and ROSS G S (1997), 'Antisense apple ACC-oxidase RNA reduces ethylene production in transgenic tomato fruit'. *Plant Science* 122, 91–99.

BRADFORD K J, CHEN F, COOLEY M B, DAHAL P, DOWNIE B, FUKUNAGA . K, GEE O H, GURUSINGHE S, MELLA R A, NONOGAKI H, WU C-T, YANG H and KIM K-O (2000), 'Gene expression prior to radicle emergence in imbibed tomato seeds' in M Black, K J Bradford, J Vazquez-Ramos (eds) *Seed Biology: Advances and Applications*, 231–251.

BROMMONSCHENKEL S H, FRARY A, FRARY A and TANKSLEY S (2000), 'The broad-spectrum tospovirus resistance gene *Sw-5* of tomato is a homolog of the root-knot nematode resistance gene *Mi*'. *Molecular Plant-Microbe Interactions* 13, 1130–1138.

BRUEGGEMANN W, BEYEL V, BRODKA M, POTH H, WEIL M and STOCKHAUS J (1999), 'Antioxidants and antioxidative enzymes in wild type and transgenic *Lycopersicon* genotypes of different chilling tolerance'. *Plant Science* 140, 145–154.

BRUMMELL D A, HALL B D and BENNETT A B (1999a), 'Antisense suppression of tomato endo-1,4-beta-glucanase *Cel2* mRNA accumulation increases the force required to break fruit abscission zones but does not affect fruit softening'. *Plant Molecular Biology* 40, 615–622.

BRUMMELL D A, HARPSTER M H and DUNSMUIR P (1999b), 'Differential expression of expansin gene family members during growth and ripening of tomato fruit'. *Plant Molecular Biology* 39, 161–169.

BRUMMELL D A, HARPSTER M H, CIVELLO P M, PALYS, J M., BENNETT A B and DUNSMUIR P (1999c), 'Modification of expansin protein abundance in tomato fruit alters softening and cell wall polymer metabolism during ripening'. *Plant Cell* 11, 2203–2216.

BRUMMELL D A and LABAVITCH J M (1997), 'Effect of antisense suppression of endopolygalacturonase activity on polyuronide molecular weight in ripening tomato fruit and fruit homogenates'. *Plant Physiology* 115, 717–725.

BRUNETTI A, TAVAZZA M, NORIS E, TAVAZZA R, CACIAGLI P, ANCORA G, CRESPI S and ACCOTTO G P (1997), 'High expression of truncated viral rep protein confers resistance to tomato yellow leaf curl virus in transgenic tomato plants'. *Molecular Plant-Microbe Interactions* 10, 571–579.

BUTTERY R (1993), 'Quantitative and sensory aspects of flavor of tomato and other vegetables and fruits' in T E Acree and R Teranishi, (eds) *Flavor science: Sensible principles and techniques*, Washington D.C., American Chem. Soc., 259–286.

CARINGTON C M S, GREVE L C and LABAVITCH J M (1993), 'Cell wall metabolism in ripening fruit: VI. Effect of the antisense polygalacturonase gene on cell wall changes accompanying ripening in transgenic tomatoes'. *Plant Physiology* 103, 429–434.

CARRERA E and PRAT S (1998), 'Expression of the *arabidopsis abi1-1* mutant allele inhibits proteinase inhibitor wound-induction in tomato'. *Plant Journal* 15, 765–771.

CASAL J J, SANCHEZ R A, BOYLAN M, VIERSTRA R D and QUAIL P H (1995), 'Is the far-red-absorbing form of Avena phytochrome A that is present at the end of the day able to sustain stem-growth inhibition during the night in transgenic tobacco and tomato seedlings?' *Planta* 197, 225–232.

CHANDRA S, MARTIN G B and LOW P S (1996), 'The *Pto* kinase mediates a signaling pathway leading to the oxidative burst in tomato'. *Proceedings of the National Academy of Sciences of the United States of America* 93, 13393–13397.

CHEN F and BRADFORD K J (2000), 'Expression of an Expansin Is Associated with Endosperm Weakening during Tomato Seed Germination'. *Plant Physiology* 124, 1265–1274.

CHENGAPPA S, GUILLEROUX M, PHILLIPS W and SHIELDS R (1999), 'Transgenic tomato plants with decreased sucrose synthase are unaltered in starch and sugar accumulation in the fruit'. *Plant Molecular Biology* 40, 213–221.

CHETELAT R T, DEVERNA J W and BENNETT A B (1995), 'Effects of the *Lycopersicon chmielewskii* sucrose accumulator gene (*sucr*) on fruit yield and quality parameters following introgression into tomato'. *Theoretical and Applied Genetics* 91, 334–339.

CIARDI J A, TIEMAN D M, LUND S T, JONES J B, STALL R E and KLEE H J (2000), 'Response to *Xanthomonas campestris pv. vesicatoria* in tomato involves regulation of ethylene receptor gene expression'. *Plant Physiology* 123, 81–92.

CONSTABEL C P, BERGEY D R and RYAN C A (1995), 'Systemin activates synthesis of wound-inducible tomato leaf polyphenol oxidase via the octadecanoid defense signaling pathway'. *Proceedings of the National Academy of Sciences of the United States of America* 92, 407–411.

COOLEY M B and YODER J I (1998), 'Insertional inactivation of the tomato polygalacturonase gene'. *Plant Molecular Biology* 38, 521–530.

COOPER W, BOUZAYEN M, HAMILTON A, BARRY C, ROSSALL S and GRIERSON D (1998), 'Use of transgenic plants to study the role of ethylene and polygalacturonase during infection of tomato fruit by *Colletotrichum gloeosporioides*'. *Plant Pathology* 47, 308–316.

DAI N, SCHAFFER A, PETREIKOV M, SHAHAK Y, GILLER Y, RATNER K, LEVINE A and GRANOT D (1999), 'Overexpression of *Arabidopsis* hexokinase in tomato plants inhibits growth, reduces photosynthesis, and induces rapid senescence'. *Plant Cell* 11, 1253–1266.

D'AOUST M, NGUYEN-QUOC B, LE V Q. and YELLE S (1999a), 'Upstream regulatory regions from the maize *Sh1* promoter confer tissue-specific expression of the beta-glucuronidase gene in tomato'. *Plant Cell Reports* 18, 803–808.

D'AOUST M-A, YELLE S and NGUYEN-QUOC B (1999b), 'Antisense inhibition of tomato fruit sucrose synthase decreases fruit setting and the sucrose unloading capacity of young fruit'. *Plant Cell* 11, 2407–2418.

DARASELIA N D, TARCHEVSKAYA S and NARITA J O (1996), 'The promoter for tomato 3-hydroxy-3-methylglutaryl coenzyme A reductase gene 2 has unusual regulatory elements that direct high-level expression'. *Plant Physiology* 112, 727–733.

DEIKMAN J, KLINE R and FISCHER R L (1992), 'Organization of ripening and ethylene regulatory regions in a fruit-specific promoter from tomato *Lycopersicon esculentum*'. *Plant Physiology* 100, 2013–2017.

DEIKMAN J, XU R, KNEISSL M L, CIARDI J A, KIM K-N and PELAH D (1998), 'Separation of cis elements responsive to ethylene, fruit development, and ripening in the 5′-flanking region of the ripening-related *E8* gene'. *Plant Molecular Biology* 37, 1001–1011.

DOWD P F, LAGRIMINI L M and NELSEN T C (1998), 'Relative resistance of transgenic tomato tissues expressing high levels of tobacco anionic peroxidase to different insect species'. *Natural Toxins* 6, 241–249.

DU JARDIN P, HARVENGT L, KIRSCH F, LE V-Q, NGUYEN-QUOC B and YELLE S (1997), 'Sink-cell-specific activity of a potato ADP-glucose pyrophosphorylase B-subunit promoter in transgenic potato and tomato plants'. *Planta* 203, 133–139.

EL MANSOURI I, MERCADO J A, SANTIAGO-DOMENECH N, PLIEGO-ALFARO F, VALPUESTA V and QUESADA M A (1999), 'Biochemical and phenotypical characterization of transgenic tomato plants overexpressing a basic peroxidase'. *Physiologia Plantarum* 106, 355–362.

ERRINGTON N, TUCKER G A and MITCHELL J R (1998), 'Effect of genetic down-regulation of polygalacturonase and pectin esterase activity on rheology and composition of tomato juice'. *Journal of the Science of Food and Agriculture* 76, 515–519.

EYAL Y, CURIE C and MCCORMICK S (1995), 'Pollen Specificity Elements Reside in 30 bp of the Proximal Promoters of Two Pollen-Expressed Genes'. *Plant Cell* 7, 373–384.

FENG P C C, RUFF T G, RANGWALA S H and RAO S R (1997), 'Engineering plant resistance to thiazopyr herbicide via expression of a novel esterase deactivation enzyme'. *Pesticide Biochemistry and Physiology* 59, 89–103.

FENWICK K M, JARVIS M C, APPERLEY D C, SEYMOUR G B and BIRD C R (1996), 'Polymer mobility in cell walls of transgenic tomatoes with reduced polygalacturonase activity'. *Phytochemistry* 42, 301–307.

FERRIE B J, BEAUDOIN N, BURKHART W, BOWSHER C G and ROTHSTEIN S J (1994), 'The cloning of two tomato lipoxygenase genes and their differential expression during fruit ripening'. *Plant Physiology* 106, 109–118.

FICCADENTI N, SESTILI S, PANDOLFINI T, CIRILLO C, ROTINO G L and SPENA A (1999), 'Genetic engineering of parthenocarpic fruit development in tomato'. *Molecular Breeding* 5, 463–470.

FILLATTI J J, KISER J, ROSE B and COMAI L (1987), 'Efficient transformation of tomato and the introduction and expression of a gene for herbicide tolerance', in D J Nevins and R Jones (eds) *Tomato Biotechnology*, New York, Alan R. Liss, 199–210.

FRARY A, NESBITT T C, FRARY A, GRANDILLO S, VAN DER KNAAP E, CONG B, LIU J, MELLER J, ELBER R, ALPERT K B and TANKSLEY S D (2000), '*fw2.2*: a quantitative trait locus key to the evolution of tomato fruit size'. *Science* 289, 85–88.

FRASER P D, HEDDEN P, COOKE D T, BIRD C R, SCHUCH W and BRAMLEY P M (1995), 'The effect of reduced activity of phytoene synthase on isoprenoid levels in tomato pericarp during fruit development and ripening'. *Planta* 196, 321–326.

FRASER P D, KIANO J W, TRUESDALE M R, SCHUCH W and BRAMLEY P M (1999), 'Phytoene synthase-2 enzyme activity in tomato does not contribute to carotenoid synthesis in ripening fruit'. *Plant Molecular Biology* 40, 687–698.

FRAY R G and GRIERSON D (1993) 'Identification and genetic analysis of normal and mutant phytoene synthase genes of tomato by sequencing, complementation and co-suppression'. *Plant Molecular Biology* 22, 589–602.

FRAY R G, WALLACE A, FRASER P D, VALERO D, HEDDEN P, BRAMLEY P M and GRIERSON D (1995), 'Constitutive expression of a fruit phytoene synthase gene in transgenic tomatoes causes dwarfism by redirecting metabolites from the gibberellin pathway'. *Plant Journal* 8, 693–701.

FUCHS M, PROVVIDENTI R, SLIGHTOM J L and GONSALVES D (1996), 'Evaluation of transgenic tomato plants expressing the coat protein gene of cucumber mosaic virus strain WL under field conditions'. *Plant Disease* 80, 270–275.

FURUI H, INAKUMA T, ISHIGURO Y and KISO M (1998), 'Tomatine content in host and transgenic tomatoes by absorptiometric measurement'. *Bioscience Biotechnology and Biochemistry* 62, 556–557.

GISBERT C, RUS A M, BOLARIN M C, LOPEZ-CORONADO J M, ARRILLAGA I, MONTESINOS C, CARO M, SERRANO R and MORENO V (2000), 'The yeast *HAL1* gene improves salt tolerance of transgenic tomato'. *Plant Physiology* 123, 393–402.

GOOD X, KELLOGG J A, WAGONER W, LANGHOFF D, MATSUMURA W and BESTWICK R K (1994), 'Reduced ethylene synthesis by transgenic tomatoes expressing S-adenosylmethionine hydrolase'. *Plant Molecular Biology* 26, 781–790.

GRAY J, PICTON S, SHABBEER J, SCHUCH W and GRIERSON D (1992), 'Molecular biology of fruit ripening and its manipulation with antisense genes'. *Plant Molecular Biology* 19, 69–87.

GRAY J E, PICTON S, GIOVANNONI J J and GRIERSON, D (1994), 'The use of transgenic and naturally occurring mutants to understand and manipulate tomato fruit ripening'. *Plant Cell and Environment* 17, 557–571.

GRICHKO V P, FILBY B and GLICK B R (2000), 'Increased ability of transgenic plants expressing the bacterial enzyme ACC deaminase to accumulate Cd, Co, Cu, Ni, Pb, and Zn'. *Journal of Biotechnology* 81, 45–53.

GRIERSON D and FRAY R (1994), 'Control of ripening in transgenic tomatoes'. *Euphytica* 79, 251–263.

GRIERSON D, SMITH C J S, WATSON C F, MORRIS P C, TURNER A J, SCHUCH W, BIRD C R, RAY J and HAMILTON A (1990), 'Controlling Gene Expression in Transgenic Tomatoes', in A B Bennett and S D O'Neill, (eds) *Horticultural Biotechnology*, New York, Wiley-Liss, Inc., 229–236.

GRIFFITHS A, BARRY C, ALPUCHE-SOLIS A G and GRIERSON D (1999a), 'Ethylene and developmental signals regulate expression of lipoxygenase genes during tomato fruit ripening'. *Journal of Experimental Botany* 50, 793–798.

GRIFFITHS A, PRESTAGE S, LINFORTH R, ZHANG J, TAYLOR A and GRIERSON D (1999b), 'Fruit-specific lipoxygenase suppression in antisense-transgenic tomatoes'. *Postharvest Biology and Technology* 17, 163–173.

GROSSMAN K and SCHMUELLING T (1995), 'The effects of the herbicide quinclorac on shoot growth in tomato is alleviated by inhibitors of ethylene biosynthesis and by the presence of an antisense construct to the 1-aminocyclopropane-1-carboxylic acid (ACC) synthase gene in transgenic plants'. *Plant Growth Regulation* 16, 183–188.

HADFIELD K A and BENNETT A B (1998), 'Polygalacturonases: Many genes in search of a function'. *Plant Physiology* 117, 337–343.

HALDRUP A, PETERSEN S G and OKKELS F T (1998), 'The xylose isomerase gene from *Thermoanaerobacterium thermosulfurogenes* allows effective selection of transgenic plant cells using D-xylose as the selection agent'. *Plant Molecular Biology* 37, 287–296.

HAMILTON A J, BOUZAYEN M and GRIERSON D (1991), 'Identification of a tomato gene for the ethylene-forming enzyme by expression in yeast'. *Proceedings of the National Academy of Sciences of the United States of America* 88, 7434–7437.

HAMMOND-KOSACK K E, TANG S, HARRISON K and JONES J D G (1998), 'The tomato

Cf-9 disease resistance gene functions in tobacco and potato to confer responsiveness to the fungal avirulence gene product *Avr9*'. *Plant Cell* 10, 1251–1266.

HAMNER K C and MAYNARD L A (1942), *Factors influencing the nutritive value of tomato*, Washington D.C., United States Department of Agriculture.

HONEE G, MELCHERS L S, VLEESHOUWERS V G A A, VAN ROEKEL J S C and DE WIT P J G M (1995), 'Production of the *AVR9* elicitor from the fungal pathogen *Cladosporium fulvum* in transgenic tobacco and tomato plants'. *Plant Molecular Biology* 29, 909–920.

HONG S-B, SEXTON R and TUCKER M L (2000), 'Analysis of gene promoters for two tomato polygalacturonases expressed in abscission zones and the stigma'. *Plant Physiology* 123, 869–881.

HOWE G A and RYAN C A (1999), 'Suppressors of systemin signaling identify genes in the tomato wound response pathway'. *Genetics* 153, 1411–1421.

HWANG C-F, BHAKTA A V, TRUESDELL G M, PUDLO W M and WILLIAMSON V M (2000), 'Evidence for a role of the N terminus and leucine-rich repeat region of the *Mi* gene product in regulation of localized cell death'. *Plant Cell* 12, 1319–1329.

JACINTO T, FERNANDES K V S, MACHADO O L T and SIQUERIA-JUNIOR C L (1998), 'Leaves of transgenic tomato plants overexpressing prosystemin accumulate high levels of cystatin'. *Plant Science* 138, 35–42.

JACINTO T, MCGURL B, FRANCESCHI V, DELANO-FREIER J and RYAN C A (1997), 'Tomato prosystemin promoter confers wound-inducible, vascular bundle-specific expression of the beta-glucuronidase gene in transgenic tomato plants'. *Planta* 203, 406–412.

JENSEN E B and VEIERSKOV B (1998), 'Interaction between photoperiod, photosynthesis and ethylene formation in tomato plants (*Lycopersicon esculentum* cv. Ailsa Craig and ACC-oxidase antisense *pTOM13*)'. *Physiologia Plantarum* 103, 363–368.

JOHN I, DRAKE R, FARRELL A, COOPER W, LEE P, HORTON P and GRIERSON D (1995), 'Delayed leaf senescence in ethylene-deficient ACC-oxidase antisense tomato plants: Molecular and physiological analysis'. *Plant Journal* 7, 483–490.

JONGEDIJK E, TIGELAAR H, VAN ROEKEL J S C, BRES-VLOEMANS S A, DEKKER I, VAN DEN ELZEN P J M, CORNELISSEN B J C and MELCHERS L S (1995), 'Synergistic activity of chitinases and beta-1,3-glucanases enhances fungal resistance in transgenic tomato plants'. *Euphytica* 85, 173–180.

KANAYAMA Y, GRANOT D, DAI N, PETREIKOV M, SCHAFFER A, POWELL A and BENNETT A B (1998), 'Tomato fructokinases exhibit differential expression and substrate regulation'. *Plant Physiology* 117, 85–90.

KANIEWSKI W, ILARDI V, TOMASSOLI L, MITSKY T, LAYTON J and BARBA M (1999), 'Extreme resistance to cucumber mosaic virus (CMV) in transgenic tomato expressing one or two viral coat proteins'. *Molecular Breeding* 5, 111–119.

KAUSCH K D and HANDA A K (1997), 'Molecular cloning of a ripening-specific lipoxygenase and its expression during wild-type and mutant tomato fruit development'. *Plant Physiology* 113, 1041–1050.

KENDE H (1993), 'Ethylene biosynthesis', in W Briggs, *Annual Review of Plant Physiology and Plant Molecular Biology*, Palo Alto, California, Annual Reviews Inc., 283–307.

KERDNAIMONGKOL K and WOODSON W R (1999), 'Inhibition of catalase by antisense RNA increases susceptibility to oxidative stress and chilling injury in transgenic tomato plants'. *Journal of the American Society for Horticultural Science* 124, 330–336.

KESARWANI M, AZAM M, NATARAJAN K, MEHTA A and DATTA A (2000), 'Oxalate decarboxylase from *Collybia velutipes*: Molecular cloning and its overexpression to confer resistance to fungal infection in transgenic tobacco and tomato'. *Journal of Biological Chemistry* 275, 7230–7238.

KIM J W, SUN S S M and GERMAN T L (1994), 'Disease resistance in tobacco and tomato plants transformed with the tomato spotted wilt virus nucleocapsid gene'. *Plant Disease* 78, 615–621.

KLANN E M, HALL B and BENNETT A B (1996), 'Antisense acid invertase (*TIV1*) gene alters soluble sugar composition and size in transgenic tomato fruit'. *Plant Physiology* 112, 1321–1330.

KLEE H J (1993), 'Ripening physiology of fruit from transgenic tomato (*Lycopersicon esculentum*) plants with reduced ethylene synthesis'. *Plant Physiology* 102, 911–916.

KLEE H J, HAYFORD M B, KRETZMER K A, BARRY G F and KISHORE G M (1991), 'Control of ethylene synthesis by expression of a bacterial enzyme in transgenic tomato plants'. *Plant Cell* 3, 1187–1194.

KOEHLER S M, MATTERS G L, NATH P, KEMMERER E C and TUCKER M L (1996), 'The gene promoter for a bean abscission cellulase is ethylene-induced in transgenic tomato and shows high sequence conservation with a soybean abscission cellulase'. *Plant Molecular Biology* 31, 595–606.

KRAMER M, SANDERS R, BOLKAN H, WATERS C, SHEEHY R E and HIATT W R (1992), 'Postharvest evaluation of transgenic tomatoes with reduced levels of polygalacturonase: processing, firmness and disease resistance'. *Postharvest Biol. Tech.* 1, 244–255.

KUNTZ M, CHEN H C, SIMKIN A J, ROMER S, SHIPTON C A, DRAKE R, SCHUCH W and BRAMLEY P M (1998), 'Upregulation of two ripening-related genes from a non-climacteric plant (pepper) in a transgenic climacteric plant (tomato)'. *Plant Journal* 13, 351–361.

LAGRIMINI L M, VAUGHN J, ERB W A and MILLER S A (1993), 'Peroxidase overproduction in tomato: Wound-induced polyphenol deposition and disease resistance'. *HortScience* 28, 218–221.

LASHBROOK C C, GIOVANNONI J J, HALL B D, FISCHER R L and BENNETT A B (1998), 'Transgenic analysis of tomato endo-beta-1,4-glucanase gene function. Role of cel1 in floral abscission'. *Plant Journal* 13, 303–310.

LASHBROOK C C, GONZALEZ-BOSCH C and BENNETT A B (1994), 'Two divergent endo-beta-1,4-glucanase genes exhibit overlapping expression in ripening fruit and abscising flowers'. *Plant Cell* 6, 1485–1493.

LEE H I and RAIKHEL N V (1995), 'Prohevein is poorly processed but shows enhanced resistance to a chitin-binding fungus in transgenic tomato plants'.

Brazilian Journal of Medical and Biological Research 28, 743–750.

LEE K Y, BADEN C, HOWIE W J, BEDBROOK J and DUNSMUIR P (1997), 'Post-transcriptional gene silencing of ACC synthase in tomato results from cytoplasmic RNA degradation'. *Plant Journal* 12, 1127–1137.

MANDAOKAR A D, GOYAL R K, SHUKLA A, BISARIA S, BHALLA R, REDDY V S, CHAURASIA A, SHARMA R P, ALTOSAAR I and KUMAR P A (2000), 'Transgenic tomato plants resistant to fruit borer (*Helicoverpa armigera Hubner*)'. *Crop Protection* 19, 307–312.

MARTIN G B., FRARY A, WU T, BROMMONSCHENKEL S, CHUNWONGSE J, EARLE E D and TANKSLEY S D (1994), 'A Member of the Tomato *Pto* Gene Family Confers Sensitivity to Fenthion Resulting in Rapid Cell Death'. *Plant Cell* 6, 1543–1552.

MARTINEAU B M, HOUCK C M, SHEEHY R E and HIATT W R (1994), 'Fruit-specific expression of the *A. tumefaciens* isopentenyl transferase gene in tomato: effects on fruit ripening and defense-related gene expression in leaves'. *The Plant Journal* 5, 11–19.

MARTINEAU B, SUMMERFELT K R, ADAMS D F and DEVERNA J W (1995), 'Production of high solids tomatoes through molecular modification of levels of the plant growth regulator cytokinin'. *Bio-Technology* 13, 250–254.

MCGARVEY P B, HAMMOND J., DIENELT M M, HOOPER D C, FU Z F, DIETZSCHOLD B, KOPROWSKI H and MICHAELS F H (1995), 'Expression of the rabies virus glycoprotein in transgenic tomatoes'. *Bio-Technology* 13, 1484–1487.

MCGARVEY P B, MONTASSER M S and KAPER J M (1994), 'Transgenic tomato plants expressing satellite RNA are tolerant to some strains of cucumber mosaic virus'. *Journal of the American Society for Horticultural Science* 119, 642–647.

MCGURL B, OROZCO-CARDENAS M, PEARCE G and RYAN C A (1994), 'Overexpression of the prosystemin gene in transgenic tomato plants generates a systemic signal that constitutively induces proteinase inhibitor synthesis'. *Proceedings of the National Academy of Sciences of the United States of America* 91, 9799–9802.

MCGURL B, PEARCE G., OROZCO-CARDENAS M and RYAN C A (1992), 'Structure, expression, and antisense inhibition of the system in precursor gene'. *Science* 255, 1570–1573.

MEIER I, CALLAN K L, FLEMING A J and GRUISSEM W (1995), 'Organ-specific differential regulation of a promoter subfamily for the ribulose-1,5-bisphosphate carboxylase/oxygenase small subunit genes in tomato'. *Plant Physiology* 107, 1105–1118.

MICALLEF B J, HASKINS K A, VANDERVEER P J, ROH K-S, SHEWMAKER C K and SHARKEY T D (1995), 'Altered photosynthesis, flowering, and fruiting in transgenic tomato plants that have an increased capacity for sucrose synthesis'. *Planta* 196, 327–334.

MISHRA K K and HANDA A K (1998), 'Post-transcriptional silencing of pectin methylesterase gene in transgenic tomato fruits results from impaired pre-mRNA processing'. *Plant Journal* 14, 583–592.

MOHAN R, VIJAYAN P and KOLATTUKUDY P E (1993), 'Developmental and tissue-specific expression of a tomato anionic peroxidase (*tap1*) gene by a minimal promoter, with wound and pathogen induction by an additional 5′-flanking region'. *Plant Molecular Biology* 22, 475–490.

MONTGOMERY J, GOLDMAN S, DEIKMAN J, MARGOSSIAN L and FISCHER R L (1993a), 'Identification of an ethylene-responsive region in the promoter of a fruit ripening gene'. *Proceedings of the National Academy of Sciences of the United States of America* 90, 5939–5943.

MONTGOMERY J, POLLARD V, DEIKMAN J and FISCHER R L (1993b), 'Positive and negative regulatory regions control the spatial distribution of polygalacturonase transcription in tomato fruit pericarp'. *Plant Cell* 5, 1049–1062.

MONTI M M, VALANZUOLO S, CASSANI G and COLOMBO M (1999), 'Transgenic tomatoes expressing a Cucumber Mosaic Virus satellite RNA: Field testing and analysis of satellite RNA spread'. *Journal of Plant Pathology* 81, 113–122.

MOONEY M, DESNOS T, HARRISON K, JONES J, CARPENTER R and COEN E (1995), 'Altered regulation of tomato and tobacco pigmentation genes caused by the *delila* gene of *Antirrhinum*'. *Plant Journal* 7, 333–339.

MURPHY J F, SIKORA E J, SAMMONS B and KANIEWSKI W K (1998), 'Performance of transgenic tomatoes expressing cucumber mosaic virus CP gene under epidemic conditions'. *HortScience* 33, 1032–1035.

NGUYEN-QUOC B, N'TCHOBO H, FOYER C H and YELLE S (1999), 'Overexpression of sucrose phosphate synthase increases sucrose unloading in transformed tomato fruit'. *Journal of Experimental Botany* 50, 785–791.

NICHOLASS F J, SMITH C J S, SCHUCH W, BIRD C R and GRIERSON D (1995), 'High levels of ripening-specific reported gene expression directed by tomato fruit polygalacturonase gene-flanking regions'. *Plant Molecular Biology* 28, 423–435.

NONOGAKI H, GEE O H and BRADFORD K J (2000), 'A germination-specific endo-beta-mannanase gene is expressed in the micropylar endosperm cap of tomato seeds'. *Plant Physiology* 123, 1235–1245.

OELLER P W, MIN-WONG L, TAYLOR L P, PIKE D A and THEOLOGIS A (1991), 'Reversible inhibition of tomato fruit senescence by antisense RNA'. *Science* 254, 437–439.

OROZCO-CARDENAS M, MCGURL B and RYAN C A (1993), 'Expression of an antisense prosystemin gene in tomato plants reduces resistance toward *Manduca sexta* larvae'. *Proceedings of the National Academy of Sciences of the United States of America* 90, 8273–8276.

PICTON S, GRAY J E and GRIERSON D (1995), 'The manipulation and modification of tomato fruit ripening by expression of antisense RNA in transgenic plants'. *Euphytica* 85, 193–202.

PIECHULLA B, MERFORTH N and RUDOLPH B (1998), 'Identification of tomato *Lhc* promoter regions necessary for circadian expression'. *Plant Molecular Biology* 38, 655–662.

PNUELI L, HAREVEN D, ROUNSLEY S D, YANOFSKY M F and LIFSCHITZ E (1994), 'Isolation of the tomato *AGAMOUS* gene *TAG1* and analysis of its homeotic role in transgenic plants'. *Plant Cell* 6, 163–173.

PORRETTA S and POLI G (1997), 'Tomato puree quality from transgenic processing tomatoes'. *International Journal of Food Science & Technology* 32, 527–534.

PORRETTA S, POLI G and MINUTI E (1998), 'Tomato pulp quality from transgenic fruits with reduced polygalacturonase (PG)'. *Food Chemistry* 62, 283–290.

POWELL A L T, VAN KAN J, TEN HAVE A, VISSER J, GREVE L C, BENNETT A B and LABAVITCH J M (2000), 'Transgenic expression of pear PGIP in tomato limits fungal colonization'. *Molecular Plant-Microbe Interactions* 13, 942–950.

PRESTAGE S, LINFORTH R S T, TAYLOR A J, LEE E, SPEIRS J and SCHUCH W (1999), 'Volatile production in tomato fruit with modified alcohol dehydrogenase activity'. *Journal of the Science of Food and Agriculture* 79, 131–136.

PROVVIDENTI R and GONSALVES D (1995), 'Inheritance of resistance to cucumber mosaic virus in a transgenic tomato line expressing the coat protein gene of the white leaf strain'. *Journal of Heredity* 86, 85–88.

RHIM S-L (1998), 'Molecular breeding of transgenic tomato plants expressing the delta-endotoxin gene of *Bacillus thuringiensis subsp. Tenebrionis*'. *Agricultural Chemistry and Biotechnology* 41, 137–140.

ROCHANGE SF and MCQUEEN-MASON S J (2000), 'Expression of a heterologous expansin in transgenic tomato plants'. *Planta* 211, 583–586.

ROMEIS T, PIEDRAS P and JONES J D G (2000), 'Resistance gene-dependent activation of a calcium-dependent protein kinase in the plant defense response'. *Plant Cell* 12, 803–815.

ROMER S, FRASER P D, KIANO J W, SHIPTON C A, MISAWA N, SCHUCH W and BRAMLEY P M (2000) 'Elevation of the provitamin A content of transgenic tomato plants'. *Nature Biotechnology* 18, 666–669.

ROSE J K C, LEE H H and BENNETT A B (1997), 'Expression of a divergent expansin gene is fruit-specific and ripening-regulated'. *Proceedings of the National Academy of Sciences of the United States of America* 94, 5955–5960.

ROSSI M, GOGGIN F L, MILLIGAN S B, KALOSHIAN I, ULLMAN D E and WILLIAMSON V M (1998) 'The nematode resistance gene Mi of tomato confers resistance against the potato aphid'. *Proceedings of the National Academy of Sciences of the United States of America* 95, 9750–9754.

RUDAS V A, PIVEN N M, RIVKIN, M I, VERSHININ A V and INFANTE D C (1997), 'Genetic transformation on *Lycopersicon peruvianum var. dentatum* by *Agrobacterium tumefaciens* having plasmid with introduced human interferon-beta gene'. *Tsitologiya i Genetika* 31, 17–22.

RUIZ-RIVERO O J and PRAT S (1998), 'A-308 deletion of the tomato LAP promoters is able to direct flower-specific and MeJA-induced expression in transgenic plants'. *Plant Molecular Biology* 36, 639–648.

RYAN C A (2000), 'The systemin signaling pathway: Differential activation of plant defensive genes'. *Biochimica et Biophysica Acta* 1477, 112–121.

RYAN C A and PEARCE G (1998), 'SYSTEMIN: A polypeptide signal for plant defensive genes'. *Annual Review of Cell and Developmental Biology*, Palo Alto, California, Annual Reviews Inc., 1–17.

SANDHU J S, KRASNYANSKI S F, DOMIER L L, KORBAN S S, OSADJAN M D and BUETOW D E (2000), 'Oral immunization of mice with transgenic tomato fruit expressing respiratory syncytial virus-F protein induces a systemic immune response'. *Transgenic Research* 9, 127–135.

SCIPPA G S, GRIFFITHS A, CHIATANTE D and BRAY E A (2000), 'The H1 histone variant of tomato, *H1-S*, is targeted to the nucleus and accumulates in chromatin in response to water-deficit stress'. *Planta* 211, 173–181.

SESSA G, D'ASCENZO M and MARTIN G B (2000), 'Thr38 and Ser198 are *Pto* autophosphorylation sites required for the *AvrPto-Pto*-mediated hypersensitive response'. *EMBO (European Molecular Biology Organization) Journal* 19, 2257–2269.

SEYMOUR G B, FRAY R G, HILL P and TUCKER G A (1993), 'Down-regulation of two non-homologous endogenous tomato genes with a single chimaeric sense gene construct'. *Plant Molecular Biology* 23, 1–9.

SHEEHY R E, KRAMER M and HIATT W R (1988), 'Reduction of polygalacturonase activity in tomato fruit by antisense RNA'. *Proceedings of the National Academy of Sciences of the United States of America* 85, 8805–8809.

SHERF B A, BAJAR A M and KOLATTUKUDY P E (1993), 'Abolition of an inducible highly anionic peroxidase activity in transgenic tomato'. *Plant Physiology* 101, 201–208.

SITRIT Y and BENNETT A B (1998), 'Regulation of tomato fruit polygalacturonase mRNA accumulation by ethylene: A re-examination'. *Plant Physiology* 116, 1145–1150.

SITRIT Y, HADFIELD K A, BENNETT A B, BRADFORD K J and DOWNIE A B (1999), 'Expression of a polygalacturonase associated with tomato seed germination'. *Plant Physiology* 121, 419–428.

SMIGOCKI A, HEU S and BUTA G (2000), 'Analysis of insecticidal activity in transgenic plants carrying the *ipt* plant growth hormone gene'. *Acta Physiologiae Plantarum* 22, 295–299.

SMITH C J S, WATSON C F, BIRD C R, RAY J, SCHUCH W and GRIERSON D (1990a), 'Expression of a truncated tomato polygalacturonase gene inhibits expression of the endogenous gene in transgenic plants'. *Molecular & General Genetics* 224, 477–481.

SMITH C J S, WATSON C F, MORRIS P C, BIRD C R, SEYMOUR G B, GRAY J E, ARNOLD C, TUCKER G A and SCHUCH W (1990b), 'Inheritance and effect on ripening of antisense polygalacturonase genes in transgenic tomatoes'. *Plant Molecular Biology* 14, 369–380.

SMITH C J S, WATSON C F, RAY J, BIRD C R, MORRIS P C, SCHUCH W and GRIERSON D (1988), 'Antisense RNA inhibition of polygalacturonase gene expression in transgenic tomatoes'. *Nature* 334, 724–726.

SMITH D L and GROSS K C (2000), 'A family of at least seven β-galactosidase genes is expressed during tomato fruit development'. *Plant Physiol* 123, 1173–1183.

SMITH R M, MARSHALL J A, DAVEY M R, LOWE K C and POWER J B (1996), 'Comparison of volatiles and waxes in leaves of genetically engineered tomatoes'. *Phytochemistry* 43, 753–758.

SOZZI QUIROGA G O and FRASCHINA A A (1997), 'Evaluation of sensory attributes and biochemical parameters in transgenic tomato fruit with reduced polygalacturonase activity'. *Food Science and Technology International* 3, 93–102.

SPANU P, REINHARDT D and BOLLER T (1991), 'Analysis and cloning of the ethylene-forming enzyme from tomato by functional expression of its messenger RNA in *Xenopus laevis* oocytes'. *EMBO (European Molecular Biology Organization) Journal* 10, 2007–2014.

SPEIRS J, LEE E., HOLT K, YONG-DUK K, SCOTT N S, LOVEYS B and SCHUCH W (1998), 'Genetic manipulation of alcohol dehydrogenase levels in ripening tomato fruit affects the balance of some flavor aldehydes and alcohols'. *Plant Physiology* 117, 1047–1058.

SREE VIDYA C S, MANOHARAN M, RANJIT KUMAR C T, SAVITHRI H S and SITA G L (2000), '*Agrobacterium*-mediated transformation of tomato (*Lycopersicon esculentum* var. Pusa Ruby) with coat-protein gene of Physalis mottle tymovirus'. *Journal of Plant Physiology.* 156, 106–110.

STILL D W and BRADFORD K J (1997), 'Endo-beta-mannanase activity from individual tomato endosperm caps and radicle tips in relation to germination rates'. *Plant Physiology* 113, 21–29.

STILL D W, DAHAL P and BRADFORD K J (1997), 'A single-seed assay for endo-beta-mannanase activity from tomato endosperm and radicle tissues'. *Plant Physiology* 113, 13–20.

STOMMEL J R, TOUSIGNANT M E, WAI T, PASINI R and KAPER J M (1998), 'Viral satellite RNA expression in transgenic tomato confers field tolerance to cucumber mosaic virus'. *Plant Disease* 82, 391–396.

STORTI E, BOGANI P, BETTINI P, BITTINI P, GUARDIOLA M L, PELLEGRINI M G, INZE D and BUIATTI M (1994), 'Modification of competence for *in vitro* response to *Fusarium oxysporum* in tomato cells: II. Effect of the integration of *Agrobacterium tumefaciens* genes for auxin and cytokinin synthesis'. *Theoretical and Applied Genetics* 88, 89–96.

TABAEIZADEH Z, AGHARBAOUI Z, HARRAK H and POYSA V (1999), 'Transgenic tomato plants expressing a *Lycopersicon chilense* chitinase gene demonstrate improved resistance to *Verticillium dahliae* race 2'. *Plant Cell Reports* 19, 197–202.

TAI T H, DAHLBECK D, CLARK E T, GAJIWALA P, PASION R, WHALEN M C, STALL R E and STASKAWICZ, B J (1999), 'Expression of the *Bs2* pepper gene confers resistance to bacterial spot disease in tomato'. *Proceedings of the National Academy of Sciences of the United States of America* 96, 14153–14158.

TANG X, XIE M, KIM Y J, ZHOU J, KLESSIG D F and MARTIN G B (1999), 'Overexpression of *Pto* activates defense responses and confers broad resistance'. *Plant Cell* 11, 15–29.

TAYLOR J E, TUCKER G A, LASSLETT Y, SMITH C J S, ARNOLD C M, WATSON C F, SCHUCH W, GRIERSON D and ROBERTS J A (1991), 'Polygalacturonase expression during leaf abscission of normal and transgenic tomato plants'. *Planta* 183, 133–138.

THAKUR B R, SINGH R K and HANDA A K (1996a), 'Effect of an antisense pectin methylesterase gene on the chemistry of pectin in tomato (*Lycopersicon esculentum*) juice'. *Journal of Agricultural and Food Chemistry* 44, 628–630.

THAKUR B R, SINGH R K, TIEMAN D M and HANDA A K (1996b), 'Tomato product quality from transgenic fruits with reduced pectin methylesterase'. *Journal of Food Science* 61, 85–87, 108.

THEOLOGIS A, OELLER P W, WONG L-M, ROTTMANN W H and GANTZ D M (1993), 'Use of a tomato mutant constructed with reverse genetics to study fruit ripening, a complex developmental process'. *Developmental Genetics* 14, 282–295.

THEOLOGIS A, ZAREMBINSKI T I, OELLER P W, LIANG X and ABEL S (1992), 'Modification of Fruit Ripening by Suppressing Gene Expression'. *Plant Physiology* 100, 549–551.

THOMPSON A J, JACKSON A C, SYMONDS R C, MULHOLLAND B J, DADSWELL A R, BLAKE P S, BURBIDGE A and TAYLOR I B (2000), 'Ectopic expression of a tomato 9-cis-epoxycarotenoid dioxygenase gene causes over-production of abscisic acid'. *Plant Journal* 23, 363–374.

THOMZIK J E, STENZEL K, STOECKER R, SCHREIER P H, HAIN R and STAHL D J (1997), 'Synthesis of a grapevine phytoalexin in transgenic tomatoes (*Lycopersicon esculentum* Mill.) conditions resistance against *Phytophthora infestans*'. *Physiological and Molecular Plant Pathology* 51, 265–278.

TIEMAN D M and HANDA A K (1994), 'Reduction in pectin methylesterase activity modifies tissue integrity and action levels in ripening tomato (*Lycopersicon esculentum* Mill.) fruits. *Plant Physiology* 106, 429–436.

TIEMAN D M, HARRIMAN R W, RAMAMOHAN G and HANDA A K (1992), 'An antisense pectin methylesterase gene alters pectin chemistry and soluble solids in tomato fruit'. *Plant Cell* 4, 667–679.

TIEMAN D M, KAUSCH K D, SERRA D M and HANDA A K (1995), 'Field performance of transgenic tomato with reduced pectin methylesterase activity'. *Journal of the American Society for Horticultural Science* 120, 765–770.

TIEMAN D M, TAYLOR M G, CIARDI J A and KLEE H J (2000), 'The tomato ethylene receptors *NR* and *LeETR4* are negative regulators of ethylene response and exhibit functional compensation within a multigene family'. *Proceedings of the National Academy of Sciences of the United States of America* 97, 5663–5668.

TOMASSOLI L, ILARDI, V, BARBA M and KANIEWSKI W (1999), 'Resistance of transgenic tomato to cucumber mosaic cucumovirus under field conditions'. *Molecular Breeding* 5, 121–130.

TUCKER G A, SIMONS H and ERRINGTON N (1999), 'Transgenic tomato technology: Enzymic modification of pectin pastes'. *Biotechnology & Genetic Engineering Reviews*, 293–308.

ULTZEN T, GIELEN J, VENEMA F, WESTERBROEK A, DE HAAN P, TAN M-L, SCHRAM A, VAN GRINSVEN M and GOLDBACH R (1995), 'Resistance to tomato spotted wilt virus in transgenic tomato hybrids'. *Euphytica* 85, 159–168.

VAN DER SALM T, BOSCH D, HONEE G, FENG L, MUNSTERMAN E, BAKKER P, STIEKEMA W J and VISSER B (1994), 'Insect resistance of transgenic plants that express modified *Bacillus thuringiensis cryIA(b) and cryIC* genes: A resistance management strategy'. *Plant Molecular Biology* 26, 51–59.

VAN HAAREN M J J and HOUCK C M (1993), 'A functional map of the fruit-specific promoter of the tomato 2A11 gene'. *Plant Molecular Biology* 21, 625–640.

VOS P, SIMONS G, JESSE T, WIJBRANDI J, HEINEN L, HOGERS R, FRIJTERS A, GROENENDIJK J, DIERGAARDE P, REIJANS M, FIERENS-ONSTENK J, DE BOTH M, PELEMAN J, LIHARSKA T, HONTELEZ J and ZABEAU M (1998), 'The tomato *Mi-1* gene confers resistance to both root-knot nematodes and potato aphids'. *Nature Biotechnology* 16, 1365.

WANG C, CHIN C K and CHEN A (1998), 'Expression of the yeast *DELTA-9* desaturase gene in tomato enhances its resistance to powdery mildew'. *Physiological and Molecular Plant Pathology* 52, 371–383.

WANG C, CHIN C-K, HO C-T, HWANG C-F, POLASCHOCK J J and MARTIN C E (1996), 'Changes of fatty acids and fatty acid-derived flavor compounds by expressing the yeast *DELTA-9* desaturase gene in tomato'. *Journal of Agricultural and Food Chemistry* 44, 3399–3402.

WATSON C F, ZHENG L and DELLAPENNA D (1994), 'Reduction of tomato polygalacturonase beta subunit expression affects pectin solubilization and degradation during fruit ripening'. *Plant Cell* 6, 1623–1634.

WHITHAM S, MCCORMICK S and BAKER B (1996), 'The *N* gene of tobacco confers resistance to tobacco mosaic virus in transgenic tomato'. *Proceedings of the National Academy of Sciences of the United States of America* 93, 8776–8781.

WILKINSON J E, LINDSEY K and TWELL D (1998), 'Antisense-mediated suppression of transgenic expression targeted specifically to pollen'. *Journal of Experimental Botany* 49, 1481–1490.

WILKINSON J Q, LANAHAN M B, YEN H-S, GIOVANNONI J J and KLEE H J (1995), 'An ethylene-inducible component of signal transduction encoded by Never-ripe'. *Science* 270, 1807–1809.

WILLIAMSON V M (1998), 'Root-knot nematode resistance genes in tomato and their potential for future use'. *Annual Review of Phytopathology*, Palo Alto, California, Annual Reviews Inc., 277–293.

XU R, GOLDMAN S, COUPE S and DEIKMAN J (1996), 'Ethylene control of *E4* transcription during tomato fruit ripening involves two cooperative cis elements'. *Plant Molecular Biology* 31, 1117–1127.

XUE B, GONSALVES C, PROVVIDENTI R, SLIGHTOM J L, FUCHS M and GONSALVES D (1994), 'Development of transgenic tomato expressing a high level of resistance to cucumber mosaic virus strains of subgroups I and II'. *Plant Disease* 78, 1038–1041.

ZHANG, H-X and BLUMWALD, E (2001), 'Transgenic salt-tolerant tomato plants accumulate salt in foliage but not in fruit'. *Nature Biotechnology* 19, 765–768.

9

Commercial developments with transgenic potato

H.V. Davies, Scottish Crop Research Institute, Dundee

9.1 Markets and challenges

Potato is the world's fourth most important food crop behind wheat, rice and maize. Over the last three decades potato production has grown faster than any other food crop except wheat (FAO). Glennon (2000) states that, agriculturally, in the eyes of the developing countries, no other crop has more production potential, since yield potential is still largely under exploited. In developing countries potato is also seen as a candidate for resolving domestic production problems. More than one billion individuals (50% of these in developing countries) now eat potato and as little as 100 grams supplies 10% of the recommended daily calorie allowance for children. The same amount provides about 10% of essential vitamin intake (e.g. thiamine, niacin, folate) and 50% in the case of vitamin C. In 1998 global production was around 290 million tonnes with 30% of production in developing countries. In the European Union of 15 member states production totals 50 million tonnes (MT) and is dominated by Germany (11.3 MT), the United Kingdom (6.6 MT), France (6.5 MT) and the Netherlands (6.0 MT), whilst in consumption terms Ireland still has the highest intake at 140 kg per capita (Table 9.1).

What are the most important subjects demanding global attention in potatoes? Collins (2000) indicates that in developed countries the most important include:

1. disease control strategies for late blight, bacterial wilt, ring rot, nematodes and threats to the availability of appropriate chemicals to control these pests and diseases;
2. Genetically Modified Organisms (GMOs) and related issues (acceptance and use of biotechnology, ownership of intellectual property, freedom-to-operate);
3. processing and marketing;

Table 9.1 Top ten world potato production 1998 (Source FAO)

	Production (million tonnes)		Production (million tonnes)
China	45	India	19
EU 15	48	Ukraine	17.5
Russia	37	Belarus	10
Poland	26	Turkey	5.3
USA	21	Canada	4

4. seed tuber quality and health; and
5. genetic resources for future use.

Collins also points out that resources in public institutions in developed countries, which traditionally addressed these problems, are shrinking at an alarming rate. Solutions might include bilateral country collaborations on common priorities, new partnerships with the private sector, regional/global co-operation on more widespread issues. More than 40% of the world's potatoes are grown in developing countries and this is expected to increase. Within both developing and developed countries there is also a trend towards a decline in fresh consumption and a continued rise in process utilisation.

9.1.1 Global initiative on late blight (GiLB): a model for collaborative research

An example of a global initiative to resolve a key biological problem facing potato is the Global Initiative on Late Blight (GiLB). This was formed following the realisation that the magnitude of the late blight problem requires a global strategy. Late blight is caused by the fungus *Phytophthora infestans*. The pathogen is showing increasing resistance to chemicals used in developing countries (and new chemicals are subject to increasing safety regulations). It also travels easily, providing access to environments where variability can be increased through sexual hybridisations (Collins, 2000). Organisations such as the International Potato Center (CIP) are dedicated to improving the health and well being of disadvantaged populations in the developing world and appreciate that there is a role for genetic modification (GM) technology in such initiatives. CIP is playing a key role in delivering links between private and public centre research and has set up a global strategy for the uptake and deployment of genetically engineered potatoes when and where appropriate. The initiative is called potato GENE (Genetic Engineering Network) and currently involves scientists from 11 developing and three industrialised countries. Four priority areas have been focused upon, which actually reflect key issues for the developed as well as developing countries with regard to GMOs:

1. strategies for addressing public concerns;
2. seed systems and intellectual property;

3. gene flow and effects on non-target organisms; and
4. health and food safety issues.

The objective is to provide the scientific information necessary for informed choice. The GM debate in general is opening up and considerable transparency is emerging through initiatives driven by scientists themselves. The reader is directed to sites such as http://www.rikilt.wageningen-ur.nl/euprojects/euprojects.html, where information can be gathered on European Union funded research programmes such ENTRANSFOOD and GMOCARE which are aimed at improving information flow and delivering further scientific advances to assist risk assessment.

9.2 Potato breeding and a role for GM technology

Potato breeding started around the beginning of the 19th century but its complex tetraploid genetics means that targeted breeding is a time-consuming exercise. Comparatively speaking, potato has a narrow genetic base, which, at least in part, contributes to slower progress in crop improvement than with other major crop species. Attempts to introgress genes from wild relatives has met with some success for disease resistance traits but undesirable side effects from hybridisation events mean that periods of *ca.* 12 to 15 years have been required for cultivar development. Since back-crossing to remove undesirable effects is not an easy option for potato, the approach of improving existing cultivars using gene transfer or genetic engineering technology has been an attractive proposition. Realistically, this approach is not meant to replace the role of the plant breeder, but to complement it by providing additional tools to modify potato genomes for environmental and commercial benefit. From an environmental perspective GM potato poses few risks as pollen movement is extremely limited (separation distances for experimental release of nine metres are acceptable in the UK) and there are no wild relatives with which cultivated potato can outcross.

9.2.1 Potato transformation

The traditional method of potato genetic transformation employs *Agrobacterium*-mediated systems. Whilst transgenic potato is often used as a model system to assess the roles of specific gene(s), cultivars such as Desirée tend to dominate the 'transgenic' literature as transformation efficiencies are high. However, Dale and Hampson (1995) examined transformation efficiencies in 34 potato varieties using a tuber disc protocol, showing that only half of the cultivars regenerated. From those that could be regenerated all but one produced transgenic plants. Some cultivars which did not regenerate from tuber discs did so from leaf and internode segments. It follows, then, that to deliver a commercial product which is true-to-type but which has the desired level of transgene expression and trait modification may not yet be a facile operation with some genotypes.

Belknap *et al.* (1994) transformed Lemhi Russet and Russet Burbank with constructs containing GUS, ClaSP (tyrosine-rich arylphorin) or a gene encoding a bacterial lytic peptide and showed that the highest potential for deviation from typical performance occurred in yield and tuber size gradings. The frequency of off-types varied between 15 and 80%, depending on cultivar, but off-types were not always apparent until plants were grown in the field. Of the lines transformed with GUS, less than 50% produced seed tubers under field conditions. Field trialling is clearly imperative to gain a real insight into compositional, biochemical, phenotypic and agronomic performance.

9.2.2 Somaclonal variation

Potato is potentially a good model crop for selection of improved lines generated through somaclonal variation from which novel variants can arise. Sexual crosses are not always possible in potato due to sterility problems or lack of flowers and, as already stated, the genetics of tetraploid inheritance is problematic. Potato is easily regenerated in tissue culture (although as stated previously some cultivars are more recalcitrant than others) and is vegetatively propagated from tubers. However, somaclonal variation may produce undesirable effects following targeted genetic transformation events, thus modifying phenotype and agronomic performance independently of any effects induced by insertion of the target gene(s).

The *in vitro* regeneration process required to produce GM potato lines involves: (a) establishing de-differentiated cells from tissue or organ culture under defined conditions: (b) proliferation for a number of cell generations: and (c) subsequent plant regeneration under *in vitro* conditions (Karp 1990). Somaclonal variation in regenerated plants is generated during the *in vitro* culture stage and particularly during de-differentiation. This is accompanied by increased frequency of chromosomal abnormalities with time in culture. Genetic changes also occur in plant tissues and cells *in vivo* due to mutations, endoreduplication, chimeras etc (see Kumar, 1994 and references therein for a comprehensive analysis of the origins of somaclonal variation). Genetic variation in plants regenerated *in vitro* can therefore be derived from *in vivo* and *in vitro* events. The contributions of *in vivo* and *in vitro* modifications are dependent on parameters including genotypic background, culture conditions, etc.

Somaclonal variation is uncontrollable and unpredictable in nature and most variation is of no apparent use. However, stability of any useful somaclones produced may not be a problem. Morphological changes observed range from gross abnormalities to minor and more subtle modifications. There is distinct genotypic variation in the frequency of somaclonal variants that might arise. Thus selecting GM potato lines for commercialisation which have the desired impact but in which other traits are not significantly modified by the tissue culture process will require the production of several hundred independently transformed lines and full and effective field selection using criteria that breeders would normally impose. This will be in addition to the testing of a range of constructs, promoters, targeting sequences, etc. where relevant.

Compliance, in risk assessment exercises, with the need to demonstrate 'substantial equivalence' of a GM line with the parent from which it is derived should take into account compositional variation that might be induced by somaclonal variation in species such as potato, and not only from the expression or insertion of target genes. The concept of substantial equivalence is that the GM line to be marketed should be compositionally the same as the parent line from which it was derived, but with the exception of any modification expected by inserting the gene of interest. This has to be assessed using several growing sites over more than one year. From a personal perspective the value of substantial equivalence should be combined with knowledge of natural genetic variation in the compositional status of cultivars already in production and on sale. For this reason extensive databases which clearly demonstrate the ranges of metabolite concentrations that might be expected in species of crop plants will have great utility. A scenario can be envisaged in which the composition of a GM line may differ from its parent but fall well within the range expected of cultivated potato.

9.3 Commercial applications of GM potato crops

Global sales for transgenic crops were estimated at $75 million in 1995, increasing to $235 million in 1996 and $670 million in 1997 (James, 2000). In 1999 the estimated value was approximately two billion dollars (a 30-fold increase in five years). In descending order of cropping area the ranking order for GM crops is soybean (54% of total), maize (24%), cotton and canola (rapeseed; 9%), potato, squash and papaya (1% or less). The USA has the highest proportion of GM crops with 74% of the world total, followed by Argentina (15%), and Canada (10%). Herbicide-tolerant crops occupy 71% of the area grown, insect-resistant (*Bt*) crops 28%. Romania and the Ukraine have grown introductory areas of *Bt* potato (less than 1000 ha) but the majority of commercial potato crops have been grown in North America. No potato crop currently has European Union approval for commercial release into the environment.

This chapter is primarily aimed at the commercial applications of GM potato. When the chapter was conceived the only company to have this crop in the commercial market place was Monsanto, a St Louis-based agricultural biotechnology and herbicide company of which the Pharmacia Corporation now owns 85%. Monsanto markets GM potato through its NatureMark® biotechnology unit. The potato lines developed by NatureMark® will be outlined later. According to a Reuters report on 21 March 2001, the Monsanto Company stated that it would no longer be marketing genetically modified

Table 9.2 Current regulatory status of New Leaf^{TM} potato lines

	USA	Canada	Mexico	Japan	Russia	Romania	Australia
NewLeaf^{TM}	P		P	I	I	I	PU
NewLeaf^{TM} Y	P	P					U
NewLeaf^{TM} Plus	P	P					U

P – production approval
I – import approval
U – under review (expected for food import by end of 2001)

potatoes as part of a streamlining of its biotechnology crop portfolio to focus on wheat, corn, soybeans and cotton. This decision was no doubt influenced by statements from some leading processing and fast-food outlet companies that consumer concerns over GM potato products had affected their GM purchasing policy. Since the processing market is the major one for potato in North America, the GM potato business becomes non-viable as a result.

Today's reliable, cheap, potatoes are a testimony to the ingenuity of scientists, agronomists and farmers over the past 100 years, and a tremendous success story by any standards. However, even with all the advances in agricultural science some 40% of potato harvests are lost world-wide to diseases, pests and weeds, a figure which might reach 75% if it were not for crop protection products (Rhodes, 2000). NatureMark® commercialised its potato GM lines through the NewLeaf^{TM} brand name and the first-generation targets were aimed at resolving some of these key production issues. The current regulatory status of NatureMark® potato lines is shown in Table 9.2. The lines developed are described below.

9.3.1 Insect-resistant potato: NewLeaf^{TM}

NatureMark® potato lines named NewLeaf^{TM} confer resistance to Colorado potato beetle (CPB; *Leptinotarsa decemlineata* Say). Colorado beetle is a primary pest of North America and elsewhere, larvae emerging from egg masses about one week after deposition then feeding on foliage. Mature larvae leave the plant and pupate in soil to emerge as adults one week later and begin feeding on foliage yet again. NewLeaf ^{TM} potatoes were commercialised in the USA and Canada after more than ten years of development work in laboratories, glasshouses and fields across North America. The potatoes were transformed with a gene, which encodes for the Cry3A protein from the bacterium *Bacillus thuringiensis* (var. tenebrionis) using the constitutive 35S CaMV promoter. The Cry 3A protein is part of a family of proteins used for more than 30 years by organic producers, home gardeners, etc. The product has been endorsed by the World Health Organisation (WHO) and other regulatory agencies throughout the world. The protein affects directly only the target pest, Colorado potato beetle (CPB) and has no effect on other insects, mammals or wildlife. Growers have reduced insecticide usage on average by 42%. NewLeaf^{TM} was approved in the

USA in May 1995. The relevant agencies in the USA, the Food and Drug Administration, (FDA), United States Department of Agriculture (USDA) and Environmental Protection Agency (EPA) have approved the product based on substantial equivalence to other Russet Burbank potatoes. The product has also been approved by Health Canada, Agri-Food Canada and Agriculture Canada. Japan and Mexico have also approved the use of NewLeaf™ potato.

The first generation of NewLeaf™ potatoes included cultivars Atlantic, Superior and Russet Burbank and all of these NewLeaf™ varieties have been completely de-regulated by the FDA, USDA and EPA in the USA. Commercial adoption rates between 1995 and 1999 are shown in Table 9.3. A range of transgenic lines were selected with superior agronomic characteristics in addition to CPB resistance. Cultural management guidelines were also developed for growers. Out of the programme arose NewLeaf™ 6 Russet Burbank which is high yielding (earlier tuber bulking) and which produces a high percentage of quality grade potatoes classified as US and Canadian number ones. This has required optimisation of row spacings and development of appropriate fertiliser regimes. NewLeaf™ 6 also has extended dormancy and shows improved long storage characteristics. Since the potato is completely protected from CPB it appears to need no additional protection from this pest. Plants modified to express insecticidal proteins from *Bacillus thuringiensis* (referred to as *Bt*-protected plants) are believed to provide a safe and highly effective method of insect control. *Bt*-protected corn, cotton, and potato were introduced into the United States in 1995/1996 and grown on a total of approximately ten million acres in 1997, 20 million acres in 1998, and 29 million acres globally in 1999. These crops provide highly effective control of major insect pests such as the European corn borer, south-western corn borer, tobacco budworm, cotton bollworm and pink bollworm, in addition to CPB. They reduce reliance on conventional chemical pesticides and appear to provide notably higher yields in cotton and corn. The estimated total net savings to the grower using *Bt*-protected cotton in the United States was approximately $92 million in 1998. Other benefits of these crops include reduced levels of the

Table 9.3 NewLeaf™ commercial adoption rate. Varieties: Russet Burbank, Atlantic and Superior

1995	1,800 acres of commercial production. 54 M lbs of raw product used for tablestock and French fries.
1996	10,000 acres of commercial production. 300 M lbs of raw product used for tablestock and French fries.
1997	30,000 acres of commercial production. 900 M lbs of raw product used for tablestock and French fries.
1998	48,000 acres of commercial production. 1.4 B lbs of raw product used for tablestock and French fries.
1999	55,000 acres of commercial production. 1.65 B lbs of raw product used for tablestock and French fries.

fungal toxin fumonisin in corn and the opportunity for supplemental pest control by beneficial insects due to the reduced use of broad-spectrum insecticides. Insect resistance management plans are being implemented to prolong the effectiveness of these products.

Extensive testing of *Bt*-protected crops has been conducted which establishes the safety of these products to humans, animals, and the environment. Acute, sub-chronic, and chronic toxicology studies conducted over the past 40 years have established the safety of the microbial *Bt* products, including their expressed insecticidal (Cry) proteins, which are fully approved for marketing. Mammalian toxicology and digestive fate studies, which have been conducted with the proteins produced in the currently approved *Bt*-protected plant products, have confirmed that these Cry proteins are non-toxic to humans and pose no significant concern for allergenicity. Food and feed derived from *Bt*-protected crops, which have been fully approved by regulatory agencies, have been shown to be substantially equivalent to the food and feed derived from conventional crops. Non-target organisms exposed to high levels of Cry protein are virtually unaffected, except for certain insects that are closely related to the target pests. Because the Cry protein is contained within the plant (in microgram quantities), the potential for exposure to farm workers and non-target organisms is extremely low. The Cry proteins produced in *Bt*-protected crops have been shown to degrade rapidly when crop residue is incorporated into the soil. Thus the environmental impact appears to be negligible. The human and environmental safety of *Bt*-protected crops is further supported by the long history of safe use for *Bt* microbial pesticides around the world.

For several decades the primary control strategy for CPB has been the use of chemical insecticides and about 22 active ingredients are registered, in Canada for example, for this purpose. However, restrictions in the modes of actions of insecticides coupled with repeated applications have led to resistance development in most commercial production regions in Canada (Stewart, 2000). The entry of new products with novel modes of action such as *Bt* toxin and products such as cyromazine and imidacloprid give more grower flexibility and reduce the potential for resistance development. Other insecticides are under development. Insecticide with *Bt* protein as the active ingredient is primarily effective against larvae only. Crop rotation is an important control strategy for growers too, as might be biological control using insect-destroying fungi such as *Beauveria bassiana* and beneficial nematodes such as *Steinemema carpocapsae*. However, the most spectacular development has been with *Bt* transgenics. Tests on Prince Edward Island and elsewhere in North America have shown that the technology is effective against both adults and larvae. Potato plant mixtures of 70% GM and 30% non-GM appear to be as effective in controlling CPB beetle numbers as GM monocultures. Indeed, border rows of transgenics can reduce the number of colonising beetles entering a field. CPB appeared in Europe in the 1920s and since 1980 has occurred in practically the entire European continent with the exception of the UK and Scandinavia. Quarantine actions are implemented only in the UK.

Recent work (Mohammed *et al*, 2000) has evaluated potato tuber moth resistance in tubers of transgenic potato lines expressing an alternative Bt protein encoded by the *Bt-cry5* gene. The potato tuber moth is the most destructive pest of potato in tropical and sub-tropical countries. Larvae attack both foliage and tubers in the field and in storage. Moth mortality was 100% in transgenic lines of cultivar Spunta using the constitutive cauliflower mosaic virus promoter (35S CaMV).

9.3.2 Virus-resistant potato: NewLeaf™ Plus and NewLeaf™ Y

Virus resistance in potato has been developed using a range of approaches and genetic constructs which include sequences of virus coat proteins, movement proteins, replicases, untranslatable sense or antisense RNAs, proteases, defective interfering RNAs, and satellites. Expression of ribozymes, a double stranded RNA-specific ribonuclease, antiviral proteins, a plant pathogen resistance gene and 'plantibodies' have also provided virus resistance (see Kawchuk and Prufer (1999) for a review of approaches used to deliver virus resistance).

NewLeaf™ Plus produced by NatureMark® is a high yielding Russet Burbank with combined CPB (*cry*3A expression) and potato leafroll virus (PLRV) resistance (produced using the constitutive Figwort Mosaic Virus Promoter (FMV) within a construct designed to prevent virus replication). PLRV can cause yield losses of as much as 50% and nearly all commercial varieties are susceptible to infection with world-wide losses estimated at 10% (van der Wilk *et al.*, 1991). PLRV also causes net necrosis (phloem cells affected), which greatly reduces the value of tubers for fresh and processing use. Freedom from such internal necroses provides a more consistent product and better financial returns per hectare by reducing processing costs in French fry and chip (crisp) industries and by helping to deliver improved seed quality. The transgenics are capable of reducing insecticide usage by up to 100%. NewLeaf™ Plus was approved in the USA for consumption in August 1997. The FDA and EPA in the US determined that the potato was as safe to eat as any other Russet Burbank. Large-scale agronomic trials were grown in the USA in 1998.

NewLeaf™ Y cultivars Russet Burbank and Shepody have been developed with combined CPB and potato virus Y (PVY) resistances. PVY is considered one of the most damaging potato viruses because it causes economically significant yield depression. Severe infestations can reduce yield by as much as 80% (Bemster and de Boks, 1987). The PVY coat protein gene used to generate resistance is also expressed using the FMV promoter and is more effective at PVY control than any insecticide programme allowing more sustained crop protection through reduced insecticide usage. Protection against PVY has reduced seed de-certification risk for seed growers and helped to maximise yields for commercial growers. Other benefits include improved processing quality and storage and higher tuber set in cultivars such as Shepody (more uniform tuber size distributions have been claimed through improved line selections). As far as NewLeaf™ Y is concerned the FDA and Health Canada

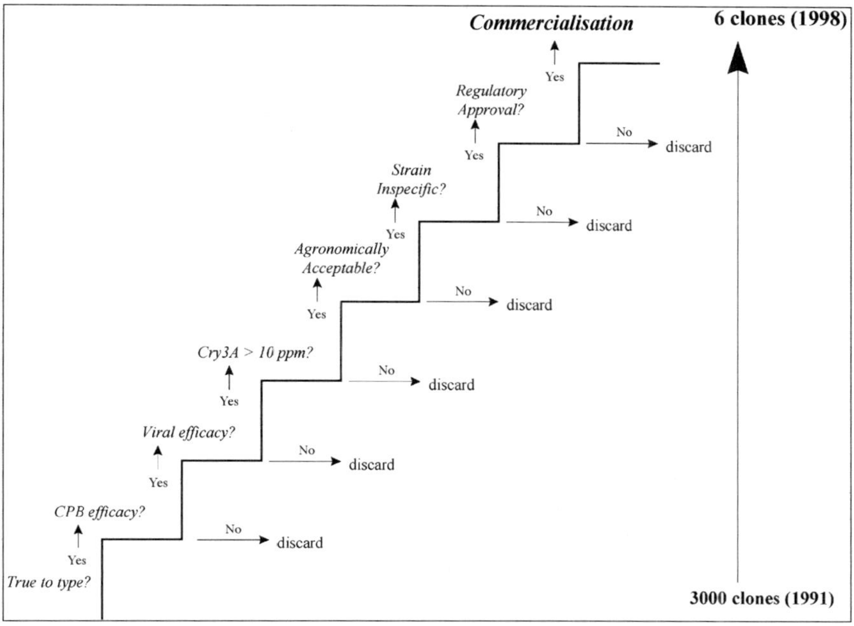

Fig. 9.1 Steps to commercialisation for NewLeaf™ Plus and New Leaf™ Y CPB and virus-resistant potato clones (from Rogan *et al.* submitted publication).

completed their review in May 1999 and agreed they were safe for human consumption. The EPA (August1997), USDA (February 1999) and Canadian Food Inspection Agency (April 1999) determined that NewLeaf ™ Y poses no concern for unreasonable effects on the environment or livestock.

The steps from initial clone production to the commercialisation of NewLeaf™ Plus and NewLeaf™ Y is shown in Fig. 9.1. The process started with *ca.* 3,000 original clones for each with the first selection occurring after transplanting to soil and removal of off types. True to type plants were then subjected to a range of tests including phenotypic analysis, efficacy tests against CPB, PLRV and PVY, susceptibility to insect, fungal and bacterial pests, evaluation of tuber yield, tuber appearance, quality and additional extensive field performance evaluation. Field evaluations took place over several years, in different environments and under various agronomic conditions with emphasis on potential environmental impact (Kaniewski and Thomas, 1999). From this process six clones were selected for commercialisation. Rogan *et al.* (2000) analysed key nutritional, quality and anti-nutritional components of NewLeaf™ Plus and NewLeaf™ Y lines to assess their substantial equivalence to the parent cultivar. Dry matter content, vitamin C, soluble sugar, soluble protein, glycoalkaloids, vitamin B_6, niacin, copper, magnesium, potassium, amino acids, fat, ash, calories, total protein and crude fibre were quantified. The data confirmed that tubers produced by insect and virus protected varieties were substantially equivalent to tubers produced by conventional varieties (Tables 9.4, 9.5, 9.6).

Table 9.4 Nutritional and quality parameters of NewLeaf™ Plus Russet Burbank (RB) clones and conventional Russet Burbank[a] (from Rogan *et al.* 2000)

Parameter	NewLeaf™ Plus RB clone number [mean (S.E.)[b]]			RB control	Literature ranges[c]
	RBMT21-129	RBMT21-350	RBMT22-082		
Total solids (% FW)	21.6 (0.42)	21.9 (0.43)	21.0 (0.42)	21.5 (0.42)	16.8–24.5[d]
% Dextrose (% FW)	0.087 (0.0073)	0.094 (0.0076)	0.113 (0.0073)	0.099 (0.0073)	0.04–0.52
% Sucrose (% FW)	0.182[e] (0.0185)	0.201 (0.0186)	0.177[e] (0.0185)	0.199 (0.0185)	0.10–0.88
Vitamin C (mg/100 g FW)	10.1 (0.66)	9.9 (0.66)	10.4 (0.66)	10.0 (0.66)	10.3–22.0
% Soluble protein (% DW)	5.0 (0.11)	5.1 (0.11)	5.0 (0.11)	5.0 (0.11)	3.4–7.3
Total glycoalkaloids (mg/100 g FW)	5.4 (0.69)	4.8 (0.71)	5.1 (0.69)	4.3 (0.69)	3.1–16.1

Notes:
[a] Samples were collected from tubers harvested in 1995 from three field locations in the United States (Echo, OR; Ephrate, WA; Pasco, WA).
[b] Numbers in parentheses are standard error of the mean.
[c] Literature ranges includes values for Russet Burbank, Atlantic, Gemchip and Norchip varieties.
[d] Literature range for total solids calculated from a conversion from specific gravity.
[e] Underlined values are statistically different from the RB control ($P \leq 0.05$).

Table 9.5 Comparison of proximate values for NewLeaf™ Plus Russet Burbank (RB) clones and conventional Russet Burbank potato tubers[a, b] (from Rogan *et al.* 2000)

Parameter	NewLeaf™ Plus RB clone number [mean (S.E.)[c]]			RB control	Literature ranges
	RBMT21-129	RBMT21-350	RBMT22-082		
Total protein	9.86 (1.041)	9.94 (1.028)	9.93 (1.028)	9.90 (1.028)	7.1–14.6
Fat	0.20[d] (0.033)	0.19 (0.032)	0.20[d] (0.032)	0.16 (0.032)	0.1–0.8
Ash	4.68 (0.218)	4.70 (0.200)	4.78 (0.200)	4.75 (0.200)	2.2–9.5
Crude fibre	1.64 (0.093)	1.55 (0.080)	1.61 (0.080)	1.68 (0.080)	0.2–3.5
Total carbohydrates	85.24 (0.095)	85.17 (0.960)	85.09 (0.960)	85.18 (0.960)	84.5 (average)
Calories	382.3 (0.99)	382.1 (0.93)	381.9 (0.93)	381.8 (0.93)	350 (average)

Notes:
[a] Samples were collected from tubers harvested in 1996 from three field trial locations in Canada (Spruce Grove, Alberta; Winkler, Manitoba; New Denmark, New Brunswick).
[b] Except for calories, reported values are in g/100 g dry weight (corrected for moisture content in the tuber powder).
[c] Numbers in parentheses are standard error of the mean.
[d] Underlined values are statistically different from the RB control ($P \leq 0.05$).

Table 9.6 Vitamin, mineral, amino acid composition of New Leaf™ Plus, New Leaf™ Y and conventional Russet Burbank (RB) tubers[a] (from Rogan et al, 2000)

Component (mg/200 g FW)	NewLeaf™ Plus RB Clone Number												RB Control			Literature range[b]
	RBMT15-101			RBMT21-129			RBMT21-350			RBMT22-082						
		Range			Range			Range			Range			Range		
	Mean	Max	Min	Mean	Max	Min	Mean	Max	Min	Mean	Max	Min	Mean	Max	Min	
Vitamin B6	0.52	0.54	0.46	0.51	0.58	0.32	0.53	0.57	0.31	0.52	0.78	0.30	0.52	0.56	0.45	0.26–0.82
Niacin	4.11	4.46	3.34	4.03	5.10	2.81	3.99	4.44	3.20	4.28	5.11	2.67	4.06	4.60	3.49	0.18–6.2
Copper	0.30	0.42	0.11	0.39	0.61	0.14	0.30	0.50	0.14	0.33	0.64	0.11	0.32	0.50	0.14	0.03–1.4
Magnesium	49.77	52.32	48.04	52.18	56.60	48.54	45.73	50.80	42.99	46.98	67.81	38.47	51.54	66.12	47.12	22.5–110
Potassium	996.59	1151.94	826.50	1072.68	1185.60	955.86	1026.63	1120.24	966.72	1038.22	1512.64	931.20	1080.74	1202.70	979.20	700–1250
Aspartic acid	1193.86	1346.40	1020.44	1279.99	1822.08	811.68	1095.46	1419.00	680.96	1296.17	1982.08	814.80	1250.12	1630.20	728.16	677–1476
Threonine	138.04	148.41	125.19	146.72	192.82	115.88	134.10	161.70	106.40	152.41	211.25	114.65	147.19	172.71	118.99	102–214
Serine	144.63	157.08	135.66	151.29	196.56	108.65	146.53	174.90	120.99	158.68	214.51	121.06	154.59	185.25	123.73	125–255
Glutamic acid	750.61	826.20	644.10	770.79	1017.12	550.02	700.62	858.00	491.26	778.67	1114.92	587.82	792.52	1054.50	515.63	583–1207
Proline	116.82	134.13	96.90	123.61	159.74	91.85	111.19	140.80	77.82	128.61	185.82	83.81	119.25	160.17	88.21	89–366
Glycine	114.23	116.84	105.73	122.12	147.89	100.39	111.76	126.50	96.06	125.45	178.00	102.43	121.27	142.50	106.56	92–195
Alanine	107.29	113.73	101.52	117.62	146.64	95.59	111.88	132.55	91.81	123.86	170.82	101.27	116.46	135.09	98.86	87–238
Cystine	59.26	64.98	54.18	61.49	66.77	55.54	59.01	62.73	55.21	63.96	78.24	56.22	62.09	69.54	57.12	96–185
Valine	208.37	224.77	192.78	216.62	248.20	170.88	198.91	245.02	162.34	237.42	374.25	191.48	217.70	284.43	175.23	196–363
Methionine	54.30	57.99	49.93	58.53	66.14	47.79	52.40	62.24	42.44	62.43	99.76	48.25	56.36	83.79	41.03	57–100
Isoleucine	130.99	141.36	118.83	139.86	162.24	108.94	123.39	150.67	100.32	147.69	229.50	116.48	139.43	177.84	116.62	119–238
Leucine	206.96	219.81	183.05	224.38	286.42	173.02	200.40	240.35	155.65	229.31	323.39	168.78	220.30	262.77	176.42	171–346
Tyrosine	121.01	133.87	99.96	127.50	161.62	96.65	128.54	150.18	110.66	149.34	208.64	121.64	143.54	177.84	116.62	114–236
Phenylalanine	157.52	166.97	144.12	166.86	202.80	133.50	156.05	184.26	130.11	179.85	267.97	140.84	167.96	208.05	132.61	138–272
Histidine	78.45	88.92	66.80	78.14	94.85	64.08	71.66	82.50	57.27	83.86	122.58	66.93	82.23	100.32	65.71	33–117
Lysine	226.92	243.39	203.04	234.76	283.92	194.38	211.39	240.90	175.10	244.62	359.90	197.88	233.34	291.27	192.99	154–342
Arginine	186.81	194.14	169.90	195.57	253.34	147.38	179.61	218.90	131.33	214.91	319.48	169.94	199.55	253.65	145.04	175–362
Tryptophan	42.53	51.64	35.65	42.75	46.11	37.52	39.17	42.30	35.87	45.14	64.35	36.67	42.38	54.38	33.51	29–70

Notes:

[a] Samples were collected from Island Falls, ME, USA and two sites in Canada (Hartland, New Brunswick; Summerside, Prince Edward Island). Plots were replicated four times at Hartland, new Brunswick. Plots were not replicated at the other two locations. Values presented represent the mean calculated from all six values

[b] Fresh weight concentration for literature range was determined assuming that potatoes are composed of approximately 75% water. All values are reported as mg/200 g fresh weight

Table 9.7 Insecticides and miticides used in commercial NewLeaf Plus™ and Russet Burbank potato field comparisons in 1998 (from Riebe and Zalewski, submitted for publication)

Location USA	Variety[1]	Insecticides and miticides used and number of applications[2]														Total
		AC	CF	DM	EF	ES	GU	HT	IC	MP	OX	PH	PM	PP	PS	
Ephrata, WA	RB		3							2				1		6
	NLP													1		1
Tri-Cities, WA	RB			1			1			2			1	1		6
	NLP															0
Boardman, OR	RB								1							1
	NLP															0
Hermiston, OR	RB		1			2				3	1			2		9
	NLP													2		2
Paterson, WA	RB		2							2				1		5
	NLP									1			1	1		3
Warden, WA	RB	1								1					1	3
	NLP													1		1
Moses Lake, WA	RB	1		1	1	1				1						5
	NLP															0
Quincy, WA	RB		3							3				1		7
	NLP													1		1
Total – all locations	RB	2	9	2	1	3	1	0	1	14	1	0	1	6	1	42
	NLP	0	0	0	0	0	0	0	0	1	0	0	1	6	0	8

Notes:
[1] RB – Russet Burbank; NLP = NewLeaf Plus™

[2] AC = aldicarb; CF = carbofuran; DM = dimethoate; EF = esfenvalerate; ES = endosulfan; GU = guthion; HT = Havoc/Tomahawk™; IC = imidocloprid; MP = methamidiphos; OX = oxamyl; PH = phorate; PM = permethrin; PP = propargite (miticides); PS = phosmet.

Table 9.8 Cost of insect control (CHEM) and damage due to net necrosis (NN) in commercial fields of NewLeaf Plus™ and Russet Burbank in 1998 (from Riebe and Zalewski)

Location	Variety	Cost ($US/ha)		
		CHEM[1]	NN[2]	Total
Ephrata, WA	RB	240	na	na
	NLP	59	na	na
Tri-Cities, WA	RB	161	417	578
	NLP	0	22	22
Boardman, OR	RB	161	141	302
	NLP	0	0	0
Hermiston, OR	RB	489	27	516
	NLP	116	17	133
Paterson, WA	RB	388	119	507
	NLP	133	0	133
Warden, WA	RB	237	49	286
	NLP	59	0	59
Moses Lake, WA	RB	287	0	287
	NLP	0	0	0
Quincy, WA	RB	289	0	289
	NLP	59	32	91
Average all locations[3]	RB	281	108	389
	NLP	53*	10	63*

Notes:
[1] Actual cost of all applied insecticides; NLP trait premium not included.
[2] Net necrosis cost estimated from USDA grade deductions using the formula:

(% by weight *damaged* × $0.33/mt) + (% *unusable* × $1.00/mt) × 56 mt/ha.

[3] NLP averages followed by an asterisk are significantly different from RB (P<0.05).

Riebe and Zalewski (submitted for publication) surveyed insecticide usage on NewLeaf™ Plus potatoes in paired field comparisons with conventional Russet Burbank. The work showed that insecticide use could be greatly reduced or eliminated with the transgenic material, providing significant environmental and economic benefit (Tables 9.7 and 9.8). Net savings to growers averaged $212 and $313 per hectare in 1998 and 1999, respectively. On the 140 ha monitored in 1998, NewLeaf™ Plus allowed growers to reduce insecticides and miticides by a total of 2,870 kg of active ingredient and 7,700 kg of formulated material as compared with an adjacent field of conventional crop. Data indicate that more than 500,000 kg of active ingredients could be eliminated annually in the Columbia Basin, USA, if NewLeaf™ Plus replaced all of the 35,000 ha of Russet Burbank grown in the region.

9.3.3 Transferring virus resistance technology to developing countries

In 1991 a transfer agreement was initiated between Monsanto, who donated virus resistance technology and transgenic know-how, and the Centre for

Research and Advanced Studies (CRAS), a public Mexican organisation. The project was brokered by ISAAA (International Service for the Acquisition of Agri-biotech Applications). Currently, varieties resistant to potato viruses X (PVX) and Y (PVY) are in seed increase and final stages of regulatory approvals. Others, involving triple resistance to PVX, PVY and PLRV are undergoing line selection prior to field testing (Kaniewski, pers. comm.). These potatoes may become the first approved transgenics that local scientists in a developing country have generated through gene transfer and product development (Qaim, 1998). Within Mexico's potato farming system the smaller, resource-poor farms purchase fewer certified, clean seed and most plant tubers harvested from a previous year. This results in a constant build up of virus in the stock which affects yields. PVX-PVY-PLRV-resistant varieties should decrease unit production costs on farms by *ca.* 13% but on smaller farms the savings may amount to *ca.* 30%. Virus resistance delivered through transgenics have obvious benefits but will need to be established within an improved framework of seed potato production and distribution within countries such as Mexico.

9.3.4 Potential risks associated with transgenic virus resistance

The development of transgenic virus resistance has raised concerns over potential interactions between viral transgenes or their products with viruses that infect the transgenic plant itself. Such interactions include genome recombination and transcapsidation (the encapsidation of one virus with coat protein of another virus; see Greene and Allison, 1994 and references therein). This has been shown to occur in some instances where several viruses infect the same plant. Risks are perceived since capsid protein influences viral transmission properties. Thomas *et al.* (1998) looked for evidence of interactions between PLRV-derived transgenes and viruses to which the transgenic potato plants were exposed and infected. Over 25,000 plants and 400 lines were transformed with 16 different coat protein constructs and *ca.* 40,000 plants and 500 lines with seven different replicase gene constructs of PLRV. Heterologous viruses found infecting the plants were screened for modifications in transmission characteristics, host range, symptoms, etc. New viruses or viruses with altered characteristics, including host range, were not detected in field-exposed or greenhouse-inoculated plants. The studies do not preclude any of the virus-virus interactions searched for, but do indicate that such interactions are rare events and that the risks of their occurrence may not be expanded by the fact that one of the genes is a transgene.

9.3.5 Herbicide resistance: NewLeaf^{TM} Roundup Ready (under development)

Roundup (glyphosate) is a broad-spectrum herbicide used for the post-emergence control of annual and perennial weeds in major crop systems. It is rapidly degraded by naturally occurring soil microbes and is not a threat to groundwater or surface water. Nor will it accumulate in the food chain. It promotes environmentally favourable tillage methods, which allow farmers to protect soil from erosion and degradation.

Several glyphosate-resistant crop species have been produced including cotton, maize and wheat. Glyphosate operates by inhibiting the action of the plant enzyme 5-enoyl-pyruvalshikimate-3-phosphate synthase (EPSPS) involved in the shikimic acid pathway which is responsible for the production of aromatic amino acids. Tolerance to glyphosate in GM crops is generated by expressing an EPSPS gene isolated from a strain of *Agrobacteriun tumefaciens*. The gene encodes an EPSPS enzyme with reduced sensitivity to glyphosate compared with the plant enzyme. The enzyme therefore functions in essential aromatic amino acid biosynthesis in the presence of glyphosate. Glyphosate-resistant potato has not reached the marketplace but, as with other glyphosate-resistant crops, could be expected to deliver more effective and precise weed control.

9.3.6 Quality traits: NewLeaf™ anti-bruise potatoes (under development)

Blackspot bruise occurs on physical impact or following damage to tubers and can cause major losses to the commercial potato processor when producing chips (crisps) and French fries. Mechanical damage initiates enzymic browning and symptoms include production of black, brown and red pigments. Reduced bruise damage will help to minimise crop rejection and waste in processing lines due to automatic discarding of blackened fries and chips (crisps). Bruise resistance is a trait important to growers and processors alike. The reaction leading to pigment production is catalysed by the enzyme polyphenol oxidase (PPO) which converts monophenols to *o*-diphenols and *o*-dihydroxyphenols to *o*-quinones. PPO activity has been reduced in potato by the company Keygene in the Netherlands which down-regulated PPO gene expression. Bachem *et al.* (1994) showed that these transgenics were less susceptible to bruising and De Both *et al.* (1996) showed in field trials that about 6% of the GM lines produced a bruise index of less than 10%. Some lines, with no detectable PPO transcript or enzyme activity, apparently show no enzymic browning. Recently, Coetzer *et al.* (2001) reported success using a tomato PPO gene in potato (co-suppression of PPO transcription).

The production of NewLeaf™ 'anti-blackspot' bruise potatoes has also involved down-regulation of a tuber-expressed PPO gene using a tuber-specific promoter. Lines of Russet Burbank that showed >70% reduction in PPO activity were again almost devoid of any bruise symptoms under field conditions (Fig. 9.2). The construct used has been optimised to increase the frequency of lines showing >70% reduction in PPO activity. This involved cloning a full length, tuber-specific PPO gene and testing three proprietary tuber-specific promoters in the cultivar Ranger Russet. All optimised constructs produced complete elimination of PPO and a corresponding commercial level of blackspot bruise resistance over three generations of replicated field trials.

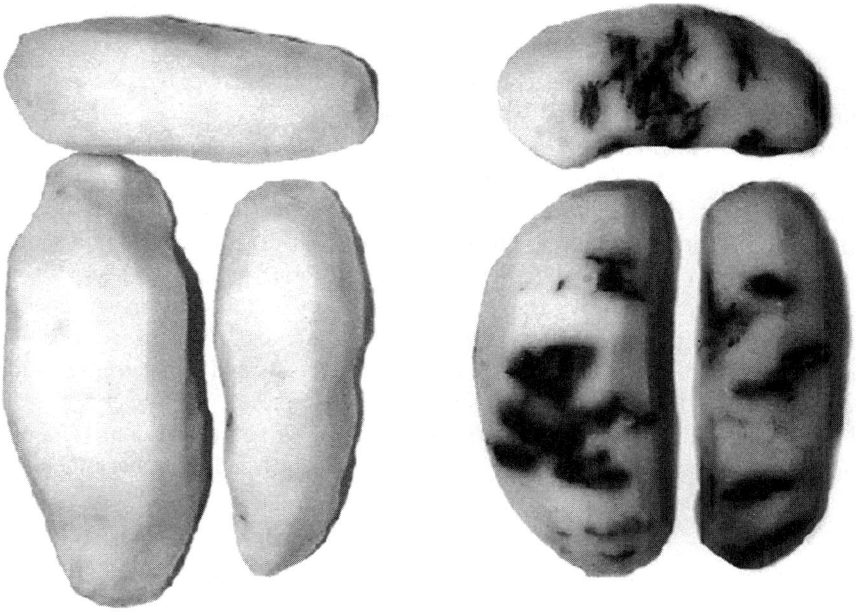

Fig. 9.2 Effect of polyphenol oxidase inhibition (antisense) on bruise damage, provided by Dr B. Krohn, Monsanto.

At the time when NatureMark®'s business was closed, commercial line selection was in progress using transgenics with combined blackspot bruise and glyphosate resistance.

9.3.7 NewLeaf™ Ultra

This line is aimed at providing growers, processors and consumers with high yielding varieties with minimal chemical, agronomic and processing inputs. To this end a 'quadra' gene construct was made to deliver resistance to CPB, PLRV, PVY and tolerance to glyphosate. Expression of all four genes has been obtained in cvs. Russet Burbank and Atlantic and field level efficacies of the target traits demonstrated (B. Krohn, Monsanto, pers. comm.). This is the first known success in stacking four genes and obtaining desired performance in the field. Conceptually, the Ultra lines could be developed further to introduce resistance to Verticillium and improved quality traits such as elevated dry matter content and bruise resistance.

9.3.8 Benefit estimates from potato agricultural biotechnology

The National Center for Food and Agricultural Policy (NCFAP) in the US published a report in January 2001 providing data on estimated benefits from commercialising GM crops (http://www.ncfap.org/pup/biotech/updated benefits.pdf). Data indicates that in 1996 only 1% of the US potato acreage was covered by GM varieties (NewLeaf™; CPB-resistant) but by 2000 this had increased to between 2 and 3%. The figures reflect combined adoption of NewLeaf™, NewLeaf™ Plus (introduced in 1999) and NewLeaf™ Y (also introduced in 1999). The report suggests that factors which contributed to low adoption rates include the need to control other pests in addition to CPB (limiting potential savings on chemical insect control), the development of a very effective conventional insecticide (imidacloprid) and, of course, lack of adoption of GM potato by large industry end-users due to consumer concerns. However, the report does indicate that by using NewLeaf™ Plus growers could save considerably in reduced losses due to net necrosis and reduced insecticide costs. Including a 'technology fee' payable to NatureMark® of $46 per acre for NewLeaf™ Plus, trials indicate an average saving of $85 per acre in 1998 and $134 per acre in 1999 (Carpenter and Giannessi, 2001 and references therein).

9.4 Current and future potential for GM potato

The following text provides examples of where future trends in GM potato production may lie, taking into account the potential for enhancing quality parameters, nutritional value and non-food uses. The examples selected are not meant to be comprehensive but illustrative.

9.4.1 Antinutritional and nutritional compounds

The USDA-ARS has developed transgenic lines with reduced glycoalkaloid content by down-regulating the expression of a gene encoding solanidine UDP-glucosyltransferase (http://www.ars.usda.gov/is/pr/1999/991115.htm). Glycoalkaloids are natural compounds which can be harmful to humans and animals when consumed in high concentration. Transgenic tubers show up to a 40% reduction in glycoalkaloid levels in field trials. This provides opportunities to rescue advanced breeding selections with excellent commercial traits but which were previously discarded due to unacceptable glycoalkaloid levels. Further benefits would accrue from reduced glycoalkaloids in potato starch wastes, as a high residue content renders the wastes unsuitable for use as fodder.

The production of transgenic crops containing proteins with improved amino acid composition should be of benefit to humans as well as to monogastric animals (pig, poultry, etc.) unable to synthesise all of the amino acids needed to sustain life. The potato is the most important non-cereal food crop in the world. However, it contains limited amounts of the essential amino acids lysine, tryptophan, methionine, and cysteine. Improvements in the nutritional value of food crops such as potato are especially important for people subsisting on a vegetarian diet in which the main source of protein comes from seeds, grains, tubers, etc., which contain limiting amounts of essential amino acids.

Chakraborty *et al.* (2000) reported improvements in the nutritive value of transgenic potato through the expression of a non-allergenic seed albumin gene (*AmA1*) from *Amaranthus hypochondriacus*. As a donor gene, the *AmA1* gene has several advantages for genetic transformation experiments. First, this seed protein has a well-balanced amino acid composition, making it nutritionally superior to other proteins recommended by the WHO. Second, the purified protein has no known allergenic properties. Finally, the protein is controlled by a single gene, which facilitates integration into other species. The team showed a five- to ten-fold increase in *AmA1* transcript levels in tubers of transgenic lines using the tuber-specific granule bound starch synthase (GBSS) promoter compared with the 35S CaMV promoter. Transgenic lines contained a significant two- to four-fold increase in lysine, methionine, cysteine, and tyrosine content in their protein amino acids. Data collected for two consecutive years revealed a 35 to 45% increase in total protein content in transgenic tubers, which corresponded to an increase in most essential amino acids. Grain amaranth is used in many foods throughout the world and amaranth forage has been used for centuries as an important component of the human diet throughout the tropics. The authors presented these facts as evidence of the non-allergenic nature of amaranth.

Fructans, or fructose-oligosaccharides, consist of short chains of fructose molecules. Inulin is a mixture of linear fructose-polymers with different chain-length and a glucose molecule at each C2-end. In over 30,000 plants (e.g. chicory, onion, asparagus, artichoke) inulin serves as a storage carbohydrate. Compounds such as inulin reduce the energy density of food and are used to enrich food with dietary fibre or to replace sugar and fat. When fructans are consumed, the undigested portion is reported to support growth of 'friendly' bacteria, such as Bifidobacteria and Lactobacillus species. Other benefits noted include increased production of beneficial short-chain fatty acids such as butyrate, increased absorption of calcium and magnesium and improved elimination of toxic compounds (van den Heuvel, 1999). Hellwege *et al.* (2000) have developed transgenic potato tubers which synthesise the full spectrum of inulin molecules naturally occurring in globe artichoke (*Cynara scolymus*). High molecular weight inulins have been produced by expressing the sucrose:sucrose 1-fructosyl transferase and the fructan:fructan 1-fructosylhydrolase genes from globe artichoke. Inulin made up 5% of the dry weight of the transgenic tuber. This approach has the potential to enhance the value of staple foods such as potato with compounds giving additional benefits.

9.4.2 Food processing and industrial uses

Starch is the primary storage compound in tubers and starchy foods are the world's most abundant staples. It is the most important source of calories in the animal and human diet and provides a starter material for the preparation of more than 500 different commercial products. The physical properties of starch vary with plant source but there are considerable opportunities to generate novel starches for use in food and non-food market sectors. Genetic engineering has already generated novel potato starch, including high amylopectin starch (with no apparent yield penalty) through the down-regulation of the granule bound starch synthase gene which controls amylose synthesis (Visser *et al.* 1991; Kossman and Lloyd, 2000 and references therein). High amylose starch is also in great demand by the starch industry for its unique functional properties, but very few high amylose crops are available. Scwall *et al.* (2000) showed that concurrent down-regulation of two starch branching enzymes, A and B, in potato tubers modifies both starch grain morphology and composition and produces a significant increase in amylose content.

Stark *et al.* (1992) increased the starch content of tubers by expressing an *E. coli* glgC16 gene which encodes for the enzyme ADPglucose pyrophosphorylase. The corresponding potato enzyme resides in the starch granule and plays a key role in starch biosynthesis. The *E. coli* enzyme is not regulated by the same fine control mechanisms which operate on the endogenous potato enzyme and is therefore able to increase the production of ADPglucose which becomes incorporated into the growing starch granule. Tuber starch content can be increased by up to 25% in some glgC16 expressing lines but the response appears to be genotype dependent. These high starch potatoes also accumulate lower levels of reducing sugars (glucose and fructose) in stored tubers which is highly relevant to the requirements of the processing sector. The processing industry requires low reducing sugar levels in tubers as these sugars are primarily responsible for non-enzymic browning through a typical Maillard reaction which occurs at the temperatures required to generate potato chips (crisps) and French fries.

Ideally, the industry would like to store tubers at low temperature (*ca.* 4°C) to minimise sprout growth and eliminate the need to use chemicals to suppress the sprouting process. However, low temperatures induce glucose and fructose accumulation. Success in minimising sugar accumulation using transgenic approaches have come from the use of the glgC16 gene and from modifying the expression of genes in pathways of primary carbohydrate metabolism, e.g., by minimising the conversion of sucrose to glucose and fructose by expressing invertase inhibitor protein (Greiner *et al.* 1999). More detail on the control of sugar accumulation starch biosynthesis and potato quality can be found in Davies and Viola (1992), Davies and Mackay (1994), Davies (1996) and Davies (1998).

9.4.3 Pharmaceutical uses

Tacket *et al.* (2000) reported a new approach for delivering vaccine antigens using inexpensive plant-based oral vaccines generated in potato. Norwalk virus capsid protein (NVCP) assembled into virus-like particles, was used as a test antigen to determine immune responses in healthy adults eating GM potato containing NVCP. Overall, 19 out of 20 volunteers developed an immune response of some kind. Similarly, Chong and Langridge (2000) demonstrated expression of bioactive antimicrobial human lactoferrin in potato plants. This was the first report of synthesis of full length biologically active hLF in edible plants. Expression was significant (up to 0.1% of total soluble protein) and antimicrobial activity against four different human pathogenic bacterial strains was detected in extracts of tuber tissues.

At the time of writing this chapter, livestock and related industries in the UK are undergoing torrid times due to a severe outbreak of foot and mouth disease. Implementation of an expensive vaccination programme has been hotly debated. It is of some interest, therefore, that Carrillo *et al.* (2001) demonstrated the induction of a virus-specific antibody response to foot and mouth disease virus using the structural protein VP1 expressed in transgenic potato plants. The group previously reported the oral and parental immunogenicity of the structural protein VP1 of foot and mouth disease virus (FMDV) expressed in different transgenic plants. Their recent report indicates that transgenic potatoes containing the VP1 gene cloned under the regulatory activity of either a single or a double copy of the 35S CaMV promoter, represents a viable strategy for increasing the level of VP1 gene expression. Furthermore, immunised animals presented a FMDV VP1 specific antibody response and showed protection against the experimental challenge. These results clearly show the potential of using plants as antigen expression systems.

9.5 Revised legislation on GM crops in Europe

The prospects for further commercial releases of GM potatoes in North America, and of any in Europe, clearly await significant consumer acceptance before consistent, large volume end-user markets emerge. The factors which will determine acceptance or otherwise are complex and beyond the scope of this chapter. However, on 14 February 2001, the European Parliament finally adopted a joint text concerning the deliberate release into the environment of genetically modified organisms. The text, which revises the European Directive 90/220/EEC (which itself lays down regulations for commercial releases of GM crops into the environment), seeks to increase the transparency and efficiency of the decision-making process on GM crops and products in the European Union. This in itself may improve public confidence. The revised directive aims to promote a harmonisation of risk assessment, and to introduce clear labelling requirements for all GMOs placed on the market. There are proposals to introduce mandatory monitoring for GM products and mandate a time limitation (renewable) of ten years, maximum, for first-time consent. There are plans to introduce compulsory monitoring of GM crops after they have been placed on the market, to provide for a common methodology to assess the risks associated with their release, and to include a mechanism to allow modification, suspension or termination of the release when new information on risks becomes available.

Other important components of the evolving legislation include: a gradual elimination of antibiotic resistance markers in commercial GMOs by the end of 2004, and by 2008 for releases into the environment for experimental purposes; a plan to bring forward a legislative proposal on environmental liability before the end of 2001, also covering damage resulting from GMOs; public registers of GMOs released into the environment for experimental purposes; introduction of general rules on traceability and labelling of GMOs and products derived from them; mandatory monitoring after GMOs are placed on the market; mandatory consultation of relevant European Scientific Committee(s); mandatory consultation of the public concerning both experimental and commercial releases; the application of the precautionary principle when implementing the Directive; the opportunity for consulting Ethics Committee(s) on issues of general nature. An excellent review of international comparisons of regulatory frameworks for food products of biotechnology has been published by MacKenzie (2000).

9.6 The future

The commercial future of transgenic potatoes will clearly depend on end user 'pull' and the demand for distinct added value that these crops might provide. Such added value will be derived from reduced production and processing costs without yield penalties, improved quality and nutritional value, novel uses (diversification) and/or proven environmental benefits. It is clear that the benefits must be transparent to consumers who have increasing expectations with regard to food safety. The demands of growers and consumers will vary with geographical location and quality of life and a key challenge will be to provide proven biotechnological advances, technologies and products to the developing world at low, or zero costs, whilst maintaining profitability (and altruism) within the industrial sector. Due to research and development costs this will be sustainable only if GM crops find their niche in global marketplaces.

At present, Europe is a desert for GM crops and market resistance is evident in North America. NatureMark® products have figured significantly in this chapter since they have been the only company to place GM potato crops on the market. The demise of NatureMark® due to market forces may be seen as an opportunity by others to fill the gap. The number of mergers between plant biotechnology companies has been significant in recent years which potentially creates the 'monopolies' that many campaigners against GM crops would not wish to see. However, in the current climate the development of GM crops and products derived from them will require the 'patient money' that only large multinationals can provide whilst satisfying shareholder demands through other components of their business. Due to the costs involved, opportunities for public sector research establishments to deliver GM crops to the market without strategic alliances with companies will be minimal to say the least. These are the realities; the industrial sector needs to make profit and the public sector needs to work closely with the industrial sector. However, this must not jeopardise independence, particularly where risk assessment is involved.

The implementation of stringent risk assessment and risk management strategies together with labelling and traceability has become an important issue with the public. The recent revision of European legislation goes some way down this road. Scientifically, one can predict an increased use in risk assessment of micro-arrays for large scale analysis of gene expression and proteomics and metabolic profiling for more detailed analysis of substantial equivalencies. Combined use of these approaches will also make major contributions to our understanding of metabolic networks and signal transduction mechanisms which govern important plant and crop traits. The explosion of sequence information delivered through genomics programmes will enable expression analysis on entire potato transcriptomes in the very near future, opening up immense research opportunities and challenges in information technology.

One can predict the development of more efficient transformation systems, advances in transgene stacking to regulate several traits simultaneously (or traits which are under polygenic control) and to provide durable, multigene resistances to pests and diseases. There is already a move towards the use of transformation vectors which are 'minimal' in the use of DNA sequences not required for the transformation event and which do not contain antibiotic resistance marker genes. The technologies are also available to sequence the regions bordering inserted sequences in the target plant. This would give information on potential disruption to important gene elements and might assist the targeting of metabolite analysis in a risk assessment exercise. However, it should be emphasised that the risk assessment procedures currently applied to GM crops are already far more rigorous than for any crops generated by traditional or mutation breeding.

9.7 Additional reading

ALEXANDRATOS N (1999), 'World food and agriculture: outlook for the medium and longer term', *Proc Nat Acad Sci*, 96, 5908–5914.

ATKINSON H J, GREEN J, COWGILL, S and LEVESLEY A (2001), 'The case for genetically modified crops with a poverty focus', *Trends in Biotech*, 19, 91–96.

BRADSHAW J E and MACKAY G R (1994), *Potato Genetics*, CAB International.

BURTON W G (1989), *The Potato Crop (third edition)*, Longman Scientific and Technical.

FULTON M, FURTAN H, GOSNELL D, GRAY R, HOBBS J, HOLZMAN J, KERR B, MCNAUGHTEN J, STEVENS J and STOVIN, D (2001), *Transforming agriculture. The benefits and costs of genetically modified crops.* Prepared for The Canadian Biotechnology Advisory Committee Project Steering Group on the Regulation on Genetically Modified Foods (http://www.cbac-ccb.ca).

INSTITUTE OF BIOLOGY (2000), *Genetically modified crops and food: scientific advice to the Government: genetically modified food*, United Kingdom, 1–23.

MORRIS J and BATE R (1999), *Fearing food: risk, health & environment.* Butterworth Heinemann, Oxford.

NUFFIELD COUNCIL ON BIOETHICS (1999), *Genetically modified crops: the ethical and social issues,* 1–165, (http://www.nuffield.org).

SHISLAIN M and GOLMIRZAIE A (1998), 'Genetic engineering for potato improvement', in Kurana S M P, Chandra R and Upadhya M D, *Comprehensive Potato Biotechnology*, Malhotra Publishing House, New Delhi, 115–162.

9.8 Acknowledgements

The author wishes to thank NatureMark® and Monsanto for providing information relevant to field performance of GM potatoes. The author also acknowledges financial support from the Scottish Executive Environment and Rural Affairs Department (SEERAD).

9.9 References

BACHEM C W B, SPECKMANN G J, VAN DER LINDE P C G, VERHEGGEN F T M, HUNT M D, STEFFENS J C and ZABEAU M (1994), 'Antisense expression of polyphenol oxidase genes inhibits enzymatic browning in potato tubers', *Bio Technology* 12, 1101–1105.

BELKNAP W R, CORSINI D, PAVEK J J, SNYDER G W, ROCKHOLD D R and VAYDA M E (1994), 'Field performance of transgenic Russet Burbank and Lemhi Russet potatoes', *Am Pot J*, 71, 285–296.

BEMSTER A AND DE BOKS J (1987), 'Survey of properties and symptoms', in *Viruses of potatoes and seed potato production*, Pudoc, Wagningen, second edition, 84–113.

CARPENTER J E and GIANESSI L P (2001), *Agricultural biotechnology: Updated benefit estimates*. National Centre for Food and Agricultural Policy (http://www.ncfap.org).

CARRILLO C, WIGDOROVITZ A, TRONO K, DUS SANTOS M J, CASTANON S, SADIR A M, ORDAS R, ESCRIBANO J M and BORCA M V (2001), 'Induction of a virus-specific antibody response to foot and mouth disease virus using the structural protein VP1 expressed in transgenic potato plants', *Viral Immunol*, 14, 49–57.

CHAKRABORTY S, CHAKRABORTY N and DATTA A (2000), 'Increased nutritive value of transgenic potato by expressing a non-allergenic seed albumin gene from *Amaranthus hypochondriacus', Proc Nat Acad Sci*, 97, 3724–3729.

CHONG D K and LANGRIDGE W H (2000), 'Expression of full length bioactive antimicrobial human lactoferrin in potato plants', *Transgenic Res*, 9, 71–78.

COETZER C, CORSINI D, LOVE S, PAVEK J and TURNER N (2001), 'Control of enzymic browning in potato (*Solanum tuberosum* L.) by sense and antisense RNA from tomato polyphenol oxidase', *J Agric Food Chem*, 49, 652–657.

COLLINS, W (2000), 'Research strategies for potatoes: a global approach', in *World Potato Congress. Proceedings of the Fourth World Potato Congress*, Amsterdam, The Netherlands, September, 209–214.

DALE P J and HAMPSON K K (1995), 'An assessment of morphogenic and transformation efficiencies in a range of varieties of potato (*Solanum tuberosum* L.)', *Euphytica*, 85, 101–108.

DAVIES H V (1996), 'Recent developments in our knowledge of potato transgenic biology', *Potato Res*, 39, 411–427.

DAVIES H V (1998), 'Prospects for manipulating carbohydrate metabolism in potato tubers,' in *Aspects Appl Biol, Protection and production of sugarbeet and potatoes*, 52, 245–254.

DAVIES H V and MACKAY G R (1994), 'Exploitation of genetic variation to improve potato quality', *Aspects Appl Biol*, 39, 45–49.

DAVIES H V and VIOLA R (1992), 'Regulation of sugar accumulation in stored potato tubers', *Post-harvest News and Information*, 3, 97–100.

DE BOTH M, VERHEGGEN F, BACHEM C, SPECKMANN G J, VAN DER LINDE P and ZABEAU M (1996), 'Genetic engineering of bruising resistance into commercial cultivars', in *Abstracts of conference papers, posters and demonstrations of the 13th EAPR Triennial Conference*, Veldhoven, The Netherlands, 19–20.

GLENNON, L (2000), 'The potato in the food business – past to present', *World Potato Congress Article*, March. http://www.potatocongress.org/articles/march00.htm.

GREENE A E and ALLISON R F (1994), 'Recombination between viral RNA and transgenic plant transcripts', *Science*, 2632, 1423–1425.

GREINER S, RAUSCH T, SONNEWALD U and HERBERS K (1999), 'Ectopic expression of a tobacco invertase inhibitor prevents cold induced sweetening of potato tubers', *Nature Biotech*, 17, 708–711.

HELLWEGE E M, CZAPLA S, JAHNKE A, WILLMITZER L and HEYER A G (2000), 'Transgenic potato (*Solanum tuberosum*) tubers synthesise the full spectrum of inulin molecules naturally occurring in globe artichoke (*Cynara scolymus*)', *Proc Nat Acad Sci*, 97, 8699–8704.

JAMES, C (2000), *Global Status of Commercialized Transgenic Crops*, ISAAA Briefs No 21, ISAAA, Ithaca, NY.

KANIEWSKI W K and THOMAS P E (1999), 'Field testing for virus resistance and agronomic performance in transgenic plants', *Molecular Biotech*, 12, 101–115.

KARP, A (1990), 'On the understanding of somaclonal variation', *Oxford Surveys of Plant Mol Biol Cell Biol*, 2, 199–234.

KAWCHUK L M and PRUFER D (1999), 'Molecular strategies for engineering resistance to potato viruses', *Can J Plant Pathol-Revue Canadienne de Phytopathologie*, 21, 231–247.

KOSSMAN J and LLOYD J (2000), 'Understanding and influencing starch biochemistry', *Crit Rev Biochem Mol Biol*, 35, 141–196.

KUMAR A (1994), 'Somaclonal variation', in Bradshaw J E and Mackay G R, *Potato Genetics,* CAB International, 197–212.

MACKENZIE D J (2000), *International Comparison of Regulatory Frameworks for Food Products of Biotechnology*. Prepared for The Canadian Biotechnology Advisory Committee Project Steering Committee on the Regulation of Genetically Modified Foods (http://cbac-cccb.ca).

MOHAMMED A, DOUCHES D S, PETT W, GRAFIUS E, COOMBS J, LISWIDOWATI W, LI W and MADKOUR M A (2000), 'Evaluation of potato tuber moth (Lepidoptera: Gelechiidae) resistance in tubers of Bt-cry5 transgenic potato lines', *J Econ Entomol*, 93, 472–476.

QAIM, M (1998), *Transgenic Virus Resistant Potatoes in Mexico. Potential socio-economic implications of North-South biotechnology transfer*, ISAAA Briefs No 7, ISAAA, Ithaca, NY.

RHODES D J (2000) 'Crop protection, environment and food safety: meeting the needs of the industry into the 21st century', in *World Potato Congress. Proceedings of the Fourth World Potato Congress*, Amsterdam, The Netherlands, September, 209–214.

RIEBE J F and ZALEWSKI J C 'Pesticide reduction and disease control with genetically modified potato'. *Environmental Biosafety Research* (submitted).

ROGAN, G J, BOOKOUT J T, DUNCAN D R, FUCHS R L, LAVRIK P B, LOVE S L, MUETH M, OLSON T, OWENS E D, RAYMOND P J and ZALEWSKI J (2000), 'Compositional analysis of tubers from insect and virus-resistant potato plants', *J Agric Food Chem*, 48, 5936–5945.

ROGAN G J, THOMAS P E, LAWSON C E, BAR-PELED M, REDING K H, ZALEWSKI J C and KANIEWSKI W K (submitted), 'From concept to commercialization of virus and insect-resistant potato plants'.

SCWALL G P, SAFFORD R, WESTCOTT R J, JEFFCOAT R, TAYAL A, SHI Y C, GIDLEY M J and JOBLING SA (2000), 'Production of very high amylose potato starch by inhibition of SBE A and B', *Nature Biotech*, 18, 551–554.

STARK D M, TIMMERMAN K P, BARRY G F, PREISS J and KISHORE G M (1992), 'Regulation of the amount of starch in plant tissues by ADPglucose pyrophosphorylase', *Science*, 258, 287–292.

STEWART, J (2000), 'Managing the Colorado Potato Beetle with Chemicals and their alternatives'. *World Potato Congress Article* March, *http://www.potatocongress.org/articles/august99.htm.*

THOMAS P E, HASSAN S, KANIEWSKI W K, LAWSON E C and ZALEWSKI J C (1998), 'A search for evidence of virus/transgene interactions in potatoes transformed with the potato leafroll virus replicase and coat protein genes', *Mol Breed*, 4, 407–417.

VAN DEN HEUVEL E G, MUYS T, VAN DOKKUM W and SCHAAFSMA G (1999), 'Oligofructose stimulates calcium absorption in adolescents', *Am J Clin Nutr*, 69, 544–548.

VAN DER WILK F, LUTKE WILLINK D (POSTHUMUS), HUISMAN M J, HUTTINGA H and GOLDBACH R (1991), 'Expression of the potato leafroll luteovirus coat protein gene in transgenic potato plants inhibits viral infection', *Plant Mol Biol* 17, 431–439.

VISSER R G F, SOMHORST I., KUIPERS G J, RUYS N J, FEENSTRA W J and JACOBSEN E (1991), 'Inhibition of the expression of the gene for granule-bound starch synthase in potato by antisense constructs', *Mol Gen Genet*, 225, 289–296.

10

Cucurbits, pepper, eggplant, legumes and other vegetables

A. Bernadac, A. Latché, J.-P. Roustan, M. Bouzayen and J.-C. Pech, Ecole nationale Supérieure Agronomique de Toulouse (INP-ENSAT/ INRA)

10.1 Introduction

This chapter is related to the biotechnology of a number of vegetables that are botanically and morphologically unrelated but that have been associated because they are generally secondary crops on the world scale as compared to the species dealt with in the previous chapters. However, some of them may have great importance in some areas as they provide an essential part of the diet and/or an important income complement for farmers.

The group includes:

- fleshy fruits of the cucurbitaceae (melon, squash, cucumber) and solaneaceae (pepper, eggplant) families that are eaten raw or after cooking
- legumes other than soybean that are eaten as vegetables such as peas and beans
- bulky organs other than potato corresponding to roots (carrots), tubers (sweet potatoes) and bulbs (Allium species such as onion and garlic)
- leafy vegetables including lettuce, spinach and brassica species (cabbage, broccoli, cauliflower, etc.).

Overall these crops have received less attention in terms of genetic engineering than other major crops such as maize, rice or soybean or than model plants such as *Arabidopsis* and tobacco, although progress in the transformation procedures in major crops or model plants has been beneficial to secondary species. Nevertheless, some of the species covered in this review remain recalcitrant to transformation and in many cases, although successful gene transfer has been claimed, the efficiency is too low for routine transformation and/or is restricted to certain cultivars.

10.2 Biotechnology of cucurbits

Cucurbits is a general term for the species of the Cucurbitaceae family. This family comprises about 130 genera and more than 900 species of which only a few are cultivated.[1,2] Melon, watermelon, pumpkins, squash, gourds and cucumber represent the most important cucurbits. The world production of these crops is in excess of 115 million metric tons, on a total area harvested of 6.9 million hectares throughout the world (FAOSTAT database, 2000). The world production and the harvested area has increased three-fold in the last three decades. Similarly, yields have been also greatly improved in the last few years, for instance from 11.6 tons/ha in 1970 for cucumber to 16.2 tons/ha today. In this chapter, we concentrate on the most widely cultivated species with emphasis on those for which biotechnology methods have been applied (Table 10.1).

10.2.1 Methods of transformation

Efficient transformation methods require good control of the regeneration step through either organogenesis or somatic embryogenesis. In cucurbits, regeneration by direct or indirect organogenesis has been achieved from various plant tissues including cotyledons, hypocotyls and leaves.[3–5] The efficiency of regeneration has been found to be highly dependent on the stage of development and the growth conditions[6] and the genotype.[7] Regeneration of cucurbits has been obtained by somatic embryogenesis from callus or different explant tissues, but development of somatic embryos into plants still remains difficult, being highly dependent on the genotype[8,9] and controlled by crossing.[10] This renders the process of plant transformation more difficult and reduces the number of cultivars susceptible to be engineered. On the other hand, the regeneration process is known to induce endo-polyploidisation[11,12] leading to reduced productivity of the plants and altered shape and smaller size of the fruit, mainly in melon. Endo-polyploidisation occurs during the development of cotyledons, and at a lower rate in leaves.[12–14] Thus, the regeneration protocols must be adapted to yield low endo-polyploid plantlets.[14]

Currently, several methods for gene transformation including *Agrobacterium*-mediated transformation and biolistic are available. A great number of melon, cucumber and squash genotypes have been transformed with different genes and in particular with genes of agronomic interest (Table 10.1). Transgenic plants currently tested in field conditions or commercialised were generated via *Agrobacterium*-mediated transformation. In cucurbits, cotyledonary explants and adventitious shoots have been widely used for generation. In these transformation systems, the *NptII* selectable marker gene is the most widely used allowing selection of transformed tissues in the presence of kanamycin.

Biolistic transformation has been used for the generation of transgenic cucumber and melon plants, respectively from highly embryogenic cell suspension cultures[15] or embryos developed on cotyledons[16] leading to a high stability of the transgene. Genetic transfer via *A. rhizogenes* in cucumber[3] or *Cucurbita*

Table 10.1 A summary of gene transfer and corresponding agricultural traits into cucurbits

Plant species	Transformation methods	Gene utilised	Agricultural traits	Inheritance	Field or greenhouse test	References
Cucumis melo	A.t.	CMV coat protein	CMV resistance			24
Cucumis melo	A.t. and Biolistic	CMV-WL coat protein	CMV resistance	R1	No	25
Cucumis melo	A.t.	ZYMV, WMV-2, CMV coat protein	ZYMV, WMV-2, CMV resistance			45
Cucumis melo	A.t.	ACC oxidase gene in antisense orientation	Slow ripening			14, 41
Cucumis melo	A.t.	HAL1 yeast	Salt tolerance	yes	No	37
Cucumis melo	A.t.	ZYMV, WMV-2, CMV coat protein	ZYMV, WMV-2, CMV resistance	yes	Field	28
Cucumis melo	A.t.	S-adenosylmethionine hydrolase	Slow ripening			39
Cucumis sativus	A.t.	CMV-C coat protein	CMV resistance	yes	No	22
Cucumis sativus	A.t.	CMV coat protein	CMV resistance	yes	Field	23
Cucumis sativus	A.t.	Rice chitinase	Resistance to *Botrytis cinerea*	yes	No	5
Cucurbita pepo	A.t.	ZYMV, WMV2 coat protein	ZYMV, WMV-2, resistance	yes	Field	29
Cucurbita pepo	A.t.	ZYMV, WMV coat protein	ZYMV, WMV resistance	yes	Greenhouse and field	45
Cucurbita pepo	A.t.	ZYMV, WMV coat protein	ZYMV, WMV resistance	yes	field	30
Cucurbita pepo	A.t.	ZYMV, WMV-2, CMV coat protein	ZYMV, WMV-2, CMV resistance	yes	Field	27

Note:
A.t.: *Agrobacterium tumefaciens*

pepo,[17,18] or via the pollen-tube pathway in watermelon[19] have been described but so far do not seem to have provided an efficient process in cucurbits.

10.2.2 Resistance to viruses

More than 30 viruses can infect cucurbits, but only twelve cause economically important losses.[20,21] The following viruses are the most commonly encountered in warm and temperate areas and capable of infecting several cucurbit species: (i) CMV (Cucumber Mosaic Virus); (ii) WMV (Watermelon Mosaic Virus); (iii) ZYMV (Zucchini Yellow Mosaic Virus); (iv) PRSV (Papaya RingSpot Virus) and (v) SqMV (Squash Mosaic Virus). The incorporation of virus resistance genes into cucurbits has been the goal of many breeding programmes. However, because resistance genes are derived from wild species, they are not simply inherited and/or are recessive, so introgression of these genes into horticulturally acceptable genotypes is not an easy task. Genetic engineering has allowed breeders rapidly to develop virus-resistant varieties by introducing dominant coat protein or replicase viral genes into inbred parents of existing commercial hybrids.

Viral coat protein-mediated protection has allowed the generation of CMV-resistant cucumber plants[22] that can exhibit a high level of CMV-resistance under field conditions.[23] Similarly, melon plants over-expressing CMV or ZYMV coat proteins have been generated.[24–26] For Clough and Hamm,[26] a significant reduction in disease incidence in the transgenic lines occurred in field conditions. Expression of the coat protein gene delayed virus disease development and the subsequent systemic spread of virus in the transgenic plants. However, because most cucurbits are susceptible to several viruses, multiple virus resistance has been sought, instead of single virus resistance.

Transgenic lines of yellow crookneck squash (*Cucurbita pepo*) containing multiple coat protein constructs of CMV, WMV and ZYMV have been generated and used for the production of hybrid varieties.[27] Field evaluation was performed for two transgenic lines: (i) CZW-30, transformed with the triple coat protein gene construct that exhibited resistance to all three viruses[28] and (ii) ZW-20, transformed with the coat protein genes of WMV and ZYMV that displayed excellent resistance to the two viruses.[29] The ZW-20 line and subsequent generations were approved for commercial distribution by Asgrow in the USA in 1995 and were the first disease resistance transgenic plants to be approved for commercialisation.[27] Similarly, two experimental transgenic summer squash hybrids, possessing resistance to ZYMV and WMV and to ZYMV, WMV and CMV exhibited outstanding resistance in field conditions to the corresponding viruses as compared to the non-transgenic virus-susceptible hybrid 'Pavo'.[30]

Another strategy to enhance virus tolerance in melon, based on the overexpression of polyribosime directed toward CMV coat protein, has been reported by Plages.[31] A further approach to reduce development of the virus in cucurbits is to use a cDNA copy of RNA1 or an altered form of the *2a replicase* gene from CMV which have been found to enhance virus tolerance in transgenic tobacco.[32, 33]

10.2.3 Resistance to fungi

The improvement of resistance to phytopathogenic fungi is one of the most crucial objectives in cucurbit cultivation. However, the problem is complex because highly resistant sources must be available for the breeding programmes. For example, *Fusarium* wilt-resistant materials have been found in genetic resources in cucumber and used in breeding programmes.[34] On the contrary, breeding materials for resistance against grey mould (*Botrytis cinerea*), one of the most serious cucumber diseases have not been found.

Transformation techniques have been used to produce resistant transgenic plants by expressing chitinase genes aimed at inhibiting fungal development in the plant. In cucumber, the response of transgenic plants expressing different chitinase genes originating from petunia, tobacco or bean to inoculation with fungal pathogens including *Alternaria radicini*, *Botrytis cinerea*, *Colletotrichum lagenarium*, and *Rhizoctonia solani* has shown that chitinase overexpression failed to reduce the development of the disease.[35] In opposition, Tabei *et al.*[5] have succeeded in obtaining *B. cinerea*-resistant lines of cucumber by expressing a rice endochitinase gene. Different responses for disease resistance were observed including inhibition of appressorium formation and penetration of hyphae as well as restriction of invasion of the infection hyphae. Furthermore, disease resistance against grey mould was confirmed to be inheritable and chitinase overexpression did not affect the morphological development of the plants. Although the effectiveness of the fungal resistance must be confirmed by field trials, this approach could be used for other cucurbits species to fight against plant diseases caused by phytopathogenic fungi. However, the extent of disease tolerance appears to be highly dependent on the type of chitinase protein expressed and the characteristics of the fungal pathogen. On the other hand, chitinase overexpression in combination with a-glucanase, PR proteins or other enzymes involved in the synthesis of antifungal compounds could provide a broader spectrum of activity against fungal pathogens.

10.2.4 Resistance to abiotic stresses

Unfavourable soil and water conditions are important constraints and lower crop yields. The adaptive response of plants to water and salt stress involves the activation of a large number of genes. The individual function of these genes is not well understood in as much as most of the genes are not specific to these types of stresses.[36] The number of plants that have been engineered for salt or drought resistance is therefore very limited. However, a yeast salt-tolerance gene encoding a water-soluble protein HAL1 has been transferred to two cultivars of *Cucumis melo* via *A. tumefaciens*. Some halotolerance has been observed in primary transformants[37] that could be transmitted to self-pollinated progeny.[38]

10.2.5 Fruit quality traits

In recent years, methods based on the genetic manipulation of fruit ripening, have been developed as an alternative strategy to increase the storage life and improve the quality of fruits. Although most studies have been carried out with the tomato as model fruit, Cantaloupe melons represent good targets due to their fast ripening rate and short postharvest life. Postharvest losses largely due to overripening have been estimated to be near 30% in the USA.[39]

In climacteric fruits like melon, the expression of many genes involved in the ripening process is stimulated by ethylene. This plant hormone represents an obvious target for controlling fruit ripening by genetic manipulation. An antisense construct of a cDNA encoding melon ACC oxidase (ACO)[40] driven by the 35S promoter has been used to generate transgenic melons of the Cantaloupe Charentais type (cv Vedrantais). Among several transformants, one line was selected that exhibited strong inhibition (over 99.5%) of ethylene production.[41] Ethylene suppression resulted in the inhibition of rind yellowing, flesh softening, climacteric respiration and peduncle detachment.[42] However, coloration of the flesh, accumulation of sugars and organic acids, and the synthesis of the ethylene precursor ACC were not affected by ethylene suppression. Ethylene-inhibited fruit also evolve far fewer aroma volatiles than wild-type (WT) fruit.[43] However, ethylene treatment of antisense ACO fruit is capable of restoring the WT phenotype.

Antisense ACO Charentais cantaloupe melons enabled an assessment of the role of ethylene in some physiological disorders. These melons, in contrast to WT fruit, do not develop the characteristic pitting and browning of the rind associated with chilling injury either when stored at low temperature (2ºC for 3 weeks) or upon rewarming to room temperature. Tolerance to chilling was clearly correlated with a lower accumulation of ethanol and acetaldehyde and a higher activity of activated oxygen scavenging enzymes in antisense ACO fruit. It appears that ethylene acts in conjunction with low temperature to induce metabolic shifts that participate in the development of chilling injury.[44]

Another strategy for reducing ethylene synthesis has been used for cantaloupes of the American type commonly referred to as muskmelon. The T3 bacteriophage gene product S-adenosylmethionine hydrolase (SAMase) catalyses the degradation of SAM, a precursor to ethylene biosynthesis. The gene was expressed under the control of a chimerical fruit-specific promoter.[39] It was confirmed that the chimerical promoter was capable of driving SAMase expression in a fruit-specific and ethylene-responsive manner. SAMase melons showed significant reduction in ethylene biosynthesis (up to 75%) both as inbred homozygous plants and as hybrids. The inhibition of ethylene production of SAMase fruit was not strong enough to dramatically alter the ripening and postharvest phenotype. However, although the onset of maturity was not significantly delayed, full maturity of transgenic fruit occurred over a shorter period. Also, the concentration of soluble sugars was frequently higher in transgenic fruit probably because fruit slip was delayed by one to three days, allowing more sugars to accumulate in the fruit before its harvest.

These examples indicate that two alternatives exist for the postharvest handling of ethylene-inhibited Cantaloupe melons: (i) use of lines and corresponding transgenic hybrids with very strong inhibition of ethylene production (in this case fruit can be stored in the cold with no risk of chilling injury and then ripened to the desired stage of maturity before distribution), or (ii) use of lines and corresponding hybrids with less reduction of ethylene production and showing delayed ripening and extended shelf-life so as to reach the consumer at the right maturity stage without ethylene treatment. A comparative evaluation of the ethylene-inhibited transgenic genotypes and already existing long-keeping varieties of melon deserves to be carried out from the points of view of quality of the fruit, storability and postharvest shelf-life.

10.3 Biotechnology of pepper

Peppers belong to the genus *Capsicum.* They are found around the world in various edible forms and they exhibit great variations in size, shape, flavour and colour as well as plant habits. Five main species have been domesticated (*C. frutescens* L., *C. chinense* Jacq. , *C. baccatum* L., *C. pubescens* R.& P., and *C. annuum* L.) among which *C. annuum* is the most widely cultivated. Two types of cultivars can be distinguished: sweet and pungent. A horticultural classification of peppers has been made by Smith[45] based on the size, shape, colour and taste (sweet or pungent) of the fruit. The transfer of characters from wild species to cultivated genotypes is hampered by interspecific incompatibility and/or hybrid sterility. Only in a few cases have these barriers been overcome through embryo culture and somatic hybridisation. A recent update on pepper *in vitro* regeneration and transformation has been made by Steinitz *et al.*,[46] and a summary of genes of interest transferred into pepper is given in Table 10.2.

10.3.1 Methods of transformation

Most of the reports on pepper regeneration deal with two main types, bell and chile. Shoot regeneration is generally dependented on organogenesis from cotyledons and hypocotyls, but the regeneration capacity is highly dependent upon the cultivar,[47,48] the developmental stage[49] and the location[50] of the plant tissue. Zhu *et al.*[51] found that leaves of a Chinese sweet pepper variety were by far the best material for successful regeneration, transformation with young leaves giving the highest rates. However, Arroyo and Revilla[52] had previously described an efficient regeneration procedure for some commercially important spanish cultivars using hypocotyls and especially cotyledons with high regeneration rates. Fari *et al.*[53] by screening a seed collection of chile peppers for *in vitro* regeneration selected plants that, after selfing, generated an inbred line n°40017-13 with a high capacity for regeneration. Regeneration from protoplasts has also been achieved[54] in one cultivar of *C. annuum* 'Dulce Italiano', while three other cultivars (Americano, Florida Gynat and Nigrum)

Table 10.2 A summary of gene transfer and corresponding agricultural traits into pepper and eggplant

Plant species	Transformation methods	Gene utilised	Agricultural traits	Inheritance	Field or greenhouse test	References
Capsicum annuum	A.t.	CMV coat protein	CMV resistance	R1	No	51
	A.t.	CMV satellite RNA	CMV resistance	R1	G	99
	A.t.	Bar gene	Herbicide resistance	R1		60
Solanum melongena	A.t.	CryIIIB	Insect resistance	R1	No	71
	A.t.	Synthetic CryIIIA	Insect resistance	R1	F	72
	A.t.	Mutagenised CryIIIB	Insect resistance	R2	F	75, 76
	A.t.	Synthetic CryIAb	Insect resistance	Not tested	No	77
	A.t.	Cysteine protease inhibitor	Insect resistance		No	70
	A.t.	IaaM from *Pseudomonas synringae*	Parthenocarpy	R2 and more	G	79, 80

Note:
A.t.: *Agrobacterium tumefaciens*

and a wild species *C. chinense* produced meristem-like structures but no shoots. Regeneration of chile peppers has also been achieved by somatic embryogenesis.[55,56]

Concerning bell peppers, an *in vitro* regeneration protocol including transformation via *A. tumefaciens* harbouring a GUS reporter gene has been defined by Liu *et al.*[57] using six cultivars and one wild accession. Regeneration of whole transgenic plants proved unsuccessful. A critical step appears to be the elongation and rooting of the shoots during kanamycin selection.[58] The virulence of *Agrobacterium* strains towards peppers seems to be variable and dependent on the cultivar,[51,57] using leaves as starting material and a specific protocol, specially for bud elongation, have succeeded in generating a Chinese sweet pepper variety harbouring a CMV coat protein.

Concerning chile peppers, Manoharan *et al.*[59] established a protocol for regeneration-transformation of a hot chile pepper variety from India using a hypervirulent *A. tumefaciens* strain (EHA 105) and involving the use of thidiazuron as a cytokinin as suggested by Szàsz *et al.*[58] for bell peppers. An effective protocol has also been established for a Korean variety of hot pepper that uses a complex cocktail of plant growth regulators and a pre-culture of cotyledons or hypocotyls in the presence of the ethylene inhibitor $AgNO_3$.[60]

Despite the elevated number of protocols published so far, their efficiency is still very low and peppers can still be considered as recalcitrant to genetic transformation. It is obvious that a good combination of an efficient protocol for regeneration-transformation, a virulent *A. tumefaciens* strain, and highly responsive cultivars is the key for the generation of transgenic peppers.

10.3.2 Transformation for herbicide resistance

Transgenic hot peppers showing resistance to bialaphos, a non-selective herbicide have been generated[60] by transfer of the *bar* gene via *A. tumefaciens*. Transgenic plants were capable of withstanding 5000 $mg.L^{-1}$ of bialaphos applied to the leaves. Sweet peppers expressing the *pat* (phosphinothricin N-acetyl transferase) gene introduced via *A. tumefaciens* exhibited tolerance to applications of 0.44% of the commercial preparation of Basta containing 20% phosphinothricin.[61]

10.3.3 Resistance to viruses

Transgenic sweet peppers harbouring a CMV satellite RNA[62] showed increased resistance to viruses. Stable inheritance of the gene was observed in the progeny and significant symptom attenuation of the offspring was confirmed upon mechanical inoculation with CMV-Y or CMV-Kor strains under greenhouse conditions. However, feasibility of satellite RNA as a biocontrol of CMV in hot peppers needs further field study experiments.

10.3.4 Peppers as a source of genes and/or promoters of interest for other crops

Peppers have been used as a model for the study of carotenoid biosynthesis. Two genes that encode major chromoplast proteins have been cloned[63] that encode a capsanthin/capsoburin synthase (*ccs*) involved in the synthesis of a red carotenoid pigment not found in tomato and a fibrillin (*fib*), a structural protein involved in carotenoid deposition in chromoplasts.[64] Both genes (*ccs* and *fib*) are strongly induced at the early stages of ripening in peppers. The two promoters of the genes, specially the *fib* promoter were exceptionally strong in expressing GUS activity in ripening tomato fruit starting at the late immature-green stage.[65] Interestingly, although ethylene is considered not to be involved in the ripening of the non-climacteric pepper fruit, it influenced *GUS* expression driven by both the *ccs* and *fib* promoters in the climacteric tomato fruit.

10.4 Biotechnology of eggplant

Eggplant and its cultivated or wild relatives, cover a wide range of *Solanum* species (subgenus *Leptostemonum*) originating from Asia and Africa. The most widely cultivated species is *S. melongena* L. Other species are cultivated in Africa such as the scarlet eggplant (*S. aethiopicum* L.) and the gboma eggplant (*S. macrocarpon* L.). Eggplant (*S. melongena* L.), with a world production of around 9 million metric tons, is an economically very important solanaceous crop in Asia and the Mediterranean basin. A list of transformation methods, genes of interest and targeted agricultural traits is given in Table 10.2.

10.4.1 Methods of transformation

Early work by Guri and Sink,[66] Rotino and Gleddie[67] and Rotino *et al.*[68] demonstrated the feasibility of transformation of eggplant via *A. tumefaciens* with a transformation efficiency of 7% in the latter case. However, Billings *et al.*[69] were unsuccessful in transforming their experimental material using one of these two methods. After studying the effects of growth regulators, they devised an improved method that included thidiazuron and gave an transformation efficiency of 20.8%. The transformation efficiency seems to depend strongly upon the genotype. For instance, La Porta *et al.*[70] were unable to transform their lines using the protocol of Billings *et al.*,[69] which proved very efficient with some other genotypes,[71,72] while they obtained a transformation frequency of 8.3% with the protocol of Rotino and Gleddie.[67] It appears therefore that eggplant is not a recalcitrant species, but care must be taken in adapting the transformation protocol to the genotype.

10.4.2 Transformation for herbicide resistance

No genetic transformation aimed at conferring herbicide resistance to eggplant has been published so far. However, eggplants resistant to atrazine have been generated by screening somatic embryos derived from EMS-mutagenised seeds for their ability to grow in the presence of 15 $mg.L^{-1}$ of atrazine.[73]

10.4.3 Resistance to insects

The commercial production of eggplants is frequently hampered in western countries by attacks by Colorado potato beetle (CPB) (*Leptinotarsa decemlineata* Say). Because the eggplant germplasm lacks an effective resistance gene to CPB, improvement of insect resistance via biotechnology represents the only alternative for the generation of insect-resistant eggplants. The insecticidal crystal protein genes (*Cry*) of *Bacillus thuringiensis* Berl. represent an important gene family for the biotechnological improvement of resistance to certain insects (mainly lepidoptera and coleoptera) in cultivated plants.[74] The native *CryIIIB* gene was first used to produce eggplants resistant to the CPB[71] following the transformation protocol described by Billings *et al.*[69] A study of the seedlings of eight independent transgenic lines revealed the absence of any significant resistance to the first and second instar larvae of the CPB. Using a modified synthetic *CryIIIA* gene[72] several lines of transgenic eggplants were generated that now showed resistance to neonate larvae and adult CPB under field conditions. These lines showed higher expression of the *CryIIIA* gene than the previously tested native *CryIIIB* gene.[71] A mutagenised version of the *CryIIIB* gene has been transferred by Arpaia *et al.*[75] to the female parent of the commercial eggplant F1 hybrid 'Rimina'. Over 150 transgenic plants were produced among which 23 showed high expression of the toxin and significant insecticidal activity on neonate CPB larvae. Further tests were done on selfed transgenic progeny showing significant resistance to CBP and higher yields under natural infestation in field conditions.[76]

In Asian countries, the most devastating insect of eggplants is a lepidopteran fruit and stem borer (*Leucinodes orbonalis* Guenee). Kumar *et al.*,[77] using the transformation method of Rotino and Gleddie,[67] have generated transgenic Brinjal type eggplants expressing a synthetic *CryIAb* gene. Plants strongly expressing the toxin showed significant insecticidal activity against the larvae in bioassay studies. Glasshouse and field evaluations remain to be performed to critically evaluate the practical usefulness of the strategy. Altogether, these results demonstrate that recovery of a high level of CPB resistance can be achieved by using modified *Cry Bt* genes rather than native ones and that biotechnology is an efficient strategy for the control of CPB in eggplants. Another type of gene for plant resistance to insects has been tested.[70] It encodes proteinase inhibitors that prevent digestion of plant proteins by some coleopteran and hemipteran insects.[78] La Porta *et al.*[70] have transformed eggplant using an optimised protocol of Rotino and Gleddie[67] with a soybean gene encoding a cysteine inhibitor of proteases, but information on the resistance to insects of the transgenic plants generated is lacking.

10.4.4 Quality traits

The absence of seeds is a desirable trait in a number of fruit crops including eggplant. Parthenocarpic mutants have been identified in several plant species, but their use for generating parthenocarpic varieties is limited by the reduction of fruit set and fruit size. The parthenocarpic trait is polygenic and it proves cumbersome in breeding programmes.[79] For these reasons biotechnology may prove to be an interesting alternative method. Transgenic eggplants expressing the coding region of the *iaaM* gene from *Pseudomonas syringae* driven by an ovule-specific promoter show parthenocarpic development.[80] The *iaaM* gene codes for an indolacetamide monoxygenase that converts tryptophan to indolacetamide, a precursor of the plant hormone auxin.[81] Transgenic plants produced seedless fruit of marketable size when the flowers were emasculated or in adverse conditions when untransformed lines where unable to set fruit. The comparison of three hybrids, transgenic for the *iaaM* gene, with untransformed hybrids and a commercial parthenocarpic cultivar in winter in an unheated greenhouse showed that the yield of the transgenic hybrids was increased by ca. 25% even in the absence of fruit set hormone treatment.[79] A 10% reduction in the cultivation costs was also observed. It is concluded that the *iaaM* gene is a powerful biotechnological tool for generating parthenocarpic eggplants that proves to be superior to the use of both agricultural practices and traditional genetic methods.

10.5 Biotechnology of legumes

Legumes or pulses comprise a large number of species among which only a few are extensively cultivated. Christou[82] has reviewed advances in grain legume biotechnology and reports on all legume species. Here we will concentrate on legumes that are mainly consumed as vegetables (Table 10.3). The two major species are common beans (*Phaseolus vulgaris* L.) and peas (*Pisum sativum* L.). Reference will also be made to other species of lesser economical importance for which biotechnological improvement has been undertaken, such as cowpea (*Vigna unguiculata* L. Walp.), mungbean (*Vigna radiata* L.), chickpea (*Cicer arietinum* L.), faba bean (*Vicia faba* L.) and lentils (*Lens culinaris* Medik)

10.5.1 Methods of transformation

Peas

Successful transformation of peas has been achieved through *A. tumefaciens*-based transformation vectors. Gene transfer through *A. rhizogenes* has also been reported but with the aim of studying expression of transgenes during hairy root development.[83] Direct DNA transfer into protoplasts using electroporation has resulted in the generation of transgenic calli that were resistant to hygromycin, but plant regeneration from these cultures was not possible.[84]

Table 10.3 A summary of gene transfer and corresponding agricultural traits into legume vegetables

Plant species	Transformation methods	Gene utilised	Agricultural traits	Inheritance	Field or greenhouse test	References
Pisum sativum.	A.t.	Bar gene	Herbicide resistance	R1	G	87
	A.t.	α-Amylase inhibitor 1	Insect resistance	R3	G and F	120–122
	Electroporation	PEMV coat protein	Virus resistance	R4	G	119
	A.t.	Proteinase inhibitor	Insect resistance	R2	G	123
Phaseolus vulgaris	Biolistic	Rep-TrAP-REn and BC1 antisense constructs	Virus resistance	R4	G	101
	Biolistic	2S-albumin	Improved amino acid content	R5	G	127
Vicia narbonensis	A.t.	2S-albumin	Improved amino acid content	R3		102
Cicer aritinum	Biolistic	CryAI(c)	Insect resistance			117

Note:
A.t.: *Agrobacterium tumefaciens*

Production of transgenic peas via *A. tumefaciens*-mediated gene transfer has been achieved by Puonti-Kaerlas *et al.*[85] Studies into the inheritance of the transgene gene showed stable transmission of the gene over two generations, but some aberrations were reported in the primary transformants including aborted flowers, limited number of seeds per pod, non-viable seeds and polyploidy.[86] The nine-month period required to recover transgenic shoots is a sign of low efficiency of the protocol and is probably responsible for the aberrations observed. More recent protocols requiring shorter times have allowed further advances to be made towards routine transformation of peas.

Schroeder *et al.*[87] have described a protocol that used longitudinal slices of immature embryos and phosphinotrocin as a selectable agent in which 1.5% to 2.5% of the starting explants gave rise to normal transformed plants. Grant *et al.*[88] have developed a method using immature cotyledons as the explant source in which both organogenesis and embryogenesis regeneration seems to occur. The method was applied to four distinct cultivars. It allowed reduction of the time from explant to seed-bearing primary regenerants. Further assessment of this method[89] demonstrated the efficiency of kanamycin in selecting transformed peas. Davies *et al.*[90] obtained transgenic plants by injecting *A. tumefaciens* into the cotyledonary node. The selection agent was kanamycin and approximately 1.4% of the injected seeds gave rise to transgenic plants. The method was further modified for greater reliability.[91] This procedure presents distinct advantages over those previously reported in that it uses dry mature seeds as starting material and cotyledonary meristems producing shoots without an intermediate callus phase. It allowed rapid generation of phenotypically normal self-fertile plants in which transgenes were inherited in a Mendelian fashion.

Beans

Although considerable efforts have been made to establish regeneration protocols by direct shoot formation from apical and axillary meristems or from embryo-derived calli, an efficient *Agrobacterium*-mediated transformation system has not yet been developed.[92] Reports claiming transformation of *Phaseolus vulgaris* are not convincing as they lack genetic and molecular analysis of the transformants to confirm stable transformation.[93–95] The slow progress in *P. vulgaris* transformation using *A. tumefaciens* could be attributed to the lack of knowledge about the sensitivity of bean genotypes to *A. tumefaciens* or factors affecting transformation itself.[95, 96]

Direct gene transfer by particle bombardment has proved more successful. Using a protocol similar to that developed for soybean with embryo axes[97, 98] generated plants that expressed GUS activity and resistance to the phosphinothricin herbicide over several generations. Kim and Minamikawa[99] also transformed beans using particle bombardment. They were able to introduce the *GUS* gene driven by the concanavalin A promoter conferring seed specificity. The protocol allowed the generation of six transgenic plants from 319 embryogenic axes. Stable integration was confirmed only in primary transformants and not in the progeny. The most obvious success of the particle bombardment process in generating transgenic beans has been reported by Aragao *et al.*[100] who generated methionine-enriched and virus resistant beans[101] in which the transgene was proved to be stably transmitted for at least three generations.

Broad bean (Vicia)

An *A. tumefaciens*-mediated transformation protocol including the generation of embryos from shoot tip calli has been devised for *Vicia narbonensis*, a close relative of the faba bean (*Vicia faba*). It has allowed the generation of stable transgenic lines producing methionine-enriched seeds.[102, 103]

Mung beans (Vigna)

The particle delivery system has been used to express the *GUS* and *NptII* genes in three *Vigna* species (*V. radiata, V. aconitifolia, and V. mungo*) at a relatively high frequency.[104] Transformed calli of *Vigna mungo* and roots of *Vigna radiata* have been obtained using *A. tumefaciens*[105] and non-disarmed *A. rhizogenes*[106] respectively, but none was capable of regenerating shoots. Similarly, attempts to transform cowpea with *Agrobacterium* failed to generate transgenic plants.[107, 108] A more successful method has been reported by Chowrira *et al.*[109] who obtained transgenic cowpea by electroporation-mediated gene transfer into intact plant tissues: the *GUS* gene was introduced in electroporated nodal axillary buds and R1 plants were recovered from seeds originating from these buds.

Lentils

Very few attempts have been made at the genetic transformation of lentils. Early work with four strains of non-disarmed *A. tumefaciens* showed that transfection was feasible,[110] but the generation of transgenic plants was not reported.[111] The possibility of transferring genes in cotyledonary nodes via particle bombardment has been explored recently by Oktem *et al.*.[112] Stable expression of the *GUS* reporter gene was achieved in regenerated shoots. Using the same electroporation method as cowpea, Chowrira *et al.*[109] obtained R2 plants expressing the *GUS* gene. However, segregation ratios in R2 populations showed a strong bias against transgene presence or expression.

Chick pea

Successful transformation and regeneration of the chickpea has been reported using embryo axes after excision of the apical meristem co-cultivated with *A. tumefaciens*.[113–115] Transgenic plants expressing *GUS* and *NPTII* genes were obtained. Altinkut *et al.*[116] used shoot primordia of mature embryos and reported an *Agrobacterium*-mediated transformation with an efficiency of 12.7%. Particle bombardment on embryo axes has been used as an alternative[117] and allowed the production of plants expressing a chimeric C*ryIA(c)* gene.

10.5.2 Transformation for herbicides resistance

The *Bar* gene, conferring resistance to Bialophos herbicide and the related compounds phosphinothricin and ammonium glufosinate, has been used as a selection marker both for pea and bean transformation.[87,91,98] The transgenic pea[87] and bean[98] plants generated showed strong resistance, in greenhouse conditions, to levels of Basta similar to those used in field practice. Stability of the transgene was proved in the progeny.

10.5.3 Resistance to viruses

Bean golden mosaic geminivirus, transmitted by the white fly (*Bemisia tabaci*, Gen) is the causal agent of a severe disease of the common bean throughout western tropical regions. Only low to moderate resistance can be found in the germplasm so highly resistant varieties cannot be generated by breeding. The potential of genetic engineering to produce geminivirus-resistant plants has been demonstrated in the tomato.[118] Antisense constructs of the *Rep-TrAP-REn* and *BC1* viral genes have been used to transform common bean[101] via the biolistic process using embryo apical meristems.[100] Two of the transgenic lines from the R3 and R4 generations exhibited both delayed and attenuated viral symptoms, but not full resistance. Another approach has been made by Chowrira *et al.*[119] with transgenic peas expressing a chimeric pea enation mosaic virus coat protein gene. R2, R3 and R4 plants were shown to display attenuated symptoms as compared to controls.

10.5.4 Resistance to insects

The stored grains of peas and other legumes such as chickpea and cowpea are susceptible to storage insect pests, mainly bruchid beetles. The presence of an α-amylase inhibitor in bean seeds confers insect resistance and the α-*AI-PV* gene is therefore a candidate gene for the improvement of other legumes. Transgenic peas expressing the α-*AI-PV* gene under a seed-specific promoter (phytohaemagglutinin) were produced[120] and seeds of the R2 generation caused increased mortality in Azuki bean weevil (*Callossobruchus chinensis*) and cowpea weevil (*C. maculatus*) larvae. Moreover, these transgenic peas showed strong resistance to the pea weevil (*Bruchus pisorum*) during seed development in the growing crop[121] and were successfully tested under field conditions.[122] Proteinase inhibitors (PI) are also involved in plant protection from insects and peas expressing a PI from *Nicotiana alata* have been obtained and were able to produce PI amounting to 0.2% of total protein.[123] Feeding trials with R2 transgenic peas displayed slowed development of *Helicoverpa armigera* larvae. Chick pea expressing a chimeric *CryAI(c)* gene from *B. thuringiensis* has been obtained by Kar *et al.*[117] through particle bombardment. Insect feeding assays indicated an inhibitory effect on larvae of *Heliothis armigera*. Further studies on the inheritabilty of the gene and field behaviour of the transgenic plants are awaited.

10.5.5 Quality traits

Although beans are rich in some essential amino acids (lysine, threonine, valine, isoleucine and leucine), their nutritional value is limited because of the small amounts of other essential amino acids methionine and cysteine.[124] A *2S-albumin* gene has been isolated from the Brazil nut (*Bertholletia excelsa* H.B.K.) encoding a methionine-rich protein (18.8% of total amino acids) which is targeted to the seed vacuoles.[125 126] This gene has been expressed in a number of laboratory plants (tobacco, *Arabidopsis*) and seed crop species like canola, sunflower and potato.[127] Expression in beans has been achieved using the biolistic method.[100] The gene driven by a double 35S promoter and AMV enhancer sequences was stable and correctly expressed in homozygous R2 to R5 seeds. In two of the five transgenic lines, the methionine content was increased by 14 and 23%. The same gene has also been expressed in *Vicia narbonensis* using either the 35S or the seed specific promoter of the legumin B4 (*LeB4*) gene from *Vicia faba*.[102] Transformation was performed via *A.tumefaciens* using the protocol of Pickard *et al.*[128] The LeB4 promoter proved to be much more efficient than the CAMV 35 promoter. Besides conferring seed-specific expression, it induced a three-fold increase in the methionine content of seed protein in one R0 line. However, these studies are hampered by the fact that Brazil nut 2S albumin has been identified as an allergen and the transgenic soybeans expressing the protein were allergenic.[129]

10.5.6 Beans as a source of genes and/or promoters of interest for other crops

Lectins are found predominantly in the seeds of legumes, where they accumulate to relatively high levels. Their role in defence against pests and pathogens has been established[130, 131] and lectin genes could be interesting candidates for agronomic improvement. Pea lectin was introduced into potato under CAMV promoter control.[132] The lectin was processed in potato leaves as in pea cotyledons, and maintained its haemagglutination activity. The cowpea trypsin inhibitor has been used as an insectidal gene in tobacco[133] and sweet potato.[134] The pea storage protein legumin has been transferred to rice to improve the amino acid composition.[135]

A number of pea storage proteins accumulate specifically in the seed. The corresponding genes exhibit seed-specific expression so that their regulatory sequences can be used to drive gene expression of heterologous proteins specifically in the seed. Such is the case for the promoters of *Phaseolus vulgaris* genes for arcelin, phaseolin, phytohaemagglutinin and a-amylase inhibitor.[120, 136–138] The sequence motifs of the pea lectin promoter that confer seed-specific expression have been characterised.[139] A *Vicia faba* legumin promoter has been used to express the 2S methionine-rich protein of Brazil nut in *Vicia narbonensis*.[102] Legumes can also be a source of other tissue-specific promoters. The promoter of the *Blec4* gene from pea confers specific gene expression to the epidermal tissue of vegetative and floral shoot apices of alfalfa.[140]

10.6 Biotechnology of bulky organs (carrots, sweet potatoes, allium species)

Roots, bulbs and other bulky organs used as vegetables belong to a wide range of horticultural species. Except for the potato, very few of them have received attention in terms of improvement via biotechnology. In this section we will review the present data available for some roots and bulbs. Most of them are related to two species.

- The carrot (*Daucus carota* L.) which is cultivated world wide, representing 3% of world vegetable production (around 14 million tons annually).
- The sweet potato (*Ipomea batatas* L.), which is one of the most important crops used for human consumption in developing countries and ranks sixth (around 115 million tons annually) among the most important crops in the world after wheat, rice, potato and barley. Sweet potatoes and carrots are both dicots.

Data on bulky monocotyledonous species are very scarce although *Allium* species (onion, garlic and leek) are important. This is mainly due to technical difficulties in transformation.

10.6.1 Methods of transformation

Carrot

Routine regeneration of carrot can be achieved through induction of somatic embryogenesis in cell or callus cultures.[141] This capacity has been used in a number of protocols for genetic transformation via *A. tumefaciens*. However, the first protocols devised[142–146] showed little efficiency for routine transformation. In addition, in many cases, transfer of the transgene to the progeny has not been demonstrated. A protocol giving a transformation efficiency of around 20% has been established.[147] It is based on the co-cultivation of the bacteria with hypocotyl segments, development of calli and cell suspension cultures and finally regeneration of plants through induction of somatic embryogenesis. Using this method, not only plants harbouring the *NptII* and the *GUS* genes have been generated[147] but also transgenic plants in which sucrose metabolism has been changed.[148, 149] Gilbert *et al.*[150] using a similar protocol found that transformation efficiency was not influenced by explant age or binary plasmid, but was significantly influenced by the *Agrobacterium* strains, co-cultivation times, and cultivars. Nevertheless, three cultivars of carrots were transformed with a maximum efficiency of 12.1% with either acidic or basic chitinase

Sweet potato

Genetic transformation of sweet potato has been attempted using various gene transfer systems including electroporation of protoplasts,[151] particle bombardment,[152] *A. rhizogenes*[153] and *A. tumefaciens.*[134, 154, 155] The generation of transgenic plants was reported only for *Agrobacterium*-mediated transformation. Previously established protocols for somatic embryogenesis (e.g. Chée and Cantliffe[156]) have been used in some transformation procedures with *A. tumefaciens*. Newell *et al.*[134] starting with root disks, obtained transgenic plants of the 'Jewel' cultivar expressing the GUS enzyme, the cowpea trypsin inhibitor and the snowdrop lectin with a transformation efficiency reaching 10%. Another protocol, also using embryogenesis but starting with apical meristems of the 'White star' cultivar, was described by Gama *et al.*[154] with a similar efficiency. More recently Moran *et al.*[155] established a protocol for the 'Jewel' cultivar using leaf disks and shoot organogenesis that gave 31.1 to 35.5% transformation efficiency, which is much higher than the protocol developed by Newell *et al.*[134] for the same cultivar. This protocol allowed the generation of plants expressing the *CryIIIA* gene of *B. thuringiensis*. The efficiency of this protocol needs to be tested on other cultivars.

Allium species

Allium species are recalcitrant to transformation by *A. tumefaciens* although some of them, such as onion (*A. cepa* L.) is a host for *Agrobacterium*.[157] In garlic (*A. sativum* L.), where gene transfer via breeding is limited by the sterility of commercial genotypes, biotechnology has great potential. In this species, an efficient method for callus culture and shoot regeneration has recently been established,[158] but genetic transformation has not been published so far. The same situation prevails with leeks (*A. ampeloprasum* L.) for which embryogenic calli were induced in cell suspension cultures[159] and regeneration-competent protoplasts obtained.[160]

10.6.2 Resistance to viruses

A programme devoted to the development of virus resistance in sweet potatoes has been initiated using the coat protein of the sweet potato potyvirus.[161]

10.6.3 Resistance to insects and fungi

The most important pests of sweet potatoes are insects, mainly the sweet potato weevil (*Cyclas formicarius*, Fab). Losses dues to insect attacks may reach 60 to 100%. The genetics of sweet potatoes is complex due to the hexaploid genome and self-incompatibility. Transfer of foreign genes via biotechnology is therefore of great interest. Moran *et al.*[155] obtained several clones of the 'Jewel' sweet potato cultivar carrying the *CryIIIA* gene that exhibited some resistance to sweet potato weevil infestation under greenhouse and field conditions as compared to control plants although the level of expression of the gene was low. Extension to other cultivar and further field experiments are needed before commercial application. The transfer of two genes capable of conferring general resistance to insects, that encode the cowpea trypsin inhibitor and a mannose-specific lectin from snowdrop has been achieved both separately and in tandem.[134]

Gilbert *et al.*[150] generated transgenic plants of three carrot cultivars (Golden State, Danvers Half Long and Nanco) with either the acidic (petunia) or basic chitinase (tobacco). Two of the cultivars (Golden State and Nanco) were evaluated for the response to inoculation of detached petioles with a number of fungi. The rate of lesions was significantly lower in transgenic plants transformed with the acidic chitinase gene for *Botrytis cinerea*, *Rhizoctonia solani*, *Sclerotinium rolfsii* but remained unchanged for *Thielaviopsis basicola* and *Alternaria radicina.*[35] These data are consistent with observations made in other plants showing that the efficacy of chitinase genes for enhancing disease tolerance in plants is variable according to the plant species, the type of chitinase and the type of pathogen.

10.6.4 Quality traits

The sugar content of carrots represents an important quality trait. Studies devoted to the elucidation of the role of enzymes involved in sugar metabolism and provision of a motive force for solute transport have been carried out. Transgenic carrots in which sucrose synthase,[148] vacuolar and cell wall invertase[149] and tonoplast H^+ATPase[162] have been repressed were generated. The transgenic plants had an altered phenotype with smaller roots and the strategy for improving the quality of the carrot (e.g. in terms of sugar content) remains to be established.

10.6.5 Resistance to abiotic stresses

The generation of plants resistant to environmental stress is of great practical interest. Modifications of heat tolerance in carrots have been achieved by constitutively expressing or down-regulating a small heat shock protein gene, *Hsp17.7.*[163] Constitutive expression resulted in an increase of heat tolerance while down-regulation resulted in lower tolerance. Practical evaluations of this strategy remain to be made.

10.6.6 Root vegetables as a source of genes and/or promoters of interest for other crops

Sweet potato was used as the source of a trypsin inhibitor gene that can be used for generating insect resistance in cauliflower.[164]

10.7 Biotechnology of leafy vegetables (cabbage, broccoli, cauliflower, lettuce, spinach) and asparagus

This section is devoted to vegetables that are eaten as fresh or cooked leaves and to asparagus (Table 10.4).

The *Brassica oleracea* (L.) species belong to the Brassicaceae family and include several important crops such as broccoli (*B. oleracea*, var *italica*), cauliflower (*B. oleracea*, var *botrytis*), cabbage (*B. oleracea*, var *capitata*), kale

Table 10.4 A summary of gene transfer and corresponding agricultural traits into roots and leafy vegetables

Plant species	Transformation methods	Gene utilised	Agricultural traits	Inheritance	Field or greenhouse test	References
Ipomea batatas	A.t	CryIIIA	Insect resistance		G and F	155
Daucus carota	A.t	Chitinase	Fungus resistance			35
Asparagus officinalis	Biolistic	Bar gene	Herbicide resistance	No		198
Lactuca sativa	A.t	Bar gene	Herbicide resistance	R3	G	199
	A.t	TSWV-BL nucleocapsid protein	Virus resistance	R3		178
	A.t	LMV coat protein	Virus resistance	R2	G	180
	A.t	Ferritine	Increased iron content			204, 205
	A.t	Chimeric nitrate reductase	Reduced nitrate concentration	R1	G	203
Chicorium intybus	A.t.	Acetolactate synthase	Herbicide resistance	R2	No	182
Brassica oleracea	A.t	CryIA(c)	Insect resistance	yes	G	171
	A.t.	Antisense ACO	Slow senescence	yes	G	174
	A.t.	Synthetic CryIC	Insect resistance			172

Note:
A.t: *Agrobacterium tumefaciens*

(*B. oleracea,* var *acephala*), Chinese kale (*B. oleracea*, var *alboglabra*) and Brussels sprouts (*B. oleracea*, var *gemmifera*). The Chinese cabbage belongs to another species *(B. campestris* ssp. *Pekinensis*).

Lettuce (*Lactuca sativa* L.) a member of the Compositae, and chicory (*Chicorium endivia* L), a member of the Asteraceae are high-value crops used in many countries as fresh leaf salads. In western Europe, Witloof chicories (*Chicorium intybus* L., var *foliosum* L.) are also used in salad as white buds, while root chicories (*Chicorium intybus* L. var *sativum*) are roasted and used as coffee surrogate.

Spinach (*Spinacia oleracea* L.) is a dioecious annual leafy vegetable of the Chenopodiaceae family.

Asparagus (*Asparagus officinalis* L.) is a monocotyledonous plant member of the Liliaceae cultivated as a herbaceous perennial and consumed as either white or green spears. Table 10.4 summarises the present state of gene transfer and corresponding agricultural traits for the above-mentionned species.

10.7.1 Methods of transformation

Brassica species

Most research efforts have been directed at developing *A. tumefaciens*-mediated transformation, with little emphasis on direct transfer methods. A common procedure for all varieties of *B. oleracea* being unavailable, the protocols generally remain genotype-specific.[165] *A. rhizogenes*-mediated transformation causes formation of hairy roots that can be induced to form shoots. However, the bacteria carry the *rol* gene that may have phenotypic effects, for instance on flowering. The method has recently been proved efficient in transferring the *NptII* and other genes in 12 vegetable brassica cultivars representing six varieties: broccoli, Brussels sprouts, cabbage, cauliflower, rapid cycling cabbage, and Chinese cabbage. However, fertility was often reduced and morphogenic changes were noted in a number of plants.[166] This method therefore needs further assessment and improvement before being used for commercial applications.

For cauliflower, a number of protocols were developed in the years 1988 to 1992 using marker genes.[165] The protocol of De Block *et al.*[167] has been modified by Bhalla *et al.*[168] using a special combination of growth hormones, starting with cotyledons rather than hypocotyls, but still using silver nitrate, an inhibitor of ethylene action. Transgenic plants of three commercial genotypes were produced harbouring an antisense *Bcp1* gene encoding a protein essential for pollen functionality driven by a pollen-specific promoter. Another protocol, also using silver nitrate but starting with hypocotyls previously treated with 2,4-dichlorophenoxyacetic acid, gave very good regeneration rates and led to the production of over 100 primary transformants putatively harbouring a trypsin inhibitor gene conferring resistance to insects.[164]

An *A. tumefaciens*-mediated transformation method has been devised for broccoli by Metz *et al.*[169] using flowering stalks. This method, derived from the protocol of Toriyama *et al.*,[170] has been used for transferring *Bacillus thuringiensis* genes into broccoli with an efficiency of transformation of about 6.4%.[171,172] An improved *A. rhizogenes*-mediated transformation of broccoli has been reported by Henzi *et al.*[173] It gave 35% and 17% efficiency in the transformation of two cultivars, Shogun and Green beauty, respectively. This protocol has been used to generate plants producing low levels of ethylene,[174] however, phenotypes altered due to the expression of the *rol* gene were often observed. This represents a serious limitation to the use of *A. rhizogenes*.

The Chinese cabbage (*B. campestris ssp pekinensis*) is considered as a recalcitrant species in plant regeneration. Procedures for the transformation and regeneration of transgenic plants have been reported by Jun *et al.*[175] for the 'Spring Flavor' genotype and by Lim *et al.*[176] for a number of other genotypes. However, the efficiency of the transformation was low and dependent upon the genotype.

Lettuce and chicory

Michelmore *et al.*[177] devised a routine protocol via *A. tumefaciens* enabling them to generate several hundred kanamycin-resistant plants, starting with cotyledon explants, that produced calli before regenerating shoots. Inheritance of the transgene was confirmed. This method has been recently used with success to generate tospovirus resistance in lettuce.[178] A method for the regeneration of lettuce from adult leaf protoplasts and a protocol for transformation by electroporation has been published by Chupeau *et al.*[179] that have not been retained in further studies. Rather, an *Agrobacterium*-mediated protocol starting with young leaves has been used for the introduction of a virus coat protein gene that led to the generation of 16 primary transformants of three different cultivars showing accumulation of the protein out of a total of 87 putative transformants.[180] Another protocol using *Agrobacterium* has been developed by Curtis *et al.*,[181] starting from the cotyledons of seven-day-old seedlings. Overall, these data show that genotype-independent transformation procedures exist for the efficient transformation of lettuce. Transformation via *A. tumefaciens* of Witloof chicory has been achieved by Vermeulen *et al.*[182] after optimisation of shoot regeneration from leaf disks in order to confer herbicide resistance.

Spinach

Although regeneration systems have been described from leaf disks, hypocotyl segments[183–185] or root sections,[186] very few reliable systems for transformation of spinach are available. Only two protocols have been recently described[187,188] using *A. tumefaciens*-mediated gene transfer which have allowed stable transformation of spinach, one of them showing high efficiency.[188] Also, expression of foreign DNA has been achieved in isolated spinach chloroplasts by electroporation,[189] but this technical advance has no immediate biotechnological application.

Asparagus

As mentioned by Conner and Abernethy,[190] asparagus has been at the forefront of biotechnology developments in monocotyledonous plants, being the first such plant to be regenerated from tissue culture and isolated protoplasts and also to be genetically transformed.[191] Transgenic asparagus plants have been generated by *A. tumefaciens*-mediated gene transfer through shoot regeneration from transformed calli[192] or through embryogenesis.[193] The efficiency of transformation was very low[190] as in other monocotyledonous plants and difficult to use for practical applications. Direct DNA uptake by asparagus protoplasts has been achieved but the recovery of transgenic plants has not been reported.[194, 195] Due to the ease of regeneration of asparagus via embryogenesis,[196] microprojectile bombardment may offer the most efficient approach for gene transfer. Li and Wolyn[197] have generated transgenic plants expressing the *NptII* and *GUS* genes and a preliminary study of *GUS* transgene inheritance was performed. Cabrera-Ponce *et al.*[198] also used the microprojectile bombardment method to transfer *hygromycin phosphotransferase, phosphinotrocin acetyl transferase* and *GUS* genes into embryogenic calli of asparagus. About 50 transgenic lines showing *GUS* expression were generated but inheritance studies are awaited.

It should be noticed that the integration of genetic engineering in an asparagus breeding programme is not easy. The most important cultivars are clonal hybrids that are genetically variable due to gene segregation among the progeny.[190]

10.7.2 Transformation for herbicide resistance

Cabbage

Putative transformants of Chinese cabbage have been shown to express the *bar* (bialaphos resistance) gene.[176] However, studies on the resistance of the transgenic plants to the herbicide and inheritability of the transgene have not been performed.

Lettuce

Resistance to bialaphos has also been introduced into lettuce of the Evola cultivar by *A. tumefaciens*-mediated transformation.[199] Resistance to glufosinate was observed in axenic conditions and in the greenhouse and stable expression was confirmed over two generations. Field tests are awaited to further assess the advantages of the transgenic lines generated.

Chicory

Transgenic Wiltloof chicories harbouring an *acetolactate synthase* gene have been generated and show resistance to the herbicide chlorsulfuron.[182] Stable transformation has been observed in two selfed progeny, but large-scale tests in the greenhouse or the field are not reported.

Asparagus

Five transgenic lines harbouring the *bar* gene and generated by particle bombardment[198] were able to withstand the prescribed application of phosphinotrocin for weed control (0.5 to 1% solution by localised application). Large-scale and/or commercial applications are awaited.

10.7.3 Resistance to viruses

Cauliflower

Transgenic cauliflower carrying the capsid gene and antisense gene *VI* of the cauliflower mosaic virus have been generated through *A. tumefaciens*-mediated transformation.[200] However, while the transcription of the transgenes was detected in all plants, the capsid protein was not present.

Chinese cabbage

The tobacco mosaic virus 35S coat protein gene has been expressed in five regenerants of the Spring Flavour cultivar of Chinese cabbage (*B. campestris, ssp pekinensis*). Stable inheritance of the gene was shown in the progeny, but virus resistance was not assessed.[175]

Lettuce

The tomato spotted wilt virus is a tospovirus transmitted mainly by the western flower thrips *Frankliniella occidentalis* to several hundred plant species, including the lettuce. Genetic sources of resistance often being limited, biotechnology has been considered as an alternative to conventional breeding. Transgenic lettuce plants, expressing the nucleocapsid protein gene of a lettuce isolate of the virus, were protected against isolates of the virus not only when the protein accumulated at high levels, but also where transgene silencing occurred with high transcription rates and low steady-state mRNA levels.[178] Confirmation of these protective effects under practical conditions and over several generations however, is still lacking.

Another virus, the lettuce mosaic potyvirus, can be destructive for lettuce crops. The coat protein of this virus has been introduced into three susceptible cultivars.[180] The progeny of five transformants showed resistance to infection not only against the strain from which the coat protein originated, but also against other strains. However, the efficiency of resistance depended on the development stage of the plant at the time of inoculation. Although some plants (13%) showed stable resistance over the growth period, late viral infection was observed at advanced stages of development for most plants. Field tests need to be performed in order to evaluate the efficiency of protection in natural growing conditions and virus inoculation by aphids.

10.7.4 Resistance to insects and fungi

Cauliflower

Insect pests represent a serious problem for cauliflower cultivation. A trypsin inhibitor from the sweet potato has been transferred to Taiwan cauliflower cultivars that gave transgenic primary transformants substantial resistance to local insects upon *in planta* feeding bioassays.[164] Progeny behaviour studies and field tests remain to be performed.

Broccoli

Metz *et al.*[171] have generated a large number of transgenic broccoli lines carrying the *Bt CryIA(c)* gene, most of them causing 100% mortality of first instar larvae of the diamond moth, a major insect pest of crucifers. However, *Cry1A*-resistant larvae were able to survive on the transgenic plants. More recently, a synthetic *Bt Cry1C* gene was introduced also using the method developed by Metz.[169] Lines producing high levels of *Cry1C* protein were protected not only from susceptible or *Cry1A* resistant diamond moth larvae, but also from larvae selected for moderate levels of resistance to *Cry1C*.[172] In addition, the *Cry1C*-transgenic broccoli were also resistant to other lepidopteran pests of crucifers such as cabbage looper and imported cabbageworm.

Cabbage

The *Cry1A(c)* gene was likewise introduced into cabbage,[171] however, the disadvantages of this gene in failing to control resistant insects is the same as already mentioned for broccoli. The introduction of other synthetic *Bt* genes is awaited in this variety.

Lettuce

Transformation of lettuce with *A. tumefaciens* harbouring a maize *Ac* transposase and *Ds*, an empty transposon donor site[201] has been used to generate mutants of lettuce that were screened for downy mildew resistance.[202] This work represents a good example of the use of T-DNA mutagenesis combined with transposon tagging and genetic mapping with the aim of isolating genes of agronomic interest.

10.7.5 Quality traits

Broccoli

Transgenic lines of broccoli containing a tomato antisense 1-amino-cyclopropane-1-carboxylic acid oxidase gene, showed significant reduction of ethylene production in the florets.[174] However, the fall in ethylene production was probably not enough, even though it sometimes reached more than 90%, to slow down senescence and preserve the quality of the florets. Higher levels of reduction are required to obtain interesting phenotypes, in particular through the use of homologous genes.

Lettuce

Since lettuce accumulates high levels of nitrate and nitrate can be harmful to human health, great interest has been shown in reducing the nitrate content of the leaves. Besides tight control of cultivation conditions, the transfer of the nitrate reductase gene has been considered as an alternative approach. Curtis *et al.*[203] have stably expressed a chimeric nitrate reductase gene of tobacco in transgenic lettuce. The level of nitrate was significantly reduced but not sufficiently to reach very low levels, especially in the older leaves. In addition, phenotypic alterations were observed such as chlorosis, dwarfing and early flowering. Further studies may render this strategy fully applicable at the commercial level.

Table 10.5 Field trials on fruits and vegetables between 1994 and 1998 (from OECD biotrack database, web site: http://www.olis.oecd.org/biotrack.nsf)

Groups of species	Species	Country	Trait wanted	Number by trait	Number by species
Fleshy fruits	Cucumber/squash (*Cucurbita pepo*)	United States	Virus resistance	55	59
		Spain	Virus resistance	2	
		Italy	Virus resistance	1	
		France	Virus resistance	1	
	Eggplant (*Solanum melongena*)	United States	Insect resistance	5	6
			Fungi resistance	1	
	Melon (*Cucumis melo*)	United States	Virus resistance	59	91
			Quality traits	21	
			Fungus resistance	2	
			Herbicide resistance	1	
		Spain	Virus resistance	4	
		France	Virus resistance	3	
		Japan	Virus resistance	1	
	Pepper (*Capsicum annum*)	United States	Virus resistance	4	8
			Quality traits	4	
	Squash (*Cucurbita texana*)	United States	Virus resistance	3	3
	Watermelon (*Citrillus lanatus*)	United States	Virus resistance	6	6
Legumes	Pea (*Pisum sativum*)	United States	Quality traits	1	3
		New Zealand	Virus resistance	1	
		Canada	Herbicide resistance	1	
Bulky vegetables	Carrot (*Daucus carota*)	United States	Fungus resistance	10	10
	Onion (*Allium cepa*)	United States	Fungi resistance	1	1
	Sweet potato (*Ipomoea batatus*)	United States	Herbicide resistance	1	1
Leafy vegetables	Broccoli (*Brassica oleracea*)	Canada	Herbicide resistance	3	6
		New Zealand	Quality traits	2	
		Japan	Herbicide tolerance	1	
	Cabbage (*Brassica oleracea*)	United States	Insect resistance	2	3
			Herbicide tolerance and quality traits	1	
	Cauliflower (*Brassica oleracea*)	Belgium	Herbicide resistance and quality traits	1	3
		Canada	Herbicide resistance	1	
		Japan	Herbicide resistance	1	

Table 10.5 Continued

Groups of species	Species	Country	Trait wanted	Number by trait	Number by species
	Chicory (*Cichorium intybus*)	Italy	Herbicide resistance and quality traits	8	27
		Belgium	Herbicide resistance and/or quality traits	7	
		Netherlands	Herbicide resistance and/or quality traits	5	
		France	Herbicide resistance and quality traits	4	
		United States	Quality traits	1	
			Herbicide resistant	1	
		United Kingdom	Quality traits and herbicide resistance	1	
	Lettuce (*Lactuca sativa*)	United States	Virus resistance	5	11
			Fungi resistance	2	
			Herbicide resistance	2	
		France	Virus resistance	1	
			Quality traits	1	

Increasing the iron content of vegetables can have health benefits. An increase in the iron content of lettuce ranging from 1.2 to 1.7 times has been achieved by expressing a cDNA of soybean ferritine in lettuce via *A. tumefaciens* transformation.[204, 205] In addition, the transgenic lettuce had higher photosynthesis and growth rates, which represent interesting agronomic characters for commercial applications.

10.8 Conclusion and future trends

Considering the relatively low economic importance of most of the crops dealt with here as compared to other major crops (soybean, potato, or tomato) and the fact that the first transgenic plants were generated less than two decades ago, it can be concluded that the research efforts reported are quite significant. In addition, the large number of field trials (Table 10.5) performed between 1994 and 1998 are a testimony to the efforts carried out mainly by private companies for the improvement of these so-called secondary or under-exploited species. It can be noticed, however, that field trials on cucurbits represent more than 60% of the total. Progress still remains to be made in a number of areas, including (i) the improvement of transformation protocols, (ii) the search and development of new genes of agronomical interest and (iii) the development of strategies to better meet with public acceptability of transgenic plants. Most of these trends are common to all transgenic plants but some have higher relevance to the less important crops.

In a number of cases, gene transfer methods have been developed that are restricted by variety or genotype. Also, sometimes, the efficiency of the protocols is too low for practical applications that require the generation of a large number of transformation events. Efforts therefore remain to be made in the improvement of the transformation protocols, by increasing the efficiency of the regeneration, choosing the right strain of *A. tumefaciens*, defining the best conditions for direct transfer via particle bombardment and using an appropriate selectable marker gene. Also, the stability of transgene expression has not always been assessed. Proof of integrative transformation must be sought in genetic (transmission to the progeny, Southern blotting) and phenotypic (expression and effects of the transgene) data.

The biotechnological approaches constitute an important supplement to conventional improvement programmes. A combination of both genetic engineering and traditional breeding techniques is necessary for the genetic improvement of vegetable species. For example, some agronomically important traits such as virus tolerance have been dealt with by biotechnology and conventional breeding. Nevertheless, although dramatic progress has been made in genetic engineering, further efforts are needed to extend the number of species that will be engineered and the number of target genes to be used for protection against a wide range of diseases. This should lead to environmentally safer agricultural practices that will use fewer pesticides and will render genetic engineering better accepted by the consumer. Likewise, improving the nutritional and sensory quality is also a major objective.

Targeting down-regulation or expression of genes at the right time and in the right tissue or organelle is one of the future challenges of biotechnology. Although some tissue-specific promoters are already available, especially for legume seeds, there is a need to develop efficient promoters that will specifically drive gene expression in the part of the plant used for food (fruit, roots, leaves, etc.) or in specific organelles such as the chloroplast. The possible dispersion of antibiotic resistance genes, although theoretical, is of great concern to consumers. Methods are now available for removing selectable marker genes.[206] However, so far they have mainly been applied to model plants. There is no doubt that, when extended to commercial products, they will contribute to overcoming public reluctance to accept genetically engineered food.

10.9 Acknowledgements

We thank the Midi-Pyrénées Regional Council (Grant No. 99009080) and the EU (FAIR CT96-1138), whose financial support has made possible our own research on the biotechnology of fruit ripening and the writing of this manuscript. We are grateful to Mireille Zaninotto, Françoise Maisonnier and Anne Konkolewski for their assistance.

10.10 References

1. JEFFREY C, 'Notes on *Curcubitaceae*, including a proposed new classification of the family'. *Kew Bulletin*, 1962, **15**: 337–371.
2. ROBINSON RW, DECKER-WALTERS DS, *Cucurbits*, Cab International, 1997.
3. TRULSON AJ, SHAHIN EA, 'In vitro plant regeneration in the genus Cucumis'. *Plant Sci*, 1986, **47**: 35–43.
4. KATHAL R, BHATNAGAR SP, BHOJWANI SS, 'Regeneration of plants from leaf explants of *Cucumis melo* cultivar Pusa Sharbati'. *Plant Cell Rep*, 1988, **7**: 449–451.
5. TABEI Y, KITADE S, NISHIZAWA Y, KIKUCHI N, KAYANO T, HIBI T, AKUTSU K, 'Transgenic cucumber plants harboring a rice chitinase gene exhibit enhanced resistance to gray mold (*Botrytis cinerea*)'. *Plant Cell Rep*, 1998, **17**: 159–164.
6. YADAV RC, SALEH MT, GRUMET R, 'High frequency shoot regeneration from leaf explants of muskmelon'. *Plant Cell Tissue Organ Cult*, 1996, **45**: 207–214.
7. FICCADENTI N, ROTINO GL, 'Genotype and medium affect shoot regeneration of melon'. *Plant Cell Tissue Organ Cult*, 1995, **40**: 293–295.
8. DEBEAUJON I, BRANCHARD M, 'Somatic embryogenesis in *Curcubitacea*'. *Plant Cell Tissue Organ Cult*, 1993, **34**: 91–100.
9. GRAY DJ, MC COLLEY DW, COMPTOM ME, 'High-frequency somatic embryogenesis from quiescent seed cotyledons of *Cucumis melo* cultivars'. *J Am Soc Hort Sci*, 1993, **118**: 425–432.
10. ORIDATE T, ATSUMI H, ITO S, ARAKI H, 'Genetic difference in somatic embryogenesis from seeds in melon (*Cucumis melo* L.)'. *Plant Cell Tissue Organ Cult*, 1992, **29**: 27–30.
11. EZURA H, AMAGAL H, YOSHIOKA K, OOSAWA K, 'Highly frequent appearance of tetraploidy in regenerated plants, a universal phenomenon, in the tissue cultures of melon (*Cucumis melo* L.)'. *Plant Sci*, 1992, **85**: 209–213.
12. COLIJN-HOOYMANS CM, HAKKERT JC, JANSEN J, CUSTER JBM, 'Competence for regeneration of cucumber cotyledons is restricted to specific developmental stages'. *Plant Cell Tissue Organ Cult*, 1994, **39**: 211–217.
13. GILISSEN LJW, VAN STAVEREN MJ, CREEMERS-MOLENAAR J, VERHOEVEN HA, 'Development of polysomaty in seedlings and plants of *Cucumis sativus* L.'. *Plant Sci*, 1993, **91**: 171–179.
14. GUIS M, BEN AMOR M, LATCHÉ A, PECH JC, ROUSTAN JP, 'A reliable system for the transformation of cantaloupe charentais melon (*Cucumis melo* L. var. *cantalupensis*) leading to a majority of diploid regenerants'. *Scientia Hort*, 1999, **84**: 91–99.
15. SCHULZE J, BALKO C, ZELLNER B, KOPREK T, HANSCH R, NERLICH A, MENDEL RR, 'Biolistic transformation of cucumber using embryogenic suspension cultures: long-term expression of reporter genes'. *Plant Sci*, 1995, **112**: 197–206.
16. GRAY DJ, HIEBERT E, KELLEY KT, COMPTON ME, GABA VP, 'Comparison of

methods to transform embryogenic cotyledons of melon'. *HortSci*, 1995, **30**: 788.
17. KATAVIC V, JELASKA S, BAKRAN-PETRICIOLI T, DAVID C, 'Host-tissue differences in transformation of pumpkin (*Cucurbita pepo* L.) by *Agrobacterium rhizogenes*'. *Plant Cell Tissue Organ Cult*, 1991, **24**: 35–42.
18. TOPPI LSD, PECCHIONI N, DURANTE M, '*Cucurbita pepo* L. can be transformed by *Agrobacterium rhizogenes*'. *Plant Cell Tissue Organ Cult*, 1997, **51**: 89–93.
19. CHEN W, CHIU C, LIU H, LEE T, CHENG J, LIN C, WU Y, CHANG H, 'Gene transfer via pollen-tube pathway for anti-*Fusarium* wilt in watermelon'. *Biochem Mol Biol Int*, 1998, **46**: 1201–1209.
20. PROVVIDENTI R, 'Viral diseases of cucurbits and sources of resistance'. *Technical Bulletin*, Food and Fertiliz Technol Center, Taipei,Taiwan, 1986, **93**.
21. ZITTER T, HOPKINS DL, THOMAS CE, (eds) *Compendium of Cucurbit Diseases*. APS Press, St. Paul, Minnesota 1996, 87pp.
22. CHEE PP, SLIGHTOM JL, 'Transfer and expression of cucumber mosaic virus coat protein gene in the genome of *Cucumis sativus*'. *J Am Soc Hort Sci*, 1991, **116**: 1098–1102.
23. GONSALVES D, CHEE P, PROVVIDENTI R, SEEM R, SLIGHTON JL, 'Comparison of coat protein-mediated and genetically-derived resistance in cucumbers to infection by cucumber mosaic virus under field conditions with natural challenge inoculations by vectors'. *Bio/Technol*, 1992, **10**: 1562–1570.
24. YOSHIOKA K, HANADA K, HARADA T, MINORE Y, OOSAWA K, 'Virus resistance in transgenic melon plants that express the cucumber mosaic virus coat protein gene and in their progeny'. *Jpn J Breed*, 1993, **43**: 629–634.
25. GONSALVES C, XUE B, YEPES M, FUCHS M, LING K, NAMBA S, CHEE P, SLIGHTOM JL, GONSALVES D, 'Transferring cucumber mosaic virus-white leaf strain coat protein gene into *Cucumis melo* L. and evaluating transgenic plants for protection against infections'. *J Am Soc Hort Sci*, 1994, **119**: 345–355.
26. CLOUGH GH, HAMM PB, 'Coat protein transgenic resistance to watermelon mosaic and zucchini yellows mosaic virus in squash and cantaloupe'. *Plant Dis*, 1995, **79**: 1107–1109.
27. TRICOLI DM, CARNEY KJ, RUSSELL PF, MCMASTER JR, GROFF DW, HADDEN KC, HIMMEL PT, HUBBARD JP, BOESHORE ML, QUEMADA HD, 'Field evaluation of transgenic squash containing single or multiple virus coat protein gene constructs for resistance to cucumber mosaic virus, watermelon mosaic virus 2, and zucchini yellow mosaic virus'. *Bio/Technol*, 1995, **13**: 1458–1465.
28. FUCHS M, MCFERSON JR, TRICOLI DM, MCMASTER JR, DENG RZ, BOESHORE ML, REYNOLDS JF, RUSSELL PF, QUEMADA HD, GONSALVES D, 'Cantaloupe line CZW-30 containing coat protein genes of cucumber mosaic virus, zucchini yellow virus, and watermelon mosaic virus-2 is resistant to these three viruses in the field.' *Molecular Breeding*, 1997, **3**: 279–290.

29. FUCHS M, GONSALVES D, 'Resistance of transgenic hybrids squash ZW-20 expressing the coat protein genes of zucchini yellow mosaic virus and watermelon mosaic virus-2 to mixed infections by both potyviruses.' *Bio/Technol*, 1995, **13**, 1466–1473.
30. ARCE-OCHOA JP, DAINELLO F, PIKE LM, DREWS D, 'Field performance comparison of two transgenic summer squash hybrids to their parental hybrid line'. *HortSci*, 1995, **30**: 492–493.
31. PLAGES JN, 'L'avenir des variétés génétiquement modifiées pour la résistance aux virus (un exemple développé par Limagrain)'. *C R Acad Agric Fr*, 1997, **83**: 161–164.
32. CANTO T, PALUKAITIS P, 'Transgenically expressed cucumber mosaic virus RNA1 simultaneously complements replication of cucumber mosaic virus RNAs2 ans 3 confers resistance to systemic infection'. *Virology*, 1998, **250**: 325–336.
33. WINTERMANTEL WM, ZAITLIN M, 'Transgene translatability increases effectiveness of replicase-mediated resistance to cucumber mosaic virus'. *J Gen Virol*, 2000, **81**: 587–595.
34. KOMADA H, EZUKA A, 'Varietal resistance to *Fusarium* wilt in cucumber'. *Bull Veg Ornament Crops Res Stn Jpn Ser*, 1974, **A1**: 233–245.
35. PUNJA ZK, RAHARJO SHT, 'Response of transgenic cucumber and carrot plants expressing different chitinase enzymes to inoculation with fungal pathogens'. *Plant Dis*, 1996, **80**: 999–1005.
36. BRAY EA, 'Plant responses to water deficit'. *Trends Plant Sci*, 1997, **2**: 48–54.
37. BORDAS M, MONTESINOS C, DABAUZA M, SALVADOR A, ROIG LA, SERRANO R, MORENO V, 'Transfer of the yeast salt tolerance gene HAL1 to *Cucumis melo* L. cultivars and in vitro evaluation of salt tolerance'. *Transgenic Res*, 1997, **6**: 41–50.
38. SERRANO R, CULIANZ-MACIA FA, MORENO V, 'Genetic engineering of salt and drought tolerance with yeast regulatory genes'. *Scientia Hort*, 1999, **78**: 261–269.
39. CLENDENNEN SK, KELLOGG JA, WOLFF KA, MATSUMURA W, PETERS S, VANWINKLE JE, COPES B, PIEPER M, KRAMER MG, 'Genetic engineering of cantaloupe to reduce ethylene biosynthesis and control ripening'. In: Kanellis AK, Klee CCH, Bleecker AB, Pech JC, Grierson D, (eds) *Biology and Biotechnology of the Plant Hormone Ethylene II*, pp. 371–379. Kluwer Academic Publishers, Dordrecht, The Netherlands 1999.
40. BALAGUÉ C, WATSON CF, TURNER AJ, ROUGÉ P, PICTON S, PECH JC, GRIERSON D, 'Isolation of a ripening and wound-induced cDNA from *Cucumis melo* L. encoding a protein with homology to the ethylene-forming enzyme'. *Eur J Biochem*, 1993, **212**: 27–34.
41. AYUB R, GUIS M, BEN AMOR M, GILLOT L, ROUSTAN JP, LATCHÉ A, BOUZAYEN M, PECH JC, 'Expression of ACC oxidase antisense gene inhibits ripening of cantaloupe melon fruits'. *Nature Biotech*, 1996, **14**: 862–866.
42. GUIS M, BOTONDI R, BEN-AMOR M, AYUB R, BOUZAYEN M, PECH JC, LATCHÉ A,

'Ripening-associated biochemical traits of Cantaloupe Charentais melons expressing an antisense ACC oxidase transgene'. *J Am Soc Hort Sci*, 1997, **122**: 748–751.

43. BAUCHOT A, MOTTRAM D, DODSON A, JOHN P, 'Effect of antisense ACC oxidase on the formation of volatile esters in Cantaloupe Charentais melons (Cv. Védrantais)'. *J Agric Food Chem*, 1998, **46**: 4787–4792.
44. BEN AMOR M, FLORES B, LATCHÉ A, BOUZAYEN M, PECH JC, ROMOJARO F, 'Inhibition of ethylene biosynthesis by antisense ACC oxidase RNA prevents chilling injury in Charentais Cantaloupe melons'. *Plant Cell Environ*, 1999, **22**: 1579–1586.
45. SMITH PG, 'Horticultural classification of peppers grown in the United States'. *HortSci*, 1987, **22**: 11–13.
46. STEINITZ B, WOLF D, MATZEVITCH-JOSEF T, ZELCER A, 'Regeneration in vitro and genetic transformation of pepper (*Capsicum* spp.): the current state of the art'. *Capsicum & Eggplant Newsletter*, 1999, **18**: 9–15.
47. OCHOA-ALEJO N, IRETA-MORENO ML, 'Cultivar differences in shoot forming capacity of hypocotyl tissues of chilli pepper (*Capsicum annuum L.)* cultured in vitro'. *Scientia Hort*, 1990, **42**: 21–28.
48. SZASZ A, NERVO G, FARI M, 'Screening for in vitro shoot-forming capacity of seedling explants in bell pepper'. *Plant Cell Rep*, 1995, **14**: 666–669.
49. MORRISON RA, KONING RE, EVANS DA, Pepper. In: Evans DA, Sharp WR, Ammirato PV (eds) *Handbook of Plant Culture*, pp. 552–573. Macmillan, New York 1986.
50. FARI M, CZAKO M, 'Relationship between position and morphogenetic response of pepper hypocotyl explants cultured in vitro'. *Scientia Hort*, 1981, **15**: 207–213.
51. ZHU Y, OUYANG W, ZHANG Y, CHEN Z, 'Transgenic sweet pepper plants from *Agrobacterium* mediated transformation'. *Plant Cell Rep*, 1996, **16**: 71–75.
52. ARROYO R, REVILLA MA, 'In vitro plant regeneration from cotyledon and hypocotyl segments in two bell pepper cultivars'. *Plant Cell Rep*, 1991, **10**: 414–416.
53. FARI M, TURI Z, CSILLAG F, 'Comparative studies on *in vitro* regeneration of seedling explants in chili pepper (*Capsicum annuum L.*)'. *Acta Hort*, 1990, **280**: 131–134.
54. DIAZ I, MORENO R, POWER JB, 'Plant regeneration from protoplasts of *Capsicum annuum*'. *Plant Cell Rep*, 1988, **7**: 210–212.
55. HARINI I, LAKSHMI SITA G, 'Direct somatic embryogenesis and plant regeneration from immature embryos of chilli (*Capsicum annuum L.*)'. *Plant Sci*, 1993, **89**: 107–112.
56. BINZEL ML, SANKHLA N, JOSHI S, SANKHLA D, 'Induction of direct somatic embryogenesis and plant regeneration in pepper (*Capsicum annuum L.*)'. *Plant Cell Rep*, 1996, **15**: 536–540.
57. LIU W, PARROTT WA, HILDEBRAND DF, COLLINS GB, WILLIAMS EG, '*Agrobacterium*-induced gall formation in bell pepper (*Capsicum annuum* L.) and formation of shoot-like structures expressing introduced genes'.

Plant Cell Rep, 1990, **9**: 360–364.
58. SZASZ A, MITYKO J, ANDRASFALVY A, FARI M, 'Methodological and genetic aspects of in vitro plant regeneration and genetic transformation of the recalcitrant pepper (*Capsicum annuum* L.)'. *Acta Hort*, 1997, **447**: 365–366.
59. MANOHARAN M, VIDYA CSS, SITA GL, '*Agrobacterium*-mediated genetic transformation in hot chilli (*Capsicum annuum* L. var. Pusa Jwala)'. *Plant Sci*, 1998, **131**: 77–83.
60. LIM H, LEE G, YOU Y, PARK E, SONG Y, YANG D, CHOI K, 'Regeneration and genetic transformation of hot pepper plants'. *Acta Hort*, 1999, **483**: 387–396.
61. TSAFTARIS A, 'The development of herbicide-tolerant transgenic crops'. *Field Crops Research*, 1996, **45**: 115–123.
62. KIM SJ, LEE SJ, KIM BD, PAEK KH, 'Satellite-RNA-mediated resistance to cucumber mosaic virus in transgenic plants of hot pepper (*Capsicum annuum* cv. Golden Tower)'. *Plant Cell Rep*, 1997, **16**: 825–830.
63. DERUÈRE J, BOUVIER F, STEPPUHN J, KLEIN A, CAMARA B, KUNTZ M, 'Structure and expression of two plant genes encoding chromoplast-specific proteins: occurrence of partially spliced transcripts'. *Biochem Biophys Res Commun*, 1994, **199**: 1144–1150.
64. DERUÈRE J, RÖMER S, D'HARLINGUE A, BACKHAUS RA, KUNTZ M, CAMARA B, 'Fibril assembly and carotenoid overaccumulation in chromoplasts: a model for supramolecular lipoprotein structures'. *Plant Cell*, 1994, **6**: 119–133.
65. KUNTZ M, CHEN HC, SIMKIN AJ, RÖMER S, SHIPTON CA, DRAKE R, SCHUCH W, BRAMLEY PM, 'Upregulation of two ripening-related genes from a non-climacteric plant (pepper) in a transgenic climacteric plant (tomato)'. *Plant J*, 1998, **13**: 351–361.
66. GURI A, SINK KC, '*Agrobacterium* transformation of eggplant'. *J Plant Physiol*, 1988, **133**: 52–55.
67. ROTINO GL, GLEDDIE S, 'Transformation of eggplant (*Solanum melongena L.*) using a binary *Agrobacterium tumefaciens* vector'. *Plant Cell Rep*, 1990, **9**: 26–29.
68. ROTINO GL, ARPAIA S, IANNACONE R, IANNAMICO V, MENNELLA G, ONOFANO V, PERRONE D, SUNSERI F, XIKE Q, SPONGA S, *Agrobacterium*-mediated transformation of *Solanum spp*. using a *Bacillus thuringiensis* gene against coleopteran. Proc. VIIIth Meeting *Genetics and Breeding on Capsicum and Eggplant*, Rome, Italy, 1992, pp. 295–300.
69. BILLINGS S, JELENKOVIC G, CHIN C-K, EBERHARDT J, 'The effect of growth regulators and antibiotics on eggplant transformation'. *J Am Soc Hort Sci*, 1997, **122**: 158–162.
70. LA PORTA N, BELLONI V, ROTINO GL, 'Regeneration of transgenic eggplants (*Solanum melongena* L.) for a cysteine proteinase inhibitor'. *Capsicum & Eggplant Newsletter*, 1998, **17**: 92–95.
71. CHEN Q, JELENKOVIC G, CHEEKOK C, BILLINGS S, EBERHARDT J, GOFFREDA JC, DAY P, 'Transfer and transcriptional expression of coleopteran *CryIIIB endotoxin* gene of *Bacillus thuringiensis* in eggplant'. *J Am Soc Hort Sci*,

1995, **120**: 921–927.

72. JELENKOVIC G, BILLINGS S, QI C, LASHOMB J, HAMILTON G, GHIDIU G, 'Transformation of eggplant with synthetic *CryIIIA* gene produces a high level of resistance to the Colorado potato beetle'. *J Am Soc Hort Sci*, 1998, **123**: 19–25.
73. ASHFAQ FAROOQUI M, RAO AV, JAYASREE T, SADANANDAM A, 'Induction of atrazine resistance and somatic embryogenesis in *Solanum melongena*'. *Theor Appl Genet*, 1997, **95**: 702–705.
74. MCGAUGHEY WH, WALON ME, 'Managing insect resistance to *Bacillus thuringiensis'*. *Science*, 1992, **128**: 1451–1455.
75. ARPAIA S, MENNELLA G, ONOFARO V, PERRI E, SUNSERI F, ROTINO GL, 'Production of transgenic eggplant (*Solanum melongena* L.) resistant to Colorado Potato Beetle (*Leptinotarsa decemlineata* Say)'. *Theor Appl Genet*, 1997, **95**: 329–334.
76. ARPAIA S, ACCIARRI N, LEO GMD, MENNELLA G, SABINO G, SUNSERI F, ROTINO GL, Field performance of Bt-expressing transgenic eggplant lines resistant to colorado potato beetle. Proceedings of the Xth EUCARPIA Meeting on Genetics and Breeding of Capsicum and Eggplant, Avignon, France 1998, pp. 191–194.
77. KUMAR PA, MANDAOKAR A, SREENIVASU K, CHAKRABARTI SK, BISARIA S, SHARMA SR, KAUR S, SHARMA RP, 'Insect-resistant transgenic brinjal plants'. *Molecular Breeding*, 1998, **4**: 33–37.
78. RYAN CA, 'Protease inhibitors in plants genes for improving defenses against insects and pathogens'. *Ann Rev Phytopathol*, 1990, **28**: 425–449.
79. DONZELLA G, SPENA A, ROTINO GL, 'Transgenic parthenocarpic eggplants: superior germplasm for increased winter production'. *Molecular Breeding*, 2000, **6**: 79–86.
80. ROTINO GL, PERRI E, ACCIARRI N, SUNSERI F, ARPAIA S, 'Development of eggplant varietal resistance to insects and diseases via plant breeding'. *Adv Hort Sci*, 1997, **11**: 193–201.
81. YAMADA T, PALM CJ, BROOKS B, KOSUGE T, 'Nucleotide sequence of the *Pseudomonas savastanoi* indoleacetic acid genes show homology with *Agrobacterium tumefaciens* T-DNA'. *Proc Natl Acad Sci USA*, 1985, **82**: 6522–6526.
82. CHRISTOU P, 'Biotechnology applied to grain legumes'. *Field Crops Research*, 1997, **53**: 83–97.
83. NICOLL SM, BRIGHAM LA, FUSHI W, HAWES MC, 'Expression of transferred genes during hairy root development in pea'. *Plant Cell Tissue Organ Cult*, 1995, **42**: 57–66.
84. PUONTI-KAERLAS J, OTTOSSON A, ERICKSSON T, 'Survival and growth of pea protoplasts after transformation by electroporation'. *Plant Cell Tissue Organ Cult*, 1992, **30**: 141–148.
85. PUONTI- KAERLAS J, ERIKSSON T, ENGSTROM P, 'Production of transgenic pea (*Pisum sativum* L.) plants by A*grobacterium tumefaciens*-mediated gene transfer'. *Theor Appl Genet*, 1990, **80**: 246–252.

86. PUONTI-KAERLAS J, ERIKSSON T, ENGSTROM P, 'Inheritance of a bacterial hygromycin phosphotransferase gene in the progeny of primary transgenic pea plants.' *Theor Appl Genet*, 1992, **84**: 443–450.
87. SCHROEDER H, SCHOTZ A, WARDLEY-RICHARDSON T, SPENCER D, HIGGINS T, 'Transformation and regeneration of two cultivars of pea (*Pisum sativum* L)'. *Plant Physiol*, 1993, **101**: 751–757.
88. GRANT JE, COOPER PA, MCARA AE, FREW TJ, 'Transformation of peas (Pisum sativum L.) using immature cotyledons'. *Plant Cell Rep*, 1995, **15**: 254–258.
89. GRANT J, PITHER-JOYCE M, FIFIELD W, COOPER P, TIMMERMAN-VAUGHAN G, 'Partial resistance to alfalfa mosaic virus in transgenic pea (*Pisum sativum* L.)'. 3rd European conference on grain legumes. Opportunities for high quality, healthy and added value crops to meet European demands, Valladolid, Spain (1998).
90. DAVIES DR, HAMILTON J, MULTINEAUX P, 'Transformation of peas'. *Plant Cell Rep*, 1993, **12**: 180–183.
91. BEAN SJ, GOODING PS, MULLINEAUX PM, DAVIES DR, 'A simple system for pea transformation'. *Plant Cell Rep*, 1997, **16**: 513–519.
92. BABAOGLU M, DAVEY MR, POWER JB, 'Genetic engineering of grain legumes: key transformation events'. *AgBiotechNet*, 2000, **2**.
93. MARIOTTI D, FONTANA GS, SANTINI L, 'Genetic transformation of grain legumes: *Phaseolus vulgaris* L. and *P. coccineus* L'. *J Genet Breed*, 1989, **43**: 77–82.
94. MCCLEAN P, CHEE P, HELD B, SIMENTAL J, DRONG RF, SLIGHTOM J, 'Susceptibility of dry bean (*Phaseolus vulgaris* L.) *Agrobacterium* infection: Transformation of cotyledonary and hypocotyl tissues'. *Plant Cell Tissue Organ Cult*, 1991, **24**: 131–138.
95. ZHANG Z, DERMOT P, COYNE, MITRA A, 'Factors affecting *Agrobacterium*-mediated transformation of common bean'. *J Am Soc Hort Sci*, 1997, **122**: 300–305.
96. LEWIS ME, BLISS FA, 'Tumor formation and β-glucuronidase expression in *Phaseolus vulgaris* inoculated with *Agrobacterium tumefaciens*'. *J Am Soc Hort Sci*, 1994, **119**: 361–366.
97. CHRISTOU P, MCCABE DE, MARTINELL BJ, SWAIN WF, 'Soybean genetic engineering – commercial production of transgenic plants'. *Trends Biotechnol*, 1990, **8**: 145–151.
98. WALLACE, KM, BATHE JH, MARTINELL BJ, MCCABE DE, 'Stable transformation of *Phaseolus vulgaris* via electric-discharge mediated particle acceleration'. *Plant Cell Rep*, 1993, **12**: 165–169.
99. KIM JW, MINAMIKAWA T, 'Transformation and regeneration of French bean plants by the particle bombardment process'. *Plant Sci*, 1996, **117**: 131–138.
100. ARAGAO FJL, BARROS LMG, BRASILEIRO ACM, RIBEIRO SG, SMITH FD, SANFORD JC, FARIA JC, RECH EL, 'Inheritance of foreign genes in transgenic bean (*Phaseolus vulgaris* L.) co-transformed via particle bombardment'. *Theor Appl Genet*, 1996, **93**: 142–150.

101. ARAGAO FJL, RIBEIRO SG, BARROS LMG, BRASILEIRO ACM, MAXWELL DP, RECH EL, FARIA JC, 'Transgenic beans (*Phaseolus vulgaris* L.) engineered to express viral antisense RNAs show delayed and attenuated symptoms to bean golden mosaic geminivirus'. *Molecular Breeding*, 1998, **4**: 491–499.
102. SAALBACH I, WADDELL D, PICKARDT T, SCHIEDER O, MUNTZ K, 'Stable expression of the sulphur-rich 2S albumin gene in transgenic *Vicia narbonensis* increases the methionine content of seeds'. *J Plant Physiol*, 1995, **145**: 674–681.
103. PICKARDT T, SAALBACH I, WADDELL D, MEIXNER MG, MUNTZ K, SCHIEDER O, 'Seed specific expression of 2S albumin gene from Brazil nut (*Bertholletia excelsa*) in transgenic Vicia'. *Molecular Breeding*, 1995, **1**: 295–301.
104. BHARGAVA SC, SMIGOCKI AC, 'Transformation of tropical grain legumes using particle bombardment.' *Current Science*, 1994, **66**: 439–442.
105. KARTHIKEYAN AS, SARMA KS, VELUTHAMBI K, '*Agrobacterium tumefaciens*-mediated transformation of *Vigna mungo* L. Hepper'. *Plant Cell Rep*, 1996, **15**: 328–331.
106. JAIWAL PK, SAUTTER C, POTRYKUS I, '*Agrobacterium rhizogenes*-mediated gene transfer in mungbean'. *Current Science*, 1998, **75**: 41–45.
107. PENZA R, LURQUIN PF, FILIPPONE E, 'Gene transfer by cocultivation of mature embryos with *Agrobacterium tumefaciens*: application to cowpea (*Vigna unguiculata* Walp.)'. *J Plant Physiol*, 1991, **138**: 39–42.
108. POTRYKUS I, 'Gene transfer to plants: assessment and perspectives'. *Physiol Plant*, 1990, **79**: 125–134.
109. CHOWRIRA GM, AKELLA V, FUERST PE, LURQUIN PF, 'Transgenic grain legumes obtained by *in planta* electroporation-mediated gene transfer'. *Molecular Biotechnology*, 1996, **5**: 85–96.
110. WARKENTIN TD, MCHUGHEN A, 'Crown gall transformation of lentil (*Lens culinaris* Medik.) with virulent strains of *Agrobacterium tumefaciens*'. *Plant Cell Rep*, 1991, **10**: 489–493.
111. WARKENTIN TD, MCHUGHEN A, '*Agrobacterium tumefaciens*-mediated beta-glucoronidase (*GUS*) gene expression in lentil (*Lens culinaris* Medik.) tissues'. *Plant Cell Rep*, 1992, **11**: 274–278.
112. OKTEM HA, MAHMOUDIAN M, EYIDODAN F, YUCEL M, 'GUS gene delivery and expression in lentil cotyledonary nodes using particle bombardment'. *Lens Newsletter*, 1999, **26**: 3–6.
113. FONTANA GS, SANTINI L, CARETTO S, FRUGIS G, MARIOTTI D, 'Genetic transformation in the grain legume *Cicer arietinum* L. (chick pea)'. *Plant Cell Rep*, 1993, **12**: 194–198.
114. KAR S, JOHNSON TM, NAYAK P, SEN SK, 'Efficient transgenic plant regeneration through *Agrobacterium*-mediated transformation of chickpea (*Cicer arietinum* L.)'. *Plant Cell Rep*, 1996, **16**: 32–37.
115. KRISHNAMURTHY KV, SUHASINI K, SAGARE AP, MEIXNER M, KATHEN AD, PICKARDT T, SCHIEDER O, '*Agrobacterium* mediated transformation of chickpea (*Cicer arietinum* L.) embryo axes'. *Plant Cell Rep*, 2000, **19**: 235–240.

116. ALTINKUT A, GOZUKIRMIZ N, BAJROVIC K, GOZUKIRMIZI N, 'High percentage of regeneration and transformation in chickpea'. *Acta Hort*, 1997, **447**: 319–320.
117. KAR S, BASU D, DAS S, RAMKRISHNAN NA, MUKHERJEE P, NAYAK P, SEN SK, 'Expression of *CryIA(c)* gene of *Bacillus thuringiensis* in transgenic chickpea plants inhibits development of podborer (*Heliothis armigera*) larvae'. *Transgenic Res*, 1997, **6**: 177–185.
118. KUNIT T, SALOMON R, ZAMIR D, NAVOR N, ZEIDAN M, MICHELSON I, GAFNI Y, CZOSNEK H, 'Transgenic tomato plants expressing the tomato yellow leaf curl virus capsid protein are resistant to the virus'. *Bio/Technol*, 1994, **12**: 500–504.
119. CHOWRIRA GM, CAVILEER TD, GUPTA SK, LURQUIN PF, BERGER PH, 'Coat protein-mediated resistance to pea enation mosaic virus in transgenic *Pisum sativum* L'. *Transgenic Res*, 1998, **7**: 265–271.
120. SHADE RE, SCHROEDER HE, PUEYO JL, TABE LM, MURDOCK LL, HIGGINS TJV, CHRISPEELS MJ, 'Transgenic pea seeds expressing the alpha-amylase inhibitor of the common bean are resistant to bruchid beetles'. *Bio/Technol*, 1994, **12**: 793–796.
121. SCHROEDER HE, GOLLASCH S, MOORE A, TABE LM, CRAIG S, HARDIE DC, CHRISPEELS MJ, SPENCER D, HIGGINS TJV, 'Bean alpha-amylase inhibitor confers resistance to the pea weevil (*Bruchus pisorum*) in transgenic peas (*Pisum sativum* L.)'. *Plant Physiol*, 1995, **107**: 1233–1239.
122. MORTON RL, SCHROEDER HE, BATEMAN KS, CHRISPEELS MJ, ARMSTRONG E, HIGGINS TJV, 'Bean α-amylase inhibitor 1 in transgenic peas (*Pisum sativum*) provides complete protection from pea weevil (*Bruchus pisorum*) under filed conditions'. *Proc Natl Acad Sci USA*, 2000, **97**: 3820–3825.
123. CHARITY JA, ANDERSON MA, BITTISNICH DJ, WHITECROSS M, HIGGINS TJV, 'Transgenic tobacco and peas expressing a proteinase inhibitor from *Nicotiana alata* have increased insect resistance'. *Molecular Breeding*, 1999, **5**: 357–365.
124. MA Y, BLISS FA, 'Seed proteins in bean'. *Crop Sci*, 1978, **18**: 431–437.
125. ALTENBACH SB, PEARSON KW, LEUNG FW, SUN SSM, 'Cloning and sequence analysis of a cDNA encoding a Brazil nut protein exceptionally rich in methionine'. *Plant Mol Biol*, 1987, **8**: 239–250.
126. GANDER ES, HOLMSTROEM KO, DE PAIVA GR, DE CASTRO LAB, CARNEIRO M, GROSSI DE SA MF, 'Isolation characterization and expression of a gene coding for a 2S albumin from *Bertholletia excelsa* (Brazil nut)'. *Plant Mol Biol*, 1991, **16**: 437–448.
127. ARAGAO FJL, BARROS LMG, SOUSA MVD, GROSSI DE SA MF, ALMEIDA ERP, GANDER ES, RECH EL, 'Expression of a methionine-rich storage albumin from the Brazil nut (*Bertholletia excelsa* H.B.K., Lecythidaceae) in transgenic bean plants (*Phaseolus vulgaris* L., Fabaceae)'. *Genet Mol Biol*, 1999, **22**: 445–449.
128. PICKARDT T, MEIXNER M, SCHADE V, SCHIEDER O, 'Transformation of *Vicia narbonensis* via *Agrobacterium*-mediated gene transfer'. *Plant Cell Rep*, 1991, **9**: 535–538.

129. NORDLEE JA, TAYLOR SL, TOWNSEND JA, THOMAS LA, BUSH RK, 'Identification of a brazil nut-allergen in transgenic soybeans'. *New England J Med*, 1996, **334**: 688–692.
130. OSBORN TC, ALEXANDER DC, SUN SSM, CARDONA C, BLISS FA, 'Insectidal activity and lectin homology of arcelin seed protein'. *Science*, 1989, **240**: 207–210.
131. PRATT RC, SINGH NK, SHADE RE, MURDOCK LL, BRESSAN RA, 'Isolation and partial characterisation of a seed lectin from tepary bean that delays bruchid beetle development'. *Plant Physiol*, 1990, **93**: 1453–1459.
132. EDWARDS GA, HEPHER A, CLERK SP, BOULTER D, 'Pea lectin is correctly processed, stable and active in leaves of transgenic potato plants'. *Plant Mol Biol*, 1991, **17**: 89–100.
133. HILDER VA, GATEHOUSE AMR, SHEERMAN SE, BARKER RF, BOULTER D, 'A novel mechanism of insect resistance engineered into tobacco'. *Nature*, 1987, **300**: 160–163.
134. NEWELL CA, LOWE JM, MERRYWEATHER A, ROOKE LM, HAMILTON WDO, 'Transformation of sweet potato (*Ipomoea batatas* (L.) Lam.) with *Agrobacterium tumefaciens* and regeneration of plants expressing cowpea trypsin inhibitor and snowdrop lectin'. *Plant Sci*, 1995, **107**: 215–227.
135. SINDHU AS, ZHENG Z, MURAI N, 'The pea storage protein legumin was synthesized, processed, and accumulated stably in transgenic rice endosperm'. *Plant Sci*, 1997, **130**: 189–196.
136. RIGGS CD, HUNT DC, LIN J, CHRISPEELS MJ, 'Utilization of luciferase fusion genes to monitor differential regulation of phytohemagglutinin and phaseolin promoters in transgenic tobacco'. *Plant Sci*, 1989, **63**: 47–57.
137. ALTABELLA T, CHRISPEELS MJ, 'Tobacco plants transformed with the bean *aai* gene express an inhibitor of insect a-amylase in their seeds'. *Plant Physiol*, 1990, **93**: 805–810.
138. GOOSSENS A, DILLEN W, CLERCQ JD, MONTAGU MV, ANGENON G, 'The *arcelin-5* gene of *Phaseolus vulgaris* directs high seed-specific expression in transgenic *Phaseolus acutifolius* and *Arabidopsis* plants'. *Plant Physiol*, 1999, **120**: 1095–1104.
139. PATER SD, PHAM K, CHUA NH, MEMELINK J, KIJNE J, 'A 22-bp fragment of the pea lectin promoter containing essential TGAC-like motifs confers seed-specific gene expression.' *Plant Cell*, 1993, **5**: 877–886.
140. MANDACI S, DOBRES MS, 'A promoter directing epidermal expression in transgenic alfafa'. *Plant Mol Biol*, 1997, **34**: 961–965.
141. DE VRIES SC, BOOIJ H, MEYERINK P, HUISMAN G, WILDE DH, THOMAS TL, KAMMEN AV, 'Acquisition of embryogenic potential in carrot cell-suspension cultures'. *Planta*, 1988, **176**: 196–204.
142. BALESTRAZZI A, CARBONERA D, CELLA R, 'Transformation of *Daucus carota* hypocotyls mediated by *Agrobacterium tumefaciens*'. *J Genet Breed*, 1991, **45**: 135–140.
143. PAWLICKI N, SANGWAN-NORREEL B, 'Factors influencing the *Agro-*

bacterium tumefaciens-mediated transformation of carrot (*Daucus carota* L.)'. *Plant Cell Tissue Organ Cult*, 1992, **31**: 129–139.

144. SCOTT R, DRAPER J, 'Transformation of carrot tissues derived from proembryogenic suspension cells: a useful model system for gene expression studies in plants'. *Plant Mol Biol*, 1987, **8**: 265–274.
145. THOMAS J, GUILTINAN M, BUSTOS S, THOMAS T, NESSLER C, 'Carrot (*Daucus carota*) hypocotyl transformation using *Agrobacterium tumefaciens*'. *Plant Cell Rep*, 1989, **8**: 354–357.
146. WURTELE E, BULKA K, 'A simple, efficient method for the *Agrobacterium*-mediated transformation of carrot callus cells'. *Plant Sci*, 1989, **61**: 253–262.
147. HARDEGGER M, STURM A, 'Transformation and regeneration of carrot (*Daucus carota* L.)'. *Molecular Breeding*, 1998, **4**: 119–127.
148. TANG GQ, STURM A, 'Antisense repression of sucrose synthase in carrot (Daucus carota L.) affects growth rather than sucrose partitioning'. *Plant Mol Biol*, 1999, **41**: 465–479.
149. TANG GQ, LUSCHER M, STURM A, 'Antisense repression of vacuolar and cell wall invertase in transgenic carrot alters early plant development and sucrose partitioning'. *Plant Cell*, 1999, **11**: 177–189.
150. GILBERT MO, ZHANG YY, PUNJA ZK, 'Introduction and expression of chitinase encoding genes in carrot following agrobacterium-mediated transformation'. *In Vitro Cell Dev Biol*, 1996, **32**: 171–178.
151. NISHIGUCHI M, UEHARA Y, KOMAKI K, 'Stable transformation of sweet potato by electroporation'. *In Vitro Cell Dev Biol*, 1992, **28**: 126.
152. PRAKASH CS, VARADARAJAN U, 'Genetic transformation of sweet potato by particle bombardment'. *Plant Cell Rep*, 1992, **11**: 53–57.
153. OTANI M, MII M, HANDA T, KAMADA H, SHIMADA T, 'Transformation of sweet potato (*Ipomoea batatas* L. Lam.) plants by *Agrobacterium rhizogenes*'. *Plant Sci*, 1993, **94**: 151–159.
154. GAMA MICS, LEITRE RP, JR., CORDEIRO AR, CANTLIFFE DJ, 'Transgenic sweet potato plants obtained by *Agrobacterium tumefaciens*-mediated transformation'. *Plant Cell Tissue Organ Cult*, 1996, **46**: 237–244.
155. MORAN R, GARCIA R, LOPEZ A, ZALDUA Z, MENA J, GARCIA M, ARMAS R, SOMONTE D, RODRIGUEZ J, GOMEZ M, PIMENTEL E, 'Transgenic sweet potato plants carrying the delta-endotoxin gene from *Bacillus thuringiensis* var. *tenebrionis*'. *Plant Sci*, 1998, **139**: 175–184.
156. CHÉE RP, CANTLIFFE DJ, 'Somatic embryony patterns and plant regeneration in *Ipomoeas batatas* Poir.'. *In Vitro Cell Dev Biol*, 1988, **24**: 955–958.
157. DOMMISSE EM, LEUNG DWM, SHAW ML, CONNER AJ, 'Onion is a monocotyledonous host for *Agrobacterium*'. *Plant Sci*, 1990, **69**: 249–257.
158. BARANDIARAN X, MARTIN N, RODRIGUEZ-CONDE MF, DI PIETRO A, MARTIN J, 'An efficient method for callus culture and shoot regeneration of garlic (*Allium sativum* L.)'. *HortSci*, 1999, **34**: 348–349.
159. BUITEVELD J, FRANSZ PF, CREEMERS-MOLENAAR J, 'Induction and characterization of embryogenic callus types for the initiation of

suspension cultures of leek (*Allium ampeloprasum* L.)'. *Plant Sci*, 1994, **100**: 195–202.

160. BUITEVELD J, CREEMERS-MOLENAAR J, 'Plant regeneration from protoplasts isolated from suspension cultures of leek (Allium ampeloprasum L.)'. *Plant Sci*, 1994, **100**: 203–210.
161. HINCHEE M, 'Development of virus-resistant sweetpotato'. In: International C (ed.) *Agricultural Biotechnology in International Development*, Wallingford, UK 1998.
162. GOGARTEN JP, FICHMANN J, BRAUN Y, MORGAN L, STYLES P, DELAPP K, TAIZ L, 'The use of antisense mRNA to inhibit the tonoplast H^+ ATPase in carrot'. *Plant Cell*, 1992, **4**: 851–864.
163. MALIK MK, SLOVIN JP, HWANG CH, ZIMMERMAN JL, 'Modified expression of a carrot small heat shock protein gene, Hsp 17.7, results in increased or decreased thermotolerance'. *Plant J*, 1999, **20**: 89–99.
164. DING LC, HU CY, 'Development of insect-resistant transgenic cauliflower plants expressing the trypsin inhibitor gene isolated from local sweet potato'. *Plant Cell Rep*, 1998, **17**: 854–860.
165. PUDDEPHAT IJ, RIGGS TJ, FENNING TM, 'Transformation of *Brassica oleracea* L. a critical review'. *Molecular Breeding*, 1996, **2**: 185–210.
166. CHRISTEY MC, SINCLAIR BK, BRAUN RH, WYKE L, 'Regeneration of transgenic vegetable brassicas (*Brassica aleracea* and B. *campestris*) via Ri-mediated transformation'. *Plant Cell Rep*, 1997, **16**: 587–593.
167. DE BLOCK M, DE BROUWER D, TENNING P, 'Transformation of *Brassica napus* and *Brassica aleracea* using *Agrobacterium tumefaciens* and the expression of the bar and neo genes in the transgenic plants'. *Plant Physiol*, 1989, **91**: 694–701.
168. BHALLA L, SMITH NA, 'Agrobacterium tumefaciens-mediated transformation of cauliflower, *Brassica oleracea* var. botrytis'. *Molecular Breeding*, 1998, **4**: 531–541.
169. METZ TD, DIXIT R, EARLE ED, '*Agrobacterium tumefaciens*-mediated transformation of broccoli (*Brassica oleracea* var. *italica*) and cabbage (B. *oleracea* var. *capitata*)'. *Plant Cell Rep*, 1995, **15**: 287–292.
170. TORIYAMA K, STEIN JC, NASRALLAH ME, NASRALLAH JB, 'Transformation of *Brassica oleracea* with an S-locus gene from *B. campestris* changes the self-incompatibility phenotype'. *Theor Appl Genet*, 1991, **81**: 769–776.
171. METZ TD, ROUSH RT, TANG JD, SHELTON AM, EARLE ED, 'Transgenic broccoli expressing a *Bacillus thuringiensis* insecticidal crystal protein: implications for pest resistance management strategies'. *Molecular Breeding*, 1995, **1**: 309–317.
172. CAO J, TANG JD, 'Transgenic broccoli with high levels of *Bacillus thuringiensis* Cry1C protein control diamondback moth larvae resistant to Cry1A or Cry1C'. *Molecular Breeding*, 1999, **5**: 131–141.
173. HENZI MX, CHRISTEY MC, MCNEIL DL, DAVIES KM, '*Agrobacterium rhizogenes*-mediated transformation of broccoli (*Brassica oleracea* L.

var. *italica*) with an antisense 1-aminocyclopropane-1-carboxylic acid oxidase gene'. *Plant Sci*, 1999, **143**: 55–62.

174. HENZI MX, MCNEIL DL, CHRISTEY MC, LILL RE, 'A tomato antisense 1-aminocyclopropane-1-carboxylic acid oxidase gene causes reduced ethylene production in transgenic broccoli'. *Aust J Plant Physiol*, 1999, **26**: 179–183.

175. JUN SI, KWON SY, '*Agrobacterium*-mediated transformation and regeneration of fertile transgenic plants of chinese cabbage (Brassica campestris ssp. pekinensis cv. "Spring Flavor"'. *Plant Cell Rep*, 1995, **14**: 620–625.

176. LIM HT, YOU YS, PARK EJ, SONG YN, 'High plant regeneration genetic stability of regenerants and genetic transformation of herbicide resistance gene (*Bar*) in chinese cabbage (*Brassica campestris* ssp. *pekinensis)*'. *Acta Hort*, 1998, **459**: 199–208.

177. MICHELMORE R, MARSH E, SEELY S, LANDRY B, 'Transformation of lettuce (*Lactuca sativa*) mediated by *Agrobacterium tumefaciens*'. *Plant Cell Rep*, 1987, **6**: 439–442.

178. PANG SZ, JAN FJ, CARNEY K, STOUT J, TRICOLI DM, QUEMADA HD, GONSALVES D, 'Post-transcriptional transgene silencing and consequent tospovirus resistance in transgenic lettuce are affected by transgene dosage and plant development'. *Plant J*, 1996, **9**: 899–909.

179. CHUPEAU MC, BELLINI C, GUERCHE P, MAISONNEUVE B, VASTRA G, CHUPEAU Y, 'Transgenic plants of lettuce (*Lactuca sativa*) obtained through electroporation of protoplasts'. *Bio/Technol*, 1989, **7**: 503–508.

180. DINANT S, MAISONNEUVE B, ALBOUY J, CHUPEAU Y, CHUPEAU MC, BELLEC Y, GAUDEFROY F, KUSIAK C, SOUCHE S, ROBAGLIA C, Lot H, 'Coat protein gene-mediated protection in *lactuca sativa* against lettuce mosaïc potyvirus strains'. *Molecular Breeding*, 1997, **3**: 75–86.

181. CURTIS IS, POWER JB, BLACKHALL NW, DE LAAT AMM, DAVEY MR, 'Genotype-independent transformation of lettuce using *Agrobacterium tumefaciens*'. *J Exp Bot*, 1994, **45**: 1441–1449.

182. VERMEULEN A, VAUCHERET H, PAUTOT V, CHUPEAU Y, '*Agrobacterium* mediated transfer of a mutant *Arabidopsis acetolactate synthase* gene confers resistance to chlorsulfuron in chicory (*Cichorium intybus* L.)'. *Plant Cell Rep*, 1992, **11**: 243–247.

183. AL-KHAYRI JM, HUANG FH, MORELOCK TE, 'Regeneration of spinach from leaf callus'. *HortSci*, 1992, **26**: 913–914.

184. XIAO XG, BRANCHARD M, 'Embryogenesis and plant regeneration of spinach (*Spinacia oleracea* L.) from hypocotyl segments'. *Plant Cell Rep*, 1993, **13**: 69–71.

185. MOLVIG L, ROSE RJ, 'A regeneration protocol for *Spinacia oleracea* using gibberellic acid'. *Aust J Bot*, 1994, **42**: 763–769.

186. KNOLL KA, SHORT KC, CURTIS IS, POWER JB, DAVEY MR, 'Shoot regeneration from cultured root explants of spinach (*Spinacia oleracea* L.): a system for *Agrobacterium* transformation'. *Plant Cell Rep*, 1997, **17**: 96–101.

187. YANG YM, AL-KHAYRI JM, ANDERSON EJ, 'Transgenic spinach plants expressing the coat protein of cucumber mosaic viruis'. *In Vitro Cell Dev Biol Plant*, 1997, **33**: 200–204.
188. ZHANG HX, ZEEVAART JAD, 'An efficient *Agrobacterium tumefaciens*-mediated transformation and regeneration system for cotyledons of spinach (*Spinacia oleracea* L.)'. *Plant Cell Rep*, 1999, **18**: 640–645.
189. TO KY, CHENG MC, CHEN LFO, CHEN SCG, 'Introduction and expression of foreign DNA in isolated spinach chloroplasts by electroporation'. *Plant J*, 1996, **10**: 737–743.
190. CONNER AJ, ABERNETHY DJ, 'Genetic engineering of asparagus: assessment of methods, field testing and safety considerations'. *Acta Hort*, 1996, **415**: 51–58.
191. BYTEBIER B, DEBOECK F, DE GREVE H, VAN MONTAGU M, HERNALSTEENS JP, 'T-DNA organization in tumor cultures and transgenic plants of the monocotyledon *Asparagus officinalis*'. *Proc Natl Acad Sci USA*, 1987, **84**: 5345–5349.
192. HERNALSTEENS JP, THIA-TOONG L, SCHELL J, VAN MONTAGU M, 'An *Agrobacterium*-transformed cell culture from the monocot *Asparagus officinalis*'. *Embo J*, 1984, **3**: 3039–3041.
193. DELBREIL B, GUERCHE P, JULLIEN M, '*Agrobacterium*-mediated transformation of *Asparagus officinalis* L. Long-term embryogenic callus and regeneration of transgenic plants'. *Plant Cell Rep*, 1993, **12**: 129–132.
194. MUKHOPADHYAY S, DESJARDINS Y, 'Direct gene transfer to protoplasts of two genotypes of *asparagus officinalis* L. by electroporation'. *Plant Cell Rep*, 1994, **13**: 421–424.
195. GUANGYU C, CONNER AJ, FAUTRIER AG, FIELD RJ, 'A shoot protoplast system for genetic manipulation of asparagus'. *Asparagus Research Newsletter*, 1996, **13**: 18–25.
196. SAITO T, NISHIZAWA S, NISHIMURA S, 'Improved culture conditions for somatic embryogenesis from *Asparagus officinalis* L. using an aseptic ventilative filter'. *Plant Cell Rep*, 1991, **10**: 230–234.
197. LI B, WOLYN DJ, 'Recovery of transgenic asparagus plants by particle gun bombardment of somatic cells'. *Plant Sci*, 1997, **126**: 59–68.
198. CABRERA-PONCE JL, LOPEZ L, ASSAD-GARCIA N, MEDINA-AREVALO C, BAILEY AM, HERRERA-ESTRELLA L, 'An efficient particle bombardment system for the genetic transformation of asparagus (*Asparagus officinalis* L.)'. *Plant Cell Rep*, 1997, **16**: 255–260.
199. MCCABE MS, SCHEPERS F, VAN DER AREND A, MOHAPATRA U, DE LAAT AMM, POWER JB, DAVEY MR, 'Increased stable inheritance of herbicide resistance in transgenic lettuce carrying a petE promoter-bar gene compared with a CaMV 35S-bar gene'. *Theor Appl Genet*, 1999, **99**: 587–592.
200. PASSELÈGUE E, KERLAN C, 'Transformation of cauliflower (*Brassica oleracea* var. *botrytis*) by transfer of cauliflower mosaic virus genes through combined cocultivation with virulent and avirulent strains of *Agrobacterium*'. *Plant Sci*, 1996, **113**: 79–89.

201. YANG CH, CAROLL B, SCOFIELD S, JONES J, MICHELMORE R, 'Transactivation of Ds elements in plants of lettuce (*Lactuca sativa*)'. *Mol Gen Genet*, 1993, **241**: 389–398.
202. OKUBARA PA, ARROYO G, R., SHEN KA, MAZIER M, MEYERS BC, OCHOA OE, KIM S, YANG CH, MICHELMORE RW, 'A transgenic mutant of *lactuca sativa* (lettuce) with a T-DNA tightly linked to loss of downy mildew resistance'. *Molecular Plant Microbe Interactions*, 1997, **10**: 970–977.
203. CURTIS IS, POWER JB, DE LAAT AMM, CABOCHE M, DAVEY MR, 'Expression of a chimeric nitrate reductase gene in transgenic lettuce reduces nitrate in leaves'. *Plant Cell Rep*, 1999, **18**: 889–896.
204. GOTO F, YOSHIHARA T, SAIKI H, TAKAIWA F, SHIGEMOTO N, 'Iron accumulation in transgenic plants expressing the soybean ferritin gene'. *Acta Hort*, 2000, **521**: 101–109.
205. GOTO F, YOSHIHARA T, SAIKI H, 'Iron accumulation and enhanced growth in transgenic lettuce plants expressing the iron-binding protein ferritin'. *Theor Appl Genet*, 2000, **5**: 658–664.
206. PUCHTA H, 'Removing selectable marker genes: taking the shortcut'. *Trends Plant Sci*, 2000, **5**: 273–274.

Part III

Consumer's attitudes and risk assessment

11

Consumer's attitudes

L. J. Frewer, Institute of Food Research, Norwich

11.1 Plant biotechnology and public attitudes

People will not consume foods that they believe are risky, or associate with some other negative attribute. Consumer concerns focus on different issues depending on the type of food under consideration and may include, for example, beliefs that there is potential for negative environmental impact associated with production processes or agricultural practices, perceptions that there is uncertainty associated with unintended human or animal health effects, or even that these unintended effects are completely unpredictable and unknown. Finally, people may believe that there are potential consequences for the way in which society is organised, (for example, people may perceive that changes in technology may shift local agricultural production to globalised systems increasing people's dependence on multi-national companies), which may result in further concern (Frewer, in press). For this reason, considerable effort has been directed towards understanding people's attitudes towards emerging food technologies generally, and genetically modified foods specifically. There has been concern within the scientific and policy community that people might potentially reject consumer products which have been introduced into the market place as a result of new developments in the biosciences. Indeed, since the early 1980s, an enormous amount of research has been conducted by social scientists directed towards understanding public perceptions of genetically modified foods (Zechendorf, 1994). Approaches adopted have ranged from simple 'opinion poll' methodologies which tend to focus on items relating to acceptance or rejection of genetic technologies (for example, Eurobarometer, 1997), to more scientific and thoughtful attempts to develop causal models explaining the interrelationships between, for example, the extent to which people trust institutions responsible for regulation and technology development, and perceptions of risk and benefits associated with gene technology (see, for example, Siegrist 1999; Frewer, Scholderer, and Bredahl, in press).

Despite increased understanding of what is driving public concern, consumer acceptance of novel foods produced using emerging bioscience techniques such as genetic modification has been rather low, perhaps because the complexity of the interrelationship between science and society has been not properly understood. The politics of technology acceptance must be considered when developing new products and processes.

> Risk managers have become keenly aware that in democratic countries [public] perceptions of a technology's risks and benefits are important components of the ... political decision process, from initial decisions to developing a technology or product, to the acceptance of management approaches to risk mitigation (Siegrist *et al.*, 2000, p. 353).

There has also been a shift in emphasis linked to *why* researchers are attempting to understand peoples attitude's towards agricultural and plant biotechnology, and associated novel foods which these processes produce. Fifteen years ago the emphasis of risk perception research and communication efforts linked to the use of technology in the food chain was technology acceptance. At the time of writing, there is much more debate about how to increase transparency in risk-management processes and greater involvement of the public in deciding how to manage and regulate technology innovation. In particular, there has been considerable emphasis in recent times on communicating information relevant to people's concerns, (for example, ethical considerations in the development and implementation of technology), as well as developing strategies to convey information to the public about probabilistic risk-assessment processes.

More recent research has implied that trust in science and risk regulators, and public confidence in scientific advice, has powerful explanatory power in the context of how people respond to and interpret information. Recent theoretical stances have developed the idea that distrust of institutions (partly through perceived exclusion from the decision-making machinery linked to government and science) represents a key driver in creating and fuelling public negativity to scientific innovation and risk management practices (HM Government, 2001). Efforts to understand the psychological determinants of trust (in information sources and regulatory institutions) laid the groundwork for subsequent analysis of how complex risk information is processed and transmitted (Cvetkovich and Löfstedt, 1999). However, the need for explicit public involvement in risk-management policy has emerged as a key driver in initiatives to increase public confidence in technological risk-management itself (Rowe and Frewer, 2000), with emphasis on how the output of consultation can explicitly, as opposed to implicitly, be used in policy (Frewer and Salter, in press).

Public attitudes towards emerging biosciences such as genetic modification must be understood if effective communication about both the associated risks and benefits is to be developed. However, research into attitudes is also relevant to understanding how the relationship between science and society, public trust and governance of technology might democratically evolve in the future. A useful first stage for the purposes of the current discussion is to define what is meant by the term 'attitude', and to describe some of the theoretical contexts in which attitude research is embedded.

11.2 What is meant by the term 'attitude?'

In very general terms, the term 'attitude' is used to describe 'a psychological tendency that is expressed by evaluating a particular entity with some degree of favour or disfavour' (Eagly and Chaiken, 1993, p. 1). Here, psychological tendency refers to a state of mind that is internal to an individual, whereas 'evaluating' refers to cognitive, affective or behavioural responses that result from the attitude. Thus 'psychological tendency' might be thought of as a psychological bias that predisposes the individual towards positive or negative evaluative responses.

Attitudes can be used to explain why some people support particular social policies, or ideologies, while others oppose them. A person who favours a particular policy is said to hold a positive attitude towards it, whereas someone who opposes it would hold a negative attitude. Attitudes are not directly observable but can be inferred from observable responses (MacCorquodal and Meehl, 1948), such as responses to questionnaires or interviewers. Evaluative responses are those that express approval or disapproval, liking or disliking, approach or avoidance, attraction or aversion, and so forth. Evaluative responses and the psychological tendencies that are assumed to underlie them differ not only in terms of direction (positive or negative) but also intensity (a very positive evaluation is likely to have a very different impact on behaviour compared to a slightly positive one). Thus social scientists usually measure attitudes along a bipolar continuum that ranges from extremely positive to extremely negative, and includes a neutral reference point.

Historically, of greatest interest to social science has been study of people's attitudes towards social policies (for example, siting of nuclear power stations, technology application and development), although attitudes towards relatively abstract or end states of human existence (for example, human equality, sovereignty of nature) have also been of interest, and are normally termed 'values'. People's responses and behaviours to events, political situations and products will depend both on their attitudes and values.

The term 'attitude object' refers to the entity, object or event about which people make their evaluations. In very broad terms, people who evaluate an attitude object in a favourable way are likely to associate it with positive attributes and are unlikely to associate it with negative attributes. Conversely, people who evaluate an attitude object unfavourably are more likely to associate it with negative attributes than positive ones (Eagly and Chaiken, 1993, p. 11).

11.3 Changes in attitudes

In order to change a person's attitude, it is necessary to provide some additional information that will influence either the extent of the attitude's strength, or its direction. There is some evidence that attitudes may be changed through direct exposure to an attitude object. In the case of foods produced with novel technologies, a person who has a very negative attitude towards the technology may change their attitude following experience with products made with the technology. Most usually, however, the attitudinal effects of direct exposure are difficult to detect because the effects are masked by information about the attitude object, beliefs about the motives of information sources, or other contextual factors associated with exposure. Exposure to an attitude object in itself may provide information about the characteristics of the object, but will produce attitude change only if the new information is very different from what is already known about the object by an individual (Stroebe *et al.*, 1988).

A second approach to attitude change relies on persuasive communication, which is theoretically underpinned by theories of social influence. A position is advocated by an information source, and different messages are presented by the information source to support this position. Persuasion frequently focuses on counter attitudinal-communication (advocating a position that does not align with attitudes already held by the individual receiving the information). Theories of persuasion use the processes or variables that mediate the impact of communication on attitudes and beliefs. Hovland (1959) has argued that, if people do not attend a message, (i.e. if they are not motivated to so do because the information is not useful or interesting), attitude change will not occur. Similarly, if they are unable to comprehend it despite being highly motivated to think about the information, then their attitudes will not change. People may also use two different routes to processing information (Petty and Cacioppo, 1986). The first of these is the central route to persuasion when people spend a considerable amount of effort on critically evaluating the message content. However, if people are loath to expend this amount of effort on processing message contents, they adopt peripheral routes to persuasion – for example, information sources which they trust or like may result in more positive evaluation of the message which the source provides (Eagly and Chaiken, 1984). Petty and Cacioppo (1986) have proposed that attitude change will be the result of both the amount of message-relevant thinking as well as the extent to which people agree with the information. For messages that elicit favourable thoughts, increased elaboration (i.e. increased thoughts about the message content) should increase persuasion. In contrast, if messages elicit mainly negative thoughts, increased processing should reduce persuasion. From this, one might extrapolate that simply telling people that products are safe is unlikely to provide the reassurance that results in public support of genetically modified foods, unless this belief is already strongly held by those people receiving the message. People who are already rather concerned about genetically modified foods are unlikely to internalise messages which have an opposite valence to those which they already hold. To some extent, this is supported by the empirical evidence available, and will be discussed in a later section.

11.4 Risk perception and impact on attitudes

Research that has been directed towards understanding public perceptions associated with potential food hazards have largely, but not exclusively, focused on issues associated with risk and benefit. How the public defines risk and benefit, and how the experts define the same issues, may be very different. This is not to say that non-experts should be viewed as irrational, but rather that public views should inform the debate about the strategic development of genetic modification. Research has demonstrated that risk perception is 'socially constructed' – that is, the way that people psychologically represent risks is a more important determinant of the way in which people react to risks relative to probabilistic risk assessments. Risk perception research has demonstrated that risks which are perceived as involuntary and unnatural are viewed as more threatening than those over which people perceive they have a choice, even if the probability of occurrence of the involuntary risk is very low (Slovic, 1993). Furthermore, specific concerns may be linked with particular hazards. For example, genetic modification of foods is associated with beliefs that the technology is ethically wrong, and representative of 'tampering with nature', that the long-term impact is unknown or unforeseen, (particularly with respect to effects on future generations), and that people have no choice over whether or not to consume them (Miles and Frewer, 2001).

That attitudes influence people's behaviours regarding their choices about whether or not to eat genetically modified products can be demonstrated by reference to real-world events. Public concerns about involuntary exposure to potential hazards are demonstrated by their reactions to situations where genetically modified ingredients are introduced into the marketplace without labelling and tracebility mechanisms. The removal of consumer choice is a key driver of consumer negativity towards genetically modified foods.

Genetically modified soya developed by Monsanto was not labelled when it was first imported into Europe from the United States. As a result, European consumers perceived that they had little choice about whether to consume genetically modified foods, creating problems for the European industry through heightened consumer concern and distrust in manufacturers and regulatory institutions. In addition, failure to label genetically modified ingredients resulted in perceptions of mistrust associated with industry, as the public assumed that failure to label indicated that the real risks were being hidden in order to promote a vested interest. Finally, the European public perceived that the benefits of genetically modified Soya accrued to American producers and to industry, but that European consumers experienced the risks. Combining all these factors, it is not surprising that acceptance of genetically modified Soya by the European public was so low. In contrast, the public accepted tomato paste produced by Zeneca, introduced to the British consumer a year earlier. This was because the product was clearly labelled as genetically modified, had a tangible consumer benefit, (reduced cost) and was consumed on a voluntary basis by the public.

11.5 Case study: impact of media reporting on public attitudes towards genetically modified foods

Genetic modification of food has been associated with a great deal of media attention in the UK and Europe, particularly in the spring of 1999 (Frewer, Miles, and Marsh, in press). In late 1998, articles written about Dr Arpad Pusztai of the Rowett Research Institute reported his (at the time, unpublished) research as providing support for the potential of negative health effects for humans as a result of consuming genetically modified potatoes (which contained chemicals designed to protect against pests). Later that year, the media reported the findings of an independent analysis, conducted by the Royal Society, which criticised the research as flawed and argued that no conclusions should be drawn from the work (The Royal Society, 1999). Spring of 1999 was associated with extensive media reporting about the potential risks of genetically modified foods, with much debate about risk and benefit proffered by different actors in the whole debate.

1998 also saw reports of over a thousand UK schools taking genetically modified foods off their menus, and the banning of genetically modified food from restaurants and bars in the House of Commons. Beginning in 1998, and continuing in 1999, Prince Charles expressed his concerns about genetic modification, questioning the necessity of the technology and calling for a public debate on the issue. Additionally, numerous genetically modified crop trial sites were destroyed around the country. A particularly well-reported case involved the destruction of a crop site in Norfolk, where Lord Peter Melchett, executive director of Greenpeace UK, was remanded in custody. The 1999 summer crop trial destructions led to discussions about the possible secrecy of locations of future crop trials. Crop trial sites were also destroyed in the US and France in 1999. Debates about the threat of cross-contamination to non-GM crops, including organic crops, were also reported.

Another area of public debate was linked to the quality of scientific advice offered to the government by scientific advisory committees. Specifically, the potential for a conflict of interest for scientific advisors to the government about the safety of genetically modified foods was seen to be an issue of concern, as there was potential for individual advisors profiting from the development and application of this technology through industrial interests. There was also an impact on food processing and food manufacturing. Since 1998 most of the major UK supermarkets have eliminated genetically modified ingredients from their own brand products in response to consumer concern. It was paralleled by similar moves by food manufacturers and restaurant chains in the latter part of 1999.

The question remains as to whether there was any impact of increased media reporting on people's attitudes towards genetic modification and its application in food production, and if there was an effect, whether this was a permanent shift in attitude or rather a temporary change linked to the extent and duration of media reporting (Frewer, Miles, and Marsh, in press).

Attitudinal data regarding people's attitudes towards genetically modified crops were collected in spring 1998 (before the media reporting had increased) and one year later, in March 1999 when reporting was peaking. The third wave of data collection was conducted in July 2000, when the levels of media attention had considerably subsided. Whilst the experimental design was somewhat opportunistic rather than the result of planning, it was possible to analyse whether the high levels of media reporting had an impact on people's attitudes associated with genetically modified foods. Just over three hundred participants took part in each phase of the experiment, new participants being recruited for each phase. Participants were asked to rate their agreement, on a seven-point scale anchored by completely agree and completely disagree, with fifty-three attitude statements. Nineteen of these statements were based on the public's own concerns (Miles and Frewer, 2001), the remaining thirty-three items being developed from attitudinal themes identified in previous studies investigating attitudes to various hazards (e.g. Sparks and Shepherd, 1994; Fife-Schaw and Rowe, 1996; Frewer *et al.*, 1997).

Principal component analysis indicated that three factors were explaining people's attitudes. The first was composed of items associated with the risk potential of genetically modified foods, and was labelled 'Risks and Negative Effects'. The second was labelled 'Trust and Choice', as it included items relating to trust in regulators and information sources, and whether or not people thought they could avoid the risks associated with genetically modified foods. The third described the potential benefits associated with genetically modified food and was labelled 'Benefits'.

The analysis showed that perceptions of risk (and other negative potential consequences) associated with genetically modified food increased during the highest levels of reporting, but these subsequently reduced as reporting levels diminished. The increase in perceptions of risk were accompanied by decreased perceptions of benefit. However, unlike perceptions of risk, perceptions of benefit remained depressed a year after the volume of reporting had declined. This was possibly because the media debate provided the public with information about what benefits associated with genetically modified foods were currently available; and at the time of reporting, these were primarily associated with industrial or producer profitability, rather than being specifically focused on desirable advantages to consumers.

In terms of attitudes, it would seem that people's risk attitudes were reinforced – people became temporarily more concerned about the risks of genetically modified foods, but this effect subsequently declined. In terms of attitude-consistent information, one might posit that the messages about risk that people received were not inconsistent with views they already held – thus messages were not persuasive and processed in an effortful and in-depth way, and did not result in long-term attitude change. However, the media debate probably provided information about consumer benefit. If we can assume that people's attitudes were not firmly developed regarding consumer benefits of genetically modified foods, then it is possible that the messages provided resulted in long-term attitude change as a result of persuasive argumentation.

The results can also be interpreted as a 'social amplification effect'. The framework was proposed to explain why 'risk events with minor physical consequences often elicit strong public concern and produce extraordinarily severe social impacts' (Kasperson *et al.*, 1998, p. 177). Very basically, the model proposes that risk information is 'amplified' through different channels (for example, the media or different social networks) that operate in such as way as to increases peoples risk perceptions. Similarly, peoples risk perceptions are 'attenuated' or decreased through similar channels. The media reporting of 1999 might have been said to amplify and, subsequently, attenuate, risk perceptions associated with genetically modified foods although only amplification was observed for perceptions of benefit associated with the same products. If perceptions of benefit are driving food acceptance, than one might surmise that people would be unlikely to accept genetically modified foods, at least in the short term.

11.6 Communication about genetically modified foods and models of attitude change

The importance of people's perceptions regarding information sources has been shown through an experiment using the Elaboration Likelihood model described earlier (Frewer *et al.*, 1999). This research built on the observation that, in the UK, government sources have been shown to be one of the least trusted providers of information about food-related risk, and consumer organisations one of the most trusted sources, (Frewer *et al.*, 1996).

It is interesting to manipulate information so that the true effects of these differences in source characteristics in terms of their influence on attitudes can be empirically examined. A part of this process is to ensure that information is attributed to a source that might realistically produce it in the 'real world'. In the experimental work reported here, both the government and consumer organisations have produced information about genetic modification, adding to the 'ecological validity' of experiment. Perceptions of risk relevance were also manipulated as part of the experiment. This was either high (respondents were told that they were able to buy genetically modified food in shops at the time of the experiment) or low (respondents were told that genetically modified foods would not be available for many years). The persuasive strength of the information was also manipulated to be high or low. In this experiment, persuasive information was directed towards public acceptance of genetically modified foods. To this end, the experimental work was conducted in two stages.

The first of these was the pre-selection of messages of high and low persuasive strength directed towards acceptance of genetic modification in food and agriculture, using a separate group of participants. This enabled identification of the ten most, and ten least, persuasive statements. The second was the systematic examination of the interaction of perceived risk relevance, persuasive strength, and trust in information source to which the information was attributed. People's attitudes towards genetically modified foods were then assessed. A three-factor experimental design was used. Levels of the first factor (persuasive content) were either high or low. Levels of the second factor (source) meant that respondents received information, which was attributed to either a consumer organisation or to government. The third factor (risk relevance) also had two levels – respondents were provided with information that led them to believe that the products of genetic modification were immediately available, or were likely to be available only at a later date, currently far in the future. All respondents then rated the information for their perceptions of source characteristics and informational qualities. Assessments were also taken of their attitudes to genetic engineering used in food production (Frewer *et al.*, 1997; Bredahl *et al.*, 1998), as well as completing other tasks designed to assess the extent to which they had internalised the information.

It was found that the information was more trusted if it was both high in persuasive strength and attributed to the government, if irrelevant risks were presented to respondents. However, it was found that highly persuasive information from a consumer organisation, or information from government that was low in persuasive strength, was more trusted if the risks were presented as being highly relevant to respondents. However, in terms of differences in attitude towards genetic modification, there were few significant differences between conditions. Thus it seemed that information had an impact on trust or distrust, rather than trust influencing attitude change following presentation of information. This might be because people had very strong attitudes towards genetic modification, and their reactions to the information (and the information sources) were influenced by these views. Indeed, social judgement theory would predict that initial attitude is likely to be one of the most important determinants of reactions to persuasive information, particularly if it appears to be promoting what might be interpreted as a vested interest (Eagly and Chaiken, 1993).

In a further experiment, (Frewer, Scholderer, and Bredahl, in press), two kinds of information about genetically modified food were presented to participants in an intervention trial – these were 'product specific' information (which was used to present genetically modified foods in a positive light) and 'balanced' information, (which discussed the potential risks and benefits of genetic modification of foods in a very neutral, but probabilistic and technical way). The information was attributed either to a consumer organisation (shown to be highly trusted in pilot research), an industry association (highly distrusted), or the European Commission (moderately trusted) in the different experimental conditions used in the study. Attitudes towards genetically modified foods were assessed before and after the information intervention.

Data about people's perceptions of information source characteristics (for example, whether they were trusted or not) were also collected. The results indicated that the extent to which people trusted information sources had little impact on attitudes towards genetically modified products or product acceptance. Prior attitudes towards genetically modified foods accounted for almost 95 and 90 per cent of the variance in perceived benefit and perceived risk respectively. Contrary to what might have been predicted, trust had negligible impact on these risk-related attitudes. The extent to which participants trusted the information sources was predominantly determined by already existing attitudes held by participants towards genetically modified foods. Attitudes were not influenced by perceptions of source characteristics. In other words, independent of the type of information provided, information provision in itself had little effect on people's attitudes towards genetically modified foods. Perceptions regarding information source characteristics did not contribute to attitude change, nor did the type of information strategy adopted have an impact on post-intervention attitudes.

Of greatest concern to industry and other institutions with an interest in information dissemination was the finding that the extent to which people trusted the information sources was driven by people's attitudes to genetically modified foods. Trust did not influence the way that people reacted to the information. On the other hand, attitudes were used to define people's perceptions regarding the motivation of the source providing the information. This perhaps is understandable in the case of the product-specific information, which was very positive about genetic modification, focusing only on benefits associated with novel products. People who favour the use of genetic modification are more likely to trust a source promoting its benefits. On the other hand, people who do not support the use of genetic modification in food production are more likely to distrust this same source providing the positive information because it does not align with their strongly held views. This does not explain why the same effect was observed in the case of the 'balanced' information strategy. The reason may be because of the way in which the information strategies were developed in the first place from the opinions of experts in the area of biotechnology, who proposed a 'rationalistic' approach to technology communication issues. Expert views regarding what is salient to risk communication may be very different from what is considered important by the public.

11.7 Approaches to communication

Communication scientists have developed a large number of communication models (Gutteling and Wiegman, 1996). Two popular risk communication models are known as the 'technical view' and the 'democratic view' (Rowan, 1994). The technical view of risk communication is based on the premise that the public needs accurate information and scientific expertise, and comprises a one-way, expert to lay-public information flow. The failure of the public to agree with this view is often attributed to 'misunderstanding' on the part of the public that can be informed or persuaded away by providing technical information in a form which can be understood by the public, or which is very influential in terms of attitude change. The basic premise is that, if only people could understand the technical risks, then they would also accept exposure to hazards where technical risk probabilities are very low. The democratic view, on the other hand, assumes that all stakeholders have maximum participation and decision-making power. In other words, persuasion is inappropriate because the aim of communication should be mutual understanding and not the exertion of power of one group over another. From this perspective, recourse to psychological models of persuasion is likely to be deemed inappropriate, although some would argue that there is an ethically sustainable role for such models in health psychology in general, for example, increasing fruit and vegetable intake in the diet in order to reduce the risks of cancer.

11.8 'Democratic' approaches

Democratic approaches to risk communication have involved facilitating public inputs into regulatory and strategic decision-making processes. Many different types of public participation methodology have been identified in the literature (e.g. Fiorino, 1990; Renn, 1995). These range from those which elicit input in the form of opinions (e.g. public opinion surveys and focus groups) to those that elicit judgements and decisions from which actual policy might be derived, and which are essentially deliberative in nature (e.g. consensus conferences and citizens' juries).

Space does not permit a substantive review of the different methodologies, and the interested reader is referred to Rowe and Frewer (2000), for a more detailed review of methodological approaches in this area. However, it is interesting to note that the practice of public participation has increased across all areas of policy development in recent years, although issues of 'best practice' are disputed. Rowe and Frewer (2000) have specified some theoretical criteria for benchmarking the effectiveness of public participation exercises, which aim to provide a framework for evaluating different approaches. Broadly speaking, evaluative criteria fall into one of two categories – those related to public acceptance of a procedure (that is, 'acceptance criteria'), and those related to the effective construction and implementation of a procedure, which refer to the procedural issues associated with the participation exercise itself ('process criteria'). These criteria, and the process of validation of these criteria, are described in greater detail elsewhere (Rowe and Frewer, 2000; Frewer, 2001). However, the potential effectiveness of public consultation may be compromised by failure to evaluate not only the process but also the substantive impact of the process on policy.

Frewer and Salter (in press) have argued that the inclusion of recommendations for best practice regarding public consultation and public involvement must include the explicit assessment of both scientific advice and public consultation on policy development if public confidence in science and risk management is not to be further eroded. There may also be scope for the use of deliberative methods in the identification of specific products (for example, novel fruits and vegetables produced with the use of biotechnology) that are acceptable and indeed desired by consumers, and will be purchased when they reach the market place.

11.9 Fruit and vegetable biotechnology – consumer issues for the future

Chronic diseases that are potentially preventable through appropriate dietary choices (such as many cancers, coronary heart disease, diabetes) will continue to represent a major public health problem in the future, particularly given demographic changes in the population, for example, ageing and increased obesity. If action is not taken to change people's dietary choices, the future of the health services will be compromised due to spiralling healthcare costs. Post-genomic technologies will provide a unique opportunity for future research to examine the relationships between diet and health and, in particular, to distinguish between genetic and dietary causes of disease. Increasing knowledge about how public health may be improved through more effective dietary choice, possibly tailored to particular genotypes, may not have a positive impact on quality of life if the public do not change their diet in line with increased knowledge. Nutrition in public health will also focus on optimal nutrition for patient groups, for example, diabetics. In the general area of fruit and vegetable biotechnology, the development of functional foods through biotechnology may improve public health in a very explicit way.

11.10 Functional foods and consumer issues – implications for fruit and vegetable biotechnology

The benefits to the consumer from consumption of so-called 'functional' foods are potentially very wide ranging. Possible benefits include, for example, an improvement in cognitive or physical performance, improvements in psychological well-being or reduction of the risks of, or prevention of, certain diseases arising.

New advances in food technology are part of a 'long-term shift in the market' associated with the development of foods which promote good health. In the earlier parts of the 20th century, foods were produced which prevented nutritional deficiencies. At the beginning of the 21st century, the focus of development is on foods that are capable of reducing risks of specific chronic diseases. However, these more recent developments 'ha[ve] complicated the task of communicating with the public about health-related benefits and risks' (Greenberg and Graham, 2000). This is partly because not all foods will benefit all individuals, but rather will improve the health of some subgroups within the population. A second factor is that these novel foods may be produced using technologies that some people see as problematic.

It is arguable that the development of 'functional foods' represents a natural progression from foods that were nutrient enhanced to compensate for vitamin deficiencies. The relationship between diet and health has later focused on relationships between food choice and diseases like cancer and heart disease (Lambert, 2001). Research has indicated that foods may play an important part in disease prevention, or slowing the progress of diseases. Greenberg and Graham (2000) note that simultaneous developments in human genetics and plant biotechnology have introduced the possibility that consumers can choose from a variety of foods aimed at preventing specific diseases for which they are at risk. This suggests that a long-term shift towards preventative therapies in health care may deliver benefits to individual subgroups in the population, but may further complicate the already uneasy relationship between science and society through the introduction of further concerns about the effectiveness and safety of novel foods.

Complicating the issue of consumer attitude is the problem of increasingly ambiguous categorisation of what is considered a food and what is considered a medicine. For example, how should fruits and vegetables genetically modified to fight diseases such as cancer, or deliver edible vaccines in the form of fruit, be regulated? Furthermore, post-genomic research is developing at a rate that is difficult to comprehend by the lay person. If the advantages of novel foods with direct and concrete benefits to health are to be realised in terms of improvements in the quality of life that people experience, an effective and appropriate strategy for enhanced public consultation and public understanding of post-genomic nutrition must be developed and operationalised.

This policy should be based on an understanding of what is driving public concern, and be linked to an effective communication strategy regarding the development and regulation of research and its applications. For example, people's concerns may be more concrete, relating to privacy, (to what extent should human genetic databases be anonymised, how easy is it to identify even an anonymous individual from their genetic data), personal economic consequences (those who are identified as at risk from particular diseases will be uninsurable, unemployable and unable to raise the finance to buy property, for example), as well as moral concerns (human genetics is morally wrong or represents tampering with nature) and risk perceptions (the technology is unnatural and the long-term consequences are unknown).

It is important that communication strategies are developed to ensure that consumers are forming attitudes based on the best and most up-to-date information available, and take these concerns into account. Information should acknowledge the uncertainties inherent in risk-analysis processes as well as the potential benefits of novel products if consumers are both to trust and believe the information source and the information that it provides. Failure to achieve this will result in increased public concerns and opposition to technological progress. Public distrust in science, scientific institutions, and the institutional mechanisms through which the governance of science is operationalised will increase.

A fruitful area for research into consumer attitudes will be to develop theoretical models which link perceptions, attitudes, and values to human decision-making. Risk communication and health education initiatives might usefully be directed towards individuals with specific dietary needs, but it cannot be assumed that, because someone is 'at risk' from a particular illness, they will automatically make food choice decisions to offset the risk. Other barriers to dietary change can readily be identified, and may prevent people responding to information about the benefits of a particular product. For example, 'optimistic bias', where people consider themselves to be at less risk from a given hazard compared to other people in similar circumstances, may result in people denying their own health risks, and thus failing to take preventative action to protect against ill-health (Miles and Scaife, in press).

A second example of barriers to dietary change can be identified through consideration of current health information campaigns associated with increased fruit and vegetable consumption, and the prevention of cancer. Some barriers to increased fruit and vegetable consumption cannot be linked to demographic factors (Havas *et al.*, 1998). However, there are some general demographic trends that make a contribution and are clearly linked to intake (Krebs-Smith *et al.*, 1995; Subar *et al.*, 1995; Johansson and Andersen 1998; Billson *et al.*, 1999). For example, more affluent and better-educated individuals are generally more health conscious and thus may be more motivated to process complex diet and health messages.

Women are generally more health conscious than males. Perhaps most obviously, people who do not like the taste of fruit and vegetables will not eat them, and if one has been brought up eating fruit and vegetables one is more likely to acquire their taste and continue eating these foods throughout life. Furthermore, Dibsdall *et al.* (in press) have reported that, in low-income groups, it is not knowledge about what constitutes a healthy diet that prevents healthy food choices, but rather that other lifestyle concerns and activities take priority in groups where financial resources are limited.

The food industry will need to predict what kind of products will be acceptable as well as beneficial to consumers, particularly in the area of 'preventative nutrition therapy'. For example, sensory properties of foods are likely to be as important as functional- or health-related factors in determining whether consumers accept novel fruit and vegetables produced with the aid of biotechnology.

11.11 Conclusions

Advances in science and technology (including in the area of fruit and vegetable biotechnology) mean that people in the developed world are living longer and healthier lives than ever before. However, people also have increased expectations regarding the extent and effectiveness of regulatory protection. The public's increased anxiety regarding risk is likely to be linked to changes in the processes that generate hazards (increasing globalisation, with potential for negative impact on food security, large-scale production of commodities, and uncertain technological implications regarding safety and ecological impact). This has resulted in a need to change institutional terms of reference and procedures, in particular to broaden the base of public consultation and dialogue on risk issues.

There are a variety of practical and ethical reasons for policy-making bodies to involve lay people in decision making on issues in which the public has a stake. Political theorists and ethicists discuss concepts such as democracy, procedural justice, and human rights, in providing the moral basis for involvement; but in a practical and expedient sense, making decisions without knowledge of the views of the public majority, or without public support, is liable to lead to confrontation, dispute, disruption, boycott, unrest, distrust, and simple public dissatisfaction. This need for public involvement would seem particularly evident in the food domain, as the food we eat, its taste, safety, price, and so on, are of fundamental, unavoidable and everyday interest to all members of society. The development of novel fruits and vegetables *must* take consumer attitudes as a starting point (implying that it is important to understand what consumers want in terms of novel products). Consumers should not be regarded as a potential barrier for technology innovation, but rather as partners in developing science that can improve the quality of life of people all over the world.

11.12 References

BILLSON, H., PRYER, J.A. and NICHOLS, R. (1999). 'Variation in fruit and vegetable consumption among adults in Britain'. *European Journal of Clinical Nutrition*, 57, 946–952.

BREDAHL, L., GRUNERT, G. and FREWER, L. J. (1998). 'Consumer attitudes and decision making with regard to genetically engineered food products – a review of the literature and a presentation of models for future research'. *Journal of Consumer Policy*, 21, 251–277.

CVETKOVICH, G. and LÖFSTEDT, R.E. (eds) (1999). *Social Trust and the Management of Risk.* London, Earthscan Publications Ltd.

DIBSDALL, L., LAMBERT, N. and FREWER, L.J. (in press). 'Using interpretive phenomenology to understand barriers to healthy eating in low-income groups'. *Journal of Nuitrition Education.*

EAGLY, A.H. and CHAIKEN, S. (1984). 'Cognitive theories of persuasion'. In L. Berkowitz (ed), *Advances in Experimental Social Psychology*, 17, 297–359.

EAGLY, A.H. and CHAIKEN, S. (1993). *The Psychology of Attitudes.* New York: Brace Jovanovich.

EUROBAROMETER (1997). European Commission Directorate General XII. *Biotechnology. Opinions of Europeans on Modern Biotechnology. Eurobarometer 46.1.* European Commission, Brussels-Luxembourg.

FIFE-SCHAW, C. and ROWE, G. (1996). 'Public Perceptions of Everyday Food Hazards: A Psychometric Study'. *Risk Analysis* 16 (4), 487–500.

FIORINO, D.J. (1990) 'Citizen participation and environmental risk: A survey of institutional mechanisms'. *Science, Technology and Human Values*, 1, 226–243.

FREWER, L. J. (2001). 'Environmental risk, public trust and perceived exclusion from risk management'. *Research in Social Problems and Public Policy*, 25, 3–29.

FREWER, L. J., HOWARD, C., HEDDERLEY, D. and SHEPHERD, R. (1996). 'What determines trust in information about food-related risks? Underlying psychological constructs'. *Risk Analysis*, 16, 473–486.

FREWER, L. J., HOWARD, C., HEDDERLEY, D. and SHEPHERD, R. (1997). 'The use of the elaboration likelihood model in developing effective food risk communication'. *Risk Analysis*, 17, 6, 269–281.

FREWER, L. J., HOWARD, C. HEDDERLEY, D. and SHEPHERD, R. (1999). 'Reactions to information about genetic engineering: impact of source credibility, perceived risk immediacy and persuasive content'. *Public Understanding of Science*, 8, 35–50.

FREWER, L. J., MILES. S. and MARSH, R. (in press). 'The GM foods controversy. A test of the social amplification of risk model'. *Risk Analysis.*

FREWER, L.J. and SALTER, B. (in press). 'Public attitudes, scientific advice and the politics of regulatory policy: the case of BSE'. *Science and Public Policy.*

FREWER, L. J., SCHOLDERER, J. and BREDAHL, L. (in press). 'Communicating about the risks and benefits of genetically modified foods: Effects of different information strategies'. *Risk Analysis.*

GREENBERG, D. and GRAHAM, M. (2000) 'Improving Communication About New Food Technologies'. *Issues in food technology online.* http://www.nap.edu/issues/16.4/greenberg.htm

GUTTELING, J.M. and WIEGMAN, O. (1996). *Exploring Risk Communication. Advances in Natural and Technological Hazards Research.* Kluwer, Dordrecht.

HM GOVERNMENT (2001) *The Interim Response to the Report of the BSE Inquiry by HM Goverment in Consultation with the Devolved Administrations.* The Stationery Office, London.

HAVAS, S., TREIMAN, K., LONGENBERG, P., BALLESTEROS, M., DAMRON and FELDMAN, R. (1998). 'Factors associated with fruit and vegetable consumption among women participating in WIC'. *Journal of the American Dietic Association*, 98, 1141–1148.

HOVLAND, C.I. (1959). 'Reconciling conflicting results derived from experimental and survey studies of attitude change'. *American Psychologist*, 14, 8–17.

JOHANSSON, L. and ANDERSEN, L.F. (1998) 'Who eats 5 a day? Intake of fruits and vegetables among Norwegians in relation to gender and lifestyle'. *Journal of the American Dietic Association* 98, 689–691.

KASPERSON, R.E., RENN, O., SLOVIC, P., BROWN, H.S., EMEL, J., GOLBE, R., KASPERSON, J.X. and RATLICK, S. (1988). 'The Social Amplification of Risk; A Conceptual Framework'. *Risk Analysis*. 8, No. 2, 177–187.

KREBS-SMITH, S.M., HEIMENDINGER, J., PATTERSON, B.H., SUBAR, A.F., KESSLER, R. and PIVONKA, E. (1995). 'Psychological factors associated with fruit and vegetable consumption'. *American Journal of Health Promotion*, 10, 98–104.

LAMBERT, N. (2001). 'Food Choice, Phytochemicals and Cancer Prevention'. In L. Frewer, E. Risvik and H. Schifferstein (eds). *Food Choice in Europe*. Springer-Verlag, Munich, 131–154.

MACCORQUODAL, K. and MEEHL, P.E. (1948). 'On a distinction between hypothetical constructs and intervening variables'. *Psychological Review*, 55, 95–107.

MILES, S. and FREWER, L. J. (2001). 'Investigating specific concerns about different food hazards – higher and lower order attributes'. *Food Quality and Preference*, 12, 47–61.

MILES, S. and SCAIFE, V. (in press). 'Optimistic bias and food'. *Nutrition Research Reviews*.

PETTY, P.E. and CACIOPPO, J.T. (1986). *Communication and persuasion. Central and peripheral routes to attitude change*. Springer, New York.

RENN, O. (1992) 'Risk communication: Towards a rational discourse with the public'. *Journal of Hazardous Materials*, *2*. 465–519.

ROWAN, K.E. (1994). 'The technical and democratic approaches to risk situations. Their appeal, limitations, and rhetorical alternative'. *Argumentation*, 8, 391–409.

ROWE, G. and FREWER, L. J. (2000). 'Public participation methods: An evaluative review of the literature'. *Science, Technology and Human Values*, 25, 3–29.

SIEGRIST, M. (1999). 'A causal model explaining the perception and acceptance of gene technology'. *Journal of Applied Social Psychology*, 29, 2093–2106.

SIEGRIST, M., CVETKOVICH, G. and ROTH, C. (2000). 'Salient value similarity, social trust and risk/benefit perception'. *Risk Analysis*, 20, 323–361.

SLOVIC, P. (1993). 'Perceived Risk, trust and democracy'. *Risk Analysis*, 13, 675–682.

SPARKS, P. and SHEPHERD, R. (1994). 'Public perception of the potential hazards associated with food production and food consumption: An empirical study'. *Risk Analysis*, 14, 799–806.

STROEBE, W., LENKERT, A. and JONAS, K. (1988). 'Familiarity may breed contempt: The impact of student exchange on national stereotypes and attitudes'. In W. Stroebe, A. Kruglanski, D. Bar-Tal and M. Hewstone (eds). *The Social Psychology of Intergroup Conflict: Theory, Research and applications*, Springer, New York.

SUBAR, A.S., HEIMENDINGER, J., KREBS-SMITH, S.N., PATTERSON, B.H., KESSLER, R. and PIVORKA, E. (1995). 'Fruit and vegetable intake in the US: The baseline survey for the 5-a-day for better health programme'. *Am. J. Health Prom.* 9, 352–360.

THE ROYAL SOCIETY (1999). *Review of data on possible toxicity of GM potatoes.* Royal Society Press Office, London.

ZECHENDORF, B. (1994). 'What the public think about biotechnology'. *Bio/ Technology*, 12, September, 870–875.

12

Risk assessment

W. Cooper, formerly National Institute of Agricultural Botany, Cambridge and J. B. Sweet, National Institute of Agricultural Botany, Cambridge

12.1 Introduction

12.1.1 Status of GM crop development

Since the first field trials of transgenic crops were conducted in the USA and France in 1986, there has been a rapid growth in activity with field trials being carried out globally (Table 12.1) involving at least 56 different crop species (Table 12.2). In 1999 the acreage of GM crop plants grown for commercial purposes world-wide was expected to reach 73 million acres, with crops grown mainly in the USA and Canada.

While Europe has led the way in terms of GM crop development and evaluation, the commercial situation, in the UK in particular, is very different.

Table 12.1 Releases of genetically modified organisms per country 1998

Country	%	Country	%
USA	70.45	Sweden	0.37
Canada	11.83	New Zealand	0.34
France	4.72	Denmark	0.31
Belgium	2.02	Brazil	0.28
UK	1.84	South Africa	0.17
Italy	1.71	Finland	0.11
Holland	1.47	Portugal	0.06
Spain	1.20	Russia	0.06
Japan	1.17	Bulgaria	0.05
Germany	0.89	Austria	0.03
Australia	0.88	Switzerland	0.03

Table 12.2 Genetically modified plant species (OECD figures, 1998)

African violet (*Saintpaulia ionantha*)
Alfalfa (*Medicago sativa*)
American Chestnut (*Castonea dentata*)
Apple (*Malus domestica*)
Asparagus (*Asparagus officianalus*)
Barley (*Hordeum vulgare*)
Beet (*Beta vulgaris*)
Belladonna (*Astropa belladonna*)
Broccoli, cauliflower and cabbage (*Brassica oleracea*)
Forage rape (*B. oleracea* var. *acephala*)
Kale rape (*B. oleracea* var. *biennis*)
Brown mustard (*Brassica nigra*)
Carnation (*Dianthus carophyllatus*)
Carrot (*Daucus carotta*)
European Chestnut (*Castanea sativa*)
Chicory (*Cichorium intybus*)
Chrysanthemum (*Chrysanthemum morifolium*)
Cotton (*Gossypium hirsutum*) (7%)
Cranberry, European (*Vaccinium oxycoccus*)
Creeping bentgrass (*Agrostis stolonifera*)
Cucumber (*Cucumis sativus*)
Cucurbita texana
Cucurbita pepo
Currant (*Rubus idaeus*)
Eggplant (*Solanum melonogea*)
Ethiopian mustard (*Brassica carinata*)
Eucalyptus (*Eucalyptus camaldulensis*)
Flax (*Linum usitatissium*)
Gladiolus sp.
Grape (*Vitis vinifera*)
Kentucky Bluegrass (*Poa patensis*)
Kiwi fruit (*Actinidia deliciosa* var. *deliciosa*)
Lettuce (*Lactua sativa*)
Lisianthus (*Eustoma grandiflorum*)
Lupin (*Lupinus angustifolius*)
Maize (*Zea mays*) (38%)
Marigold (*Tagetes sp.*)
Melon (*Cucumis melo*)
Mustard (*Brassica juncea*)
Oat (*Avena sativa*)
Oilseed rape (*Brassica napus*) (13%)
Onion (*Allium cepa*)
Orange (*Citrus sp.*)
Papaya (*Carica papaya*)
Pea (*Pisum sativum*)
Peanut (*Arachis hypogaea*)
Pelargonium sp.
Pepper (*Capsicum annuum*)
Pine (*Pinus sp.*)
Pineapple (*Ananas comosus*)
Poplar (*Populus sp.*)
Potato (*Solanum tuberosum*) (12%)
Rice (*Oryza sativa*)
Rose (*Rosa hybrida*)
Silver Birch (*Betula pendula*)
Spruce *Picea sp.*
Spruce, Norway *Picea abies*
Sorghum (*Sorghum bicolor*)
Sugar beet (*Beta vulgaris*) (2%)
Sugar cane (*Saccharum officinarum*)
Sunflower (*Helianthus annuum*)
Sweet potato (*Ipomoea batatas*)
Sweetgum (*Liguidambar sp.*)
Tamarillo (*Cyphomandra betacea*)
Thale cress (*Arabidposis thaliana*)
Tobacco (*Nicotiana benthamiana*)
Tobacco (*Nicotiana tabacum*) (5%)
Tomato (*Lycopersicon esculentum*) (10%)
Turnip rape (*Brassica rapa*)
Walnut (*Juglans sp.*)
Watermelon (*Citrullus lanatus*)
Wheat (*Triticum aestivum*)
White mustard (*Sinapsis alba*)

Those species comprising the majority of releases are indicated by the relevant percentage of releases within the OECD.

By 2000 the UK had approved 135 applications for release, but for research purposes only. Whilst there are an increasing number undergoing experimental and performance trials, no consents for release for commercial purposes have yet been granted. Commercialisation of the first GM variety is under review in response to mounting public opposition and demands for a five-year freeze until further experimental analysis satisfies concerns about GM crop safety.

The state of GM crop development in the UK can be summarised as follows: oilseed rape and maize are nearest to commercialisation; modifications include varieties tolerant to the herbicides glufosinate ammonium (Challenge variety) and glyphosate (Roundup variety). In addition oilseed rape varieties modified for expression of improved oil quality such as those expressing a high lauric acid content are also close to the marketplace.

A wide range of GM crops are currently in experimental trial including spring wheat (disease resistance), sugar beet (herbicide tolerant and altered carbohydrate metabolism), potato (altered carbohydrate, virus resistance) and maize (herbicide tolerant and insect resistance). Genetic engineering has also enabled higher yielding hybrid systems to be produced by the development of GM male sterile plants, a number of which are currently being tested for yield and overall performance. Cultivars of spring and winter oilseed rape, sugar beet, fodder beet and forage maize are currently being assessed in the UK's statutory National List (NL) trials. Inclusion of a variety onto the NL and the EC common catalogue is an essential precursor to commercialisation.

Crop development in the future is likely to continue with the production of varieties with improved pest and disease resistance. These developments are also likely to include plants compatible with effective weed control and environmentally friendly farming methods, and crops with tolerance to salinity, drought or frost. There is also likely to be an increasing emphasis on the development of varieties that are bred for processing purposes such as the production of novel oils, starches and high-value pharmaceutical compounds, for example, vaccines. GM crops are likely to be an important future development in providing a substitute for fossil fuels. There is good potential for utilising genetic engineering for the synthesis of plant-based alternatives to fossil fuels using a range of widely grown oilseed crops like oilseed rape to produce economically viable quantities of oil. GM crops may also provide substitutes for other non-renewable resources from which we derive, for example, many industrial oils used in the manufacture of plastics, detergents, inks and lubricants. One example is the isolation of the genes encoding the enzymes responsible for the synthesis of petroselinic acid, a fatty acid with potential for use in making detergents and nylon polymers. Further emphasis will also be placed on the development of designer health crops, resulting in better tasting, nutritionally enhanced or 'healthy option' crops. One example is potatoes with altered starch metabolism which have reduced oil uptake during cooking and, therefore, offer a healthy alternative to the traditional potato chip.

12.1.2 Concerns surrounding GM crops

The concept, let alone commercial reality, of genetically modified (GM) crops continues to be the cause of considerable concern to the public. Amongst the most frequently quoted concerns are the fears of gene escape to wild relatives leading to what has been termed gene pollution, the contamination of organically grown crops and the breakdown of disease/pest resistance in GM varieties. Other concerns relate to the safety of ingestion of GM ingredients by humans, for example the potential for developing allergenicity in a crop which was otherwise allergen free. GM ingredients containing antibiotic resistance genes have also met with public resistance due to fears associated with the potential for antibiotic resistant strains of bacteria developing via gene transfer in the gut of animals or even humans. Biotechnology companies and research organisations are responding to public pressure by developing GM varieties that no longer contain antibiotic resistance markers.

There is no doubt that the recent wave of public concern surrounding the safety and ethics of GM crops has overshadowed the significant potential benefits that GM technology has to offer for those involved in all parts of the food chain from primary producers/growers to the household consumer, as well as the potential environmental benefits associated with decreased spray applications. However, it is true that many GM crops may have impacts, including some very positive, upon agriculture and the environment while in some cases there may also be implications concerning food quality and safety. Ultimately there may also be ethical concerns for some sectors of society. Where there are benefits to be gained at a known or unknown risk, the question of risk assessment and subsequent risk management arises. If the potential of GM technology is to be realised, the quality, safety, benefits and ethical integrity of this new technology must be evaluated against the risks. Where the benefits are found to outweigh the risks the potential of transgenic technology needs to be realised by management under strict regulatory procedures and effective stewardship post-market release.

12.2 Risk assessment and avoidance: general principles

12.2.1 Principles of risk assessment

As previously mentioned, GM crops have a number of potential benefits for growers, processors and eventually the consumer, but it is also recognised that there are likely to be environmental impacts and implications for food quality and safety. For example, exploitation of novel GM pest-, disease- and herbicide-resistant crops will require different (often reduced) pesticide and herbicide applications. These modified management systems will have an impact upon current agricultural systems and the agricultural environment. Such impacts are best analysed by risk assessments.

The basic concepts of risk assessment for genetically modified crops are similar to those applied to chemical pesticides where the risk is equal to the frequency and the hazard. For example no exposure (frequency) would equate to zero hazard. Risk assessments study both the severity and extent of the hazard or damage as well as the likelihood and frequency at which the damage will occur. Risk is defined as:

$$\text{Risk (impact)} = \text{Frequency (exposure)} \times \text{Hazard}$$

Clearly the ideal situation would be one of zero risk. Since in reality the likelihood of risk is always greater than zero, acceptable risk levels for GM crops must be defined, as with all new technology. What is defined as acceptable is based upon cultural values and may well differ globally. Indeed the current climate of controversy surrounding GM crops signifies strong cultural differences between European and North American consumers in what is defined as acceptable levels of risk for the utilisation of GM crops.

While there are differences in the regulatory procedures controlling the development and commercialisation of GM crops in North America and Europe, both systems apply the same broad principles to assessing the safety of GM crop usage for food, animal feed and in terms of environmental impact. The first step involves thoroughly assessing the procedure for modifying the plant tissue. In the UK, for example, the Advisory Committee on Genetic Modification (ACGM) is the regulatory authority responsible for contained use evaluation; that is, the initial experimental work 'contained' within the laboratory or glasshouse. The risk evaluation procedure must be specific to each product. Broadly drawn conclusions, for example based on inter-species comparisons, are unacceptable. Most importantly the information requested in a risk assessment must be derived scientifically, with experiments designed to provide clear, interpretable, unequivocal and reproducible results. A recent addition to the risk-assessment procedure has occurred in the UK, in response to public pressure, where there is now a move towards assessing the societal and cultural impacts of this new technology alongside the environmental and human health risks.

Risk assessment can be divided into four steps (Nickson and McKee 1998):

1. problem formulation
2. risk analysis
3. risk characterisation
4. risk management.

Problem formulation requires that all available information concerning the plant, the trait and the experimental information is gathered in the context of the most likely hazards, such as toxicity/allergenicity. Once all the data are available, they can be analysed for characterisation of the likelihood and/or severity of the risk. In the final phase of the assessment procedure, the acceptability or otherwise of the identified risk must be determined and effective plans set out for its management. The risk assessment procedure is an iterative one and must continue throughout the use of the product, including post-market monitoring.

In the case of GM crops there are a number of variables/risk types to consider including impacts on the agricultural environment, closely related species, insects and animals and human health. To analyse the consequences of GM crop impact upon the agricultural environment requires a detailed understanding of the characteristics of the GM crop in question. This involves determining which wild relative, if any, it may hybridise with and studying the management systems involved in growing the GM crop itself. It also involves recognising any potential effects on other GM or non-GM crops which are likely to be grown in rotation with the variety being assessed. As an example, GM herbicide tolerant (HT) crops will be treated with different herbicides, with different activity spectra, at different crop development stages, leading to effects on the botanical diversity in the GM-HT crop which are the product of the interaction between the GM crop and the herbicide treatment.

The nature of any hazard is dependent upon the characteristics of both the crop that is modified and of the GM trait. Risk assessments require measurement and study of the hazard or impact of both. Numerous studies have concentrated on measuring frequency phenomena such as gene flow and inter-specific hybridisation without considering the impact of the transgene when it has dispersed or introgressed into other populations or species. In addition the impact of the release of the GM plant will depend on the type and location of the environment into which it is being released. To be truly effective, risk assessments may have to be carried out for a range of locations as they are not necessarily transferable from one site, area, region or country to another.

12.2.2 Impact of plant species

Plants vary in the degree to which they are dominant or are invasive in certain environments and in their ability to disperse genes to different populations and species. They will therefore have different environmental impacts when genetically modified. For any particular country or region, plants can be classified as potentially being high, medium or low impact.

Plants in the high-impact group are generally hardy, perennial, competitive, open-pollinating and prolific having a wide range of relatives with which they hybridise and an ability to colonise a range of natural and semi-natural habitats. Examples include perennial rye grasses (*Lolium perenne*) and certain indigenous and introduced trees and shrubs that form a significant proportion of forests and woodlands, e.g. *Populus* spp. Modifications of these plants, which affect their competitiveness, could have significant impacts upon the ecology of a range of environments.

Medium-impact plants are open-pollinating, hybridise with some wild relatives, are prolific and colonise a limited range of habitats. Examples of such plants include oilseed rape, oats, sugar beet and rice, all of which have closely related wild relatives with which they hybridise and an ability to colonise disturbed ground. These plants and their close relatives rarely form climax populations except in particular environments such as coastal areas or in disturbed ground. Low-impact plants are usually annual or biennial species, are largely self-pollinating with few hybridising relatives that are poorly adapted (or not native) to the area in which they are cultivated. In the UK, examples include maize and sunflower.

It is important to appreciate that the impact of a plant species will depend upon the environment into which it is being released. Maize and potato are considered low-impact plants in England. However in Central and South America, where their centres of genetic diversity occur, their impact would be considered very high.

12.2.3 Impact of transgenes

Transgene expression in GM plants will have different impacts in different environments. Since genes often operate uniquely it is not easy to classify transgenes as having high or low impact. In addition their impact is also dependent upon the nature of the receiving environment (agricultural impact).

High-impact transgenes generally encode genetic modifications that improve the fitness of the GM plants by increasing their reproduction, competitiveness, invasiveness and/or persistence and will therefore also have the greatest environmental impact. Thus transformations that significantly increase plant productivity by overcoming constraints such as broad-spectrum pest, disease and stress tolerance will have the highest impact. Many pest- and disease-resistant genes will have effects on non-target species either directly or indirectly by altering relationships between pests and beneficial organisms. It is important that these non-target effects are thoroughly understood before commercialisation progresses.

Low-impact transgenes are genes that do not noticeably enhance the fitness of the modified plant so that the modified plant's role and behaviour in a given ecosystem is not altered. Examples would include genes that modify seed composition, e.g. high lauric acid genes in oilseed rape and high starch content genes in potato. However, in preparing a comprehensive risk assessment it would be important to confirm that low-impact genes might not, unintentionally, confer an environmental advantage. As an example, in the case of high starch content genes in potato, it would be important to assess that the transgenes do not significantly increase potato seed tuber over-wintering survival rates through enhanced frost resistance. In the case of oilseed rape, it would be important to ensure, for example, that there is no increase in the dormancy characteristics of oilseed rape which may confer enhanced soil survival characteristics.

12.2.4 Mechanisms of transgene transmission

Gene flow is an important consideration in evaluating the risks associated with growing GM crops. Transgene dispersal could lead to contamination of neighbouring crops, a particular worry since the UK organic authority amended its rules to include a zero tolerance to the presence of GM material. Transgene flow from crops to closely related wild relatives is also of concern as an environmental risk. Gene flow between different species is, however, not a new concept and has in fact been occurring between natural plant species, leading to a range of hybrids in the UK flora including amongst others the Salix, Lolium and Rumex genera (Daniels and Sheail 1999).

In order for gene transfer from one species of plant to another closely related wild relative to occur a number of barriers, both physical and genetic, must be overcome. These include dispersal (either of pollen or seed), longevity of the pollen grain, sexual compatibility, competition with other pollen sources and events post-fertilisation. Most gene dispersal occurs as a result of pollen transported either on the wind or via vectors such as bees (Ramsay *et al.* 1999) or, less commonly, by seed dispersal. The distances over which pollen dispersal occurs varies depending upon the plant species, the prevailing weather conditions, in the case of wind-borne pollen, or the insect vector (Moyes and Dale 1999). As discussed by Moyes and Dale (1999), although most studies have concentrated on the range of pollen dispersal, the survivability over time of the pollen grain is actually the most important aspect of potential gene transfer and cross-contamination.

Assuming that pollination is successful and gene transfer has occurred, the barriers to successful introgression of a gene from the original donor species to the recipient will be dependent upon what the gene might offer the recipient. If, for example, the gene induces a lethal effect, the seed of the recipient plant will die and gene introgression into the recipient species will go no further. If, however, the transgene confers a selective advantage such as cold tolerance, drought or disease resistance or the ability to thrive in low-light conditions, seed from the recipient plant will thrive. This is especially true for native species, with the greatest opportunity for transgene movement occurring within the crop-weed complex (Whitton *et al.* 1997). However, in assessing the scale of transgene movement, it is important to consider whether those plants containing genes conferring an adaptive advantage in the agricultural environment might lose that selective advantage in the differing environmental conditions outside of the farm field. If the transgene provides no selective advantage to the recipient plant, such as herbicide-resistant genes present in plants growing in an environment where herbicide spraying will not occur, the transgene will have a neutral impact upon the recipient species. There will be no increase in fitness of the population.

12.2.5 Multiple transgenes and transgene stability

One of the major issues surrounding GM crops containing multiple transgenes encoding a variety of traits is the question of stability of gene expression. Might the introduction of a second transgene affect expression of the original transgene and thus the phenotype of the GM variety? In particular, genetic homology between the two transgenes may cause down regulation of gene expression and suppression of the phenotype. How this effect is caused is complex and thought to be affected by factors such as the position of the transgene within the genome, i.e. point of insertion during the transformation procedure, transgene copy number within the genome and by other factors such as reproduction and even environmental conditions. The results may be unpredictable resulting in instability or silencing of gene expression (Senior and Dale 1996). The production of GM varieties involves evaluation of transgenic lines over a number of generations, during which any unstable lines would generally be identified and discarded. One possible exception to this would be instability arising from environmental interaction. This instability is also observed in conventionally bred varieties, providing a basis for further analysis of GM varieties (Qian *et al.* 1986).

From the perspective of risk assessment and environmental impact the most significant issue arises from gene flow between closely related species. Instability of gene expression generally leads to suppression of gene expression, in which case the phenotype of the GM variety would revert to the wild type, with no expression of the transgene. The implications for agronomic practice are significant, as suppression of gene expression would render a herbicide-tolerant GM variety susceptible to that particular herbicide, with consequent loss of yield if the farmer were to spray unwittingly. While the effect of transgene instability on the natural environment is likely to be minimal, there may be important issues at stake in the case of transgenic plants engineered to remove the synthesis of harmful toxins. In this situation suppression of gene expression arising from gene flow leading to multiple transgene insertions could prove a serious human or animal health problem if undetected.

12.3 Assessing the impact of genetically modified crops

12.3.1 Impact on agricultural systems

Genetic modification can have a range of impacts on agricultural systems and therefore will require specific agronomic management. The use of GM varieties would affect the nature of crop volunteers in subsequent crops and require alterations in volunteer management practices. The GM trait may also have an impact if it disperses to other crops and weeds through cross-pollination and seed dispersal. Low-impact genes such as herbicide tolerance, which have little impact on natural environments, become highly significant because of the changes in the herbicide usage required for their management. These herbicides will differ in the effect they have on plant and other species diversity in cropped fields.

Deployment of high-impact genes such as those encoding pest and disease resistance will result in reductions and changes in pesticide usage and thus offer opportunities to enhance diversity in cropped fields, especially if the transgene products are very specific to selected pests. However, it is important that the selection pressures they impose on pests and diseases do not encourage the development of virulent races of pests and pathogens and appropriate management systems are required in order to maintain durable resistance in the GM varieties. Long-term studies on the performance of insect and herbicide-resistant transgenic crops, such as oilseed rape, potato, maize and sugar beet, grown in 12 different habitats and monitored over a period of ten years, showed that no genetically modified plants were more invasive or more persistent than their conventional counterparts.

12.3.2 Impact on uncultivated flora

Genetically modified crops may also have impacts on uncultivated and 'natural' environments. These environments may be affected by characteristics of crop and wild species induced by novel genetic constructs and their products. Risk assessments must therefore concentrate on whether the genetically modified characteristics of a GM crop and of similarly modified hybridising wild relatives are likely to change the behaviour of the plants or dependent flora and fauna in their environment, to the extent that ecological balances are altered.

12.3.3 Impact on insects and animals

The first successful example of using a foreign plant gene to confer resistance to insects was reported in 1987 (Hilder *et al.* 1987) and involved transformation of tobacco (*Nicotianum tabacum*) with the cow pea trypsin inhibitor (CpTi) gene. Since then there have been many reports of success in insect management using transgenic crop varieties.

The bacterial endotoxins isolated from *Bacillus thuringiensis (Bt)*, comprise one of several groups of proteins which have been shown to have insecticidal properties to a range of economically important insects. Transgenic crop varieties engineered with *Bt* resistance are already in commercial use in the USA and China, while a number of plant proteins, such as inhibitors of proteases, lectins and other digestive enzymes, are being evaluated for their efficacy as insect-resistance mechanisms (Gatehouse *et al.* 1998).

It is important that genes selected for the control of insect pests have acceptably little effect on non-target insects including predators of the target pest insects, in order to maintain insect diversity in GM crops. Clearly, if there is an effect upon predators that is comparable with current control practices then little benefit will accrue from the deployment of GM crops, a point made strongly by the Royal Society for the Protection of Birds in their submissions on GMOs (1997). Impact assessments are therefore required to examine the effects on non-target organisms in the crop environment.

Studies to evaluate the effect of transgenic plants expressing insect resistance on non-target species have provided, at best, equivocal and often controversial results which have served only to fuel the GM debate rather than provide hard scientific facts on which to base a thorough impact assessment.

Research into the impact of potato plants expressing the snowdrop lectin GNA upon 2-spot ladybirds which feed on the aphid *Mysus persicae* demonstrated that the ladybirds were affected adversely in terms of fecundity, egg viability and longevity (Birch *et al.* 1998). However, the authors point out that the effects may either be a direct result of ladybirds preying on aphids which have digested transgenic plant material containing the lectin, or may also be due to poor nutritional quality of the aphids themselves as a food source. Other studies involving the parasitic wasp *Eulophus pennicornis* and the tomato moth *Lacanobiua oleracea* demonstrated that the parasitic wasp was not affected when it parasitised moth larvae reared on transgenic potato plants expressing the snowdrop lectin GNA (Gatehouse *et al.* 1997).

More recently Losey *et al.* (1999) published a report indicating that pollen from transgenic *Bt*-resistant maize plants had a detrimental effect on the larvae of the non-target Monarch butterfly (*Danaus plexippus*), which is considered to be a sensitive indicator of environmental disturbance in the USA. Larvae, which normally feed on the leaves of the milkweed (*Asclepias curassavica*) plant were fed on leaves that had been dusted with unquantified amounts of pollen from the transgenic *Bt* maize plants. Results indicated that larval survival rate was only 56% compared to 100% survival for larvae fed on leaves dusted with untransformed pollen. Superficially these results indicate an unacceptable environmental impact from *Bt* maize. However, closer analyses have revealed a number of serious criticisms of the research, including the use of laboratory studies only, no-choice feeding regimes, lack of stringency, lack of quantification and the use of inappropriate controls (Hodgson 1999). The experiments were not conducted in the field so no *in vivo* data were available to confirm that (a) milkweeds occur in maize fields, and (b) that Monarch butterflies occur on these milkweeds bearing in mind the insecticide programme received by conventional maize. This once again reiterates the requirement for comprehensive risk assessments based on thorough science.

Schuler *et al.* (1999) have conducted research concerning the environmental effects of *Bt*-resistant GM oilseed rape on a non-target insect. The results demonstrated that the behaviour of non-target insects can also play a part in determining how *Bt* plants will affect their populations and should be considered when trying to evaluate the environmental impact of GM crops. Their laboratory-based experiments evaluated the ecological impact of the GM crop on the diamondback moth (*Plutella xylostella*), a pest that damages the oilseed rape crop, as well as the natural bio-control agent of the diamondback moth, a parasitic wasp (*Cotesia plutellae*), which kills the moths' caterpillars by laying its eggs in them. Results demonstrated that parasitoid wasp larvae that were oviposited in *Bt*-susceptible moth larvae not surprisingly died with their hosts. In contrast wasp larvae that had been oviposited in *Bt*-resistant moth larvae feeding on transgenic plants survived and demonstrated no adverse effects of exposure to the *Bt* toxins either as adults or in the development of their own larvae.

The research group then examined the behaviour of the female parasitic wasps in the presence of GM and non-GM leaves. It is known that the female wasps locate the host diamondback moth larvae using herbivore-induced volatiles released from the damaged plants. A wind-tunnel was used to compare the flight response of the wasp towards *Bt*-susceptible and *Bt*-resistant diamondback larvae which were allowed to feed on *Bt* leaves. The flight and feeding behaviour of each wasp was then measured. In this test, 79% of the parasitoids flew to the *Bt* leaves damaged by resistant moth larvae, with only 21% choosing *Bt* leaves damaged by susceptible larvae. The apparent lack of effect on the survival or host-seeking ability of the parasitic wasp suggested that *Bt* plants may have an environmental advantage over broad-spectrum insecticides.

12.3.4 Impact on human health

The potential for transferring genes from one unrelated species to another has caused concern that allergenicity may be introduced into a food source that was previously non-allergenic. An obvious example is that of the recent research programme by Pioneer Seeds where soya bean transformed with a gene from the Brazil nut was found to have allergenic properties. The research programme was halted before any field trials took place but public concern was heightened by reports of this work. All GM foods are now routinely tested for allergenicity using serological tests involving immuno-globulin antigens for specific allergenic proteins.

Transgene instability may be an important issue in the case of transgenic plants engineered to remove the synthesis of harmful toxins. In this situation suppression of gene expression arising from gene flow leading to multiple transgene insertions could prove a serious human or animal health problem if undetected (see Section 12.2.5).

The inappropriate choice of transgenes for achieving a desired trait may also have a serious impact on human health without adequate risk assessment. Lectins are a group of proteins known to have insecticidal properties that make them attractive candidates for the development of transgenic plants with resistance to the Homoptera insects. They are thought to work by binding carbohydrate side chains present in the gut wall resulting in inhibition of food absorption. As there is the potential of toxicity to humans, it is essential that extensive risk evaluation is required to establish any potential threats of toxicity. The need for such risk assessment is reflected in work on GM tomatoes (Noteborn *et al.* 1995) and was recently highlighted by reports from the Rowett Institute in Scotland, which indicated adverse immunological and nutritional effects from enhanced lecithin in GM potatoes (Ewen and Pusztai 1999, Fenton *et al.* 1999). However, aspects of the former of these reports in particular were strongly criticised by a number of reputable scientific bodies as being unsubstantiated and have highlighted the need for agreed methodology in this field of research so that conclusive results can be acquired (Kuiper *et al.* 1999).

There has also been concern about genetically modified ingredients containing antibiotic resistance genes used to select transformed cells prior to the regeneration of transgenic plants. Use of these genes raises the potential for antibiotic-resistant strains of bacteria to develop via horizontal gene transfer in the gastro-intestinal tract of animals or even humans (Harding and Harris 1997). This possibility is not thought to be a major hazard since the antibiotic-resistant genes most often used for plant transformation themselves come from bacteria. They encode resistance against antibiotics rarely used in medicine such as kanamycin, against which a large percentage of gut microflora is already resistant.

However, given the risk, however small, of producing more antibiotic-resistant bacteria, techniques are being developed that will enable selectable markers to be removed from crop plants after the transformation process. Alternative selectable markers, not based on antibiotic selection, are also being tested, for example a mannose permease that allows the use of mannose, a sugar not normally available to plant metabolism, as a carbon source during plant regeneration.

Risk assessment methodology will also have to be adjusted for food plants which are modified to improve nutritional and other qualities, a major area for current and future research. Target traits include, for example, improving the nutritional value of proteins, increasing the concentrations of oils low in saturated fats, or fortification with micronutrients or antioxidants. Food plants modified in this way must undergo extensive toxicological and nutritional assessment with a combination of *in vitro* and *in vivo* tests, as currently required for all novel foods by the EU, for example. In the case of genetic modification, however, particular attention needs to be given to the detection and characterisation of potential unintended effects of modification. Inferences about such effects can no longer be based solely on chemical analysis of single macronutrients and micronutrients, and known crop-specific antinutrients or toxins. New methods have been developed to screen for potential alterations in the metabolism of the modified organism by such methods as:

- analysis of gene expression (monitored, for example, by microarray technology or mRNA fingerprinting)
- overall protein analysis (proteomics)
- secondary metabolite profiling.

Studies using these designs will need to be designed carefully to take account of the complexity of foods (OECD 1996, Noteborn *et al.* 2000, Van Hal *et al.* 2000).

12.4 References

BIRCH A N E, GEOGHEGAN I E, MAJERUS M, McNICOL J W, HACKETT C, GATEHOUSE A M R and GATEHOUSE J A (1998) Ecological impact on predatory two-spot ladybirds of transgenic potatoes expressing snowdrop lectin for aphid resistance, *Journal of Molecular Breeding*.

DANIELS R E and SHEAIL J (1999) Genetic pollution: concepts, concerns and transgenic crops. In 1999 BCPC Symposium Proceedings No 72: Gene flow and agriculture: relevance for transgenic crops, 65–72.

EWEN S W and PUSZTAI A (1999) Effect of diets containing genetically-modified potatoes expressing *Galanthus nivalis* lectin on rat small intestine, *The Lancet*, 354 (9187).

FENTON B, STANLEY K, FENTON S and BOLTON-SMITH (1999) Differential binding of the insectidal lectin GNA to human blood cells, *The Lancet*, 354 (9187).

GATEHOUSE A M R, DAVISON G M, NEWELL C A, MERRYWEATHER A, HAMILTON W D O, BURGESS E P J, GILBERT R J C and GATEHOUSE J A (1997) Transgenic potato plants with enhanced resistance to the tomato moth, *Lacanobia oleracea: growth room trials. Molecular Breeding*, 3, 49–63.

GATEHOUSE A M R, BROWN D P, WILKINSON H S, DOWN R E, FORD L, GATEHOUSE J A, BELL H A and EDWARDS J P (1998) The use of transgenic plants for the control of insect pests. BCPC Symposium Proceedings No 71: biotechnology in crop protection: facts and fallacies.

HARDING K and HARRIS P S (1997) Risk assessment of the release of genetically modified plants: a review, *Agro-Food-Indust Hi-Tech*, Nov/Dec, 8–13.

HILDER V A, GATEHOUSE A M R, SHERMAN S E, BARKEER R F and BOULTER D (1987) A novel mechanism for insect resistance engineered into tobacco. *Nature*, 330, 160–3.

HODGSON J (1999) Monarch Bt-corn paper questioned. *Nature Biotechnology*, 17, (7), 627.

KUIPER H A, NOTEBORN H P and PEIJNENBURG A C (1999) Adequacy of methods of testing the safety of genetically-modified foods, *The Lancet*, 354 (9187).

LOSEY J E, RAYOR L S and CARTER M E (1999) Transgenic pollen harm monarch larvae. *Nature*, 399, 214.

MOYES C L and DALE P J (1999) Organic farming and gene transfer from genetically modified crops. MAFF research report.

NICKSON T E and McKEE M J (1998) Ecological aspects of genetically modified crops. Agricultural biotechnology and environmental quality; gene escape and pest resistance. NABC report no. 10, pp. 95–104.

NOTEBORN H P, BIENENMANN-PLOUM M E and VAN DEN BERG J H (1995) Safety assessment of the *Bacillus thuringiensis* insecticidal crystal protein CryIA(b) expressed in transgenic tomatoes. In Engel K H *et al. Genetically-modified foods: safety issues*. ACS Symposium Series 605, Washington DC: 134–47.

NOTEBORN H P, LOMMEN A, VAN DER JAGT R C, WESEMAN J M and KUIPER H A (2000) Chemical fingerprinting for the evaluation of unintended secondary metabolic changes in transgenic food crops, *J. Biotechnol* (in press).

OECD (1996) *Food safety evaluation*. Paris: OECD.

QIAN C M, XU A and LIANG G (1986) Effects of low temperatures and genotypes on pollen development in wheat. *Crop Science*, 26, 43–6.

RAMSAY G, THOMPSON C E, NEILSON S and MACKAY G R (1999) Honey bees as vectors of GM oilseed rape pollen. 1999 BCPC Symposium Proceedings No 72: Gene flow and agriculture: relevance for transgenic crops, 209–14.

SCHULER T H, POTTING R P J, DENHOLM I and POPPY G M (1999) Parasitoid behaviour and *Bt* plants. *Nature*, 400, 825–6.

SENIOR I J and DALE P J (1996) Plant transgene silencing – gremlin or gift? *Chemistry and Industry*, 19 August, 604–8.

VAN HAL H N, VORST O and VAN HOUWELINGEN A M (2000) The application of DNA micro-assays in gene expression analysis, *J. Biotechnol* (in press).

WHITTON J, WOLFE D E, ARIA D M, SNOW A A and REISBERG L H (1997) The persistence of cultivar alleles in wild populations of sunflowers five generations after hybridisation. *Theoretical and Applied Genetics*, 95, 33–40.

Index